Mathematical Series 3

SPECIAL FUNCTIONS AND ORTHOGONAL POLYNOMIALS

by
REFAAT A. EL ATTAR
rea5@hotmail.com
Professor of Mathematics
College of Engineering
Alexandria University
Egypt

Lulu Press,
3131 RDU Center, Suite 210
Morrisville NC 27560, USA
http://www.lulu.com

ISBN: 1-4116-6690-9
EAN: 978-141166690-0
Printed in the United State of America

Copyright © 2006 by the author. All rights reserved. No part of this book may be reproduced, stored in a retrieval system, or transcribed in any form or by any means – electronic, mechanical, photocopying, recording, or otherwise - without the prior permission of the author.

COPYRIGHT © 2006 BY REFAAT EL ATTAR

PREFACE

This book is written to provide an easy to follow study on the subject of Special Functions and Orthogonal Polynomials. The material presented here can be covered in eight to ten 2-hour classroom lectures; however, it is also written in a way that it can be used as a self study text. Basic knowledge of calculus and differential equations is needed. The book is intended to help students in engineering, physics and applied sciences understand various aspects of Special Functions and Orthogonal Polynomials that very often occur in engineering, physics, mathematics and applied sciences.

I have collected many problems and gave numerous solved examples on the subject that might help the reader getting on-hand experience with the techniques presented in this note. It is hoped that this work will give some motivation to the reader to dig a little bit further in the subject.

The book is organized in chapters that are in a sense self contained. Chapter 1 deals with series solutions of Differential Equations. Gamma and Beta functions are studied in Chapter 2 together with other functions that are defined by integrals. Legendre Polynomials and Functions are studied in Chapter 3. Chapters 4 and 5 deal with Hermite, Laguerre and other Orthogonal Polynomials. A detailed treatise of Bessel Function in given in Chapter 6.

I express my appreciation to the staff of Mathematics in the Department of Engineering Mathematics and Physics, Faculty of Engineering, Alexandria University, for their constant encouragement during the preparation of this book.

Finally, without the understanding that my wife Rachida showed during the long hours that I stayed in front of my computer, the achievement of this book would not have been possible.

REFAAT A. EL ATTAR
rea5@hotmail.com
Alexandria, EGYPT, December, 2005

Contents

	Page
Chapter 1: Series Solutions of Differential Equations	3
1.1. Introduction	3
1.2. Taylor's Series Method	3
1.3. The Method of Undetermined Coefficients	8
1.4. The Method of Frobenius	20
1.5. Solutions for Large Values of x	51
Chapter 2: Gamma and Beta Functions and Others	57
2.1. Gamma Function	57
2.2. Beta Function	77
2.3. Other Special Functions Defined by Integrals	84
2.3.1. Incomplete Gamma Functions	84
2.3.2. The Error Functions	85
2.3.3. The Exponential Integral Functions	85
2.3.4. The Sine and Cosine Functions	87
2.3.5. The Elliptic Integral Functions	87
2.3.6. The Fresnel Sine and Cosine Integral Functions	88
2.3.7. The Debye Function	88
Chapter 3: Legendre Polynomials	93
3.1. Introduction	93
3.2. Legendre Differential Equation and Its Solutions	93
3.3. Legendre Polynomials in Descending Powers of x	101
3.4. Generating Function for Legendre Polynomials	112
3.5. Recurrence Relations for Legendre Polynomials	120
3.6. Orthogonality Properties of Legendre Polynomials	127
3.7. Integral Form of Legendre Polynomials	137
3.8. Differential Form for Legendre Polynomials (Rodrigues' Formula)	139
3.9. Schläfli's Integral for Legendre Polynomials	148
3.10. Associated Legendre Functions	149
3.11. Series of Legendre Polynomials	153
3.12. Legendre Functions of the Second Kind $Q_n(x)$	156
3.12.1. Relation between $P_n(x)$ and $Q_n(x)$	157
3.12.2. Properties of Legendre Functions of the Second Kind	159
3.13. Shifted Legendre Polynomials	162
3.14. Summary of Legendre Polynomials and Functions	163
Chapter 4: Hermite Polynomials	173
4.1. Hermite Differential Equation and its Solutions	173
4.2. Generating Function for Hemite Polynomials	178
4.3. Recurrence Relations for Hermite Polynomials	180

	Page
4.4. Rodrigues' Formula for Hermite Polynomials	181
4.5. Orthogonality Property of Hermite Polynomials	182
4.6. Weber-Hermite Functions	184
4.7. Summary of Hermite Polynomials	195
Chapter 5: Laguerre and Other Orthogonal Polynomials	201
5.1. Laguerre Differential Equation	201
5.2. Generating Function for Laguerre Polynomials	204
5.3. Differential Form of Laguerre Polynomials (Rodrigues' Formula)	205
5.4. Recurrence Relations for Laguerre Polynomials	205
5.5. Orthogonality Property of Laguerre Polynomials	208
5.6. The Associated Laguerre Differential Equation	209
5.7. Chebyshev Polynomials	212
5.7.1. Generating Function	214
5.7.2. Recurrence Relations	215
5.7.3. Orthogonality Property	215
5.7.4. Fourier-Chebyshev Series Expansion	215
5.7.5. Rodrigues' Formula	216
5.7.6. Chebyshev Determinants	216
5.8. Gegenbauer Polynomials	217
5.9. Jacobi Polynomials	219
Chapter 6: Bessel Functions	225
6.1. Introduction	225
6.2. Solution of Bessel Differential Equation	225
6.3. Independence of the Solutions $J_n(x)$ and $J_{-n}(x)$	227
6.4. Solution of Bessel's equation when n is an integer	238
6.5. Generating Function for Bessel Functions	242
6.6. Integral Formof Bessel Functions	244
6.7. Recurrence Relations for Bessel Functions	248
6.8. Integrals Involving Bessel Functions	256
6.9. Generalized Bessel Differential Equation	262
6.10. Hankel Functions	264
6.11. Modified Bessel Functions	265
6.12. Kelvin Functions	269
6.13. Orthogonality Property and Bessel Series Expansion	270
6.14. Spherical Bessel Functions	275
6.15. Modified Spherical Bessel Functions	279
6.16. Airy Functions	281
6.17. Summary of Bessel Functions	282
References	293

Chapter One

Series Solutions of Differential Equations

$$y = \sum_{k=0}^{\infty} c_k (x - x_0)^{k+\alpha}, \quad c_0 \neq 0$$

Chapter 1
Series Solutions of Differential Equations

1.1. Introduction

It may happen that a differential equation cannot be solved by any of the known classes of methods designed to find the solutions for some particular form of equations. In such situations we must find other means to express the solutions. These methods are related to power series. There are several ways in which series are used in differential equations. We consider here methods designed to obtain these power series solutions. We restrict ourselves to second order differential equations although the methods presented apply as well to higher order equations.

To get the idea, consider the differential equation

$$y'' - 5y' + 6y = 0$$

the solution of this equation can be found by usual means as

$$y = ae^{2x} + be^{3x}$$

Where a and b are two arbitrary constants. If we expand the functions e^{2x} and e^{3x} in power series about $x = 0$, then the solution of the differential equation can be written in power series form as

$$y = a\left(1 + 2x + \frac{(2x)^2}{2!} + \frac{(2x)^3}{3!} + \cdots\right) + b\left(1 + 3x + \frac{(3x)^2}{2!} + \frac{(3x)^3}{3!} + \cdots\right)$$

From this, we see that the general solution of a second order differential equation may be expressed as a linear combination of two infinite series. This suggests that we try solutions for differential equations in terms of infinite power series.

1.2. Taylor's Series Method

If $f(x)$ is defined in some interval $a < x < b$ and if x_0 is a point in this interval and if all derivatives of $f(x)$ exist at x_0, then Taylor's series of $f(x)$ is

$$f(x) = f(x_0) + (x - x_0)f'(x_0) + \frac{(x - x_0)^2}{2!}f''(x_0) + \frac{(x - x_0)^3}{3!}f'''(x_0) + \cdots + \frac{(x - x_0)^n}{n!}f^{(n)}(x_0) + \cdots$$

Special Functions and Orthogonal Polynomials

If $x_0 = 0$, we obtain MacLaurin's (**Colin Maclaurin** (1698-1746)) series

$$f(x) = f(0) + xf'(0) + \frac{x^2}{2!}f''(0) + \frac{x^3}{3!}f'''(0) + \cdots + \frac{x^n}{n!}f^{(n)}(0) + \cdots.$$

The series will converge to $f(x)$ for all x in an interval with x_0 as midpoint under appropriate hypotheses.

Examples of convergent series are

$$e^x = 1 + x + \frac{x^2}{2!} + \cdots + \frac{x^n}{n!} + \cdots \quad (\text{convergent } \forall x)$$

$$\sin x = x - \frac{x^3}{3!} + \frac{x^5}{5!} - \frac{x^7}{7!} + \cdots + \frac{(-1)^n x^{2n+1}}{(2n+1)!} + \cdots \quad (\text{convergent } \forall x)$$

$$\frac{1}{1-x} = 1 + x + x^2 + x^3 + \cdots + x^n + \cdots \quad (\text{convergent for } |x| < 1)$$

If a series converges for all x, it is convenient to say that *the radius of convergence is infinity*. If the series diverges for all $x \neq 0$, we say that the radius of convergence is zero. Each power series has a radius of convergence R and the series is convergent when $|x| < R$.

Consider the second order differential equation

$$y'' = F(x, y, y') \tag{1}$$

It is reasonable to expect a solution of (1), that is a power series in x, to contain two arbitrary constants. We can assign the values of y and y' at $x = 0$ to be these two arbitrary constants, i.e., at $x = 0$, $y(0) = A$ and $y'(0) = B$.

So, $y''(0)$ can be computed directly from Equation (1) in terms of A and B. Differentiating Equation (1) with respect to x, we get

$$y''' = \frac{d}{dx}F(x, y, y') \tag{2}$$

Again $y'''(0)$ can be computed from Equation (2) in terms of A and B. This process can be continued to compute higher derivatives of y at $x = 0$. Then using MacLaurin's expansion, the solution y is

$$y(x) = y(0) + xy'(0) + \frac{x^2}{2!}y''(0) + \frac{x^3}{3!}y'''(0) + \cdots + \frac{x^n}{n!}y^{(n)}(0) + \cdots \tag{3}$$

Series Solutions of Differential Equations

The series in Equation (3) will converge to the value of $y(x)$ in some interval about $x = 0$, if $y(x)$ is well behaved at and near $x = 0$.

Note: If the initial conditions are not at $x = 0$ but at $x = a$, then we use Taylor's expansion instead.

We give some examples for the application of Taylor's or MacLaurin's expansions to find series solutions for differential equations.

***Example* 1:** Find the first four non-zero terms in a power series solution for the differential equation: $y' = x^2 - y^2$, when $x = 1, y = 1$.

Solution: We have

$$y(1) = 1$$

$$y' = x^2 - y^2 \qquad\qquad y'(1) = 0$$

$$y'' = 2x - 2yy' \qquad\qquad y''(1) = 2$$

$$y''' = 2 - 2y'^2 - 2yy'' \qquad\qquad y'''(1) = -2$$

$$y^{iv} = -6y'y'' - 2y''' \qquad\qquad y^{iv}(1) = 4$$

Then the power series solution is

$$y = 1 + \frac{2(x-1)^2}{2!} - \frac{2(x-1)^3}{3!} + \frac{4(x-1)^4}{4!} + \cdots.\qquad\square$$

This example reveals the following points:

1. Can we write the general term for the series? sometimes tricky! and most of the time impossible.

2. Does the series converge to $y(x)$? If we can write the general term, we can use the ratio test to obtain the values of x for which the series is convergent. Hopefully, $x = 0$ is inside the interval of convergence.

3. If the series is convergent, how rapidly is this convergence?

4. If we take only the first k terms of the series as an approximation to $y(x)$, how big can the error in approximation be?

These are some questions to be asked, but in the current context, we will not be able to answer some of them.

Example 2: Find the power series solution for the initial value problem:

$$y'' - e^{-x} y' - e^{-x} y^2 + 1 = 0, \text{ when } x = 0, \ y = 1 \text{ and } y' = 1.$$

Solution: Here $y(0) = 1$ and $y'(0) = 1$, and

$$y'' = e^{-x} y' + e^{-x} y^2 - 1 \qquad\qquad y''(0) = 1$$

$$y''' = e^{-x}(y'' + 2yy' - y' - y^2) \qquad\qquad y'''(0) = 1$$

$$y^{iv} = e^{-x}(y''' + 2y'^2 + 2yy'' - 2y'' - 4yy' + y^2 + y')$$

$$y^{iv}(0) = 1$$

Substituting these values in MacLaurin's expansion, we obtain

$$y = 1 + x + \frac{x^2}{2!} + \frac{x^3}{3!} + \cdots + \frac{x^n}{n!} + \cdots$$

The form of this series suggests that $y = e^x$ is the solution of the differential equation, indeed it is. The series in this particular example converges for all finite values of x. □

Example 3: Find the power series solution for the initial value problem:

$$(x-1)y''' + y'' + (x-1)y' + y = 0, \text{ when } x = 0, \ y = y'' = 0 \text{ and } y' = 1.$$

Solution: Here $y(0) = 0$, $y'(0) = 1$ and $y''(0) = 0$.

Solving for y''' then differentiating, we get

$$y''' = -(x-1)^{-1} y'' - y' - (x-1)^{-1} y,$$

$$y^{iv} = -(x-1)^{-1} y''' + [(x-1)^{-2} - 1]y'' - (x-1)^{-1} y' + (x-1)^{-2} y$$

$$y^v = -(x-1)^{-1} y^{iv} + [2(x-1)^{-2} - 1]y''' - [2(x-1)^{-3} + (x-1)^{-1}]y''$$
$$+ 2(x-1)^{-2} y' - 2(x-1)^{-3} y$$

At $x = 0$, we have

$$y'''(0) = -1, \quad y^{iv}(0) = 0 \text{ and } y^v(0) = 1.$$

Substituting these values in MacLaurin's expansion, we obtain

Series Solutions of Differential Equations

$$y = x - \frac{1}{3!}x^3 + \frac{1}{5!}x^5 + \cdots$$

The form of this series shows that $y = \sin x$ is the solution of the differential equation. This series converges for all finite values of x. □

Exercise 1.1

Obtain the first four non-zero terms of the power solution for each of the following initial value problems:

1. $y' = x^2 y^2 + 1$, when $x = 1, y = 1$.

    Ans: $y = 1 + 2(x-1) + 3(x-1)^2 + \frac{19}{3}(x-1)^3 + \cdots$

2. $y' = \sin(xy) + x^2$, when $x = 0, y = 3$.

    Ans: $y = 3 + \frac{3}{2}x^2 + \frac{1}{3}x^3 - \frac{3}{4}x^4 + \cdots$

3. $y'' = x^2 - y^2$, when $x = 0, y = 1$ and $y' = 0$.

    Ans: $y = 1 - \frac{1}{2}x^2 + \frac{1}{6}x^4 - \frac{7}{360}x^6 + \cdots$

4. $y''' = xy + yy'$, when $x = 0, y = 0, y' = 1$ and $y'' = 2$.

    Ans: $y = x + x^2 + \frac{1}{24}x^4 + \frac{1}{15}x^5 + \cdots$

5. $y' = x + e^y$, when $x = 0, y = a$.

    Ans: $y = a + e^a x + (1 + e^{2a})\frac{x^2}{2} + (2e^{3a} + e^a)\frac{x^3}{6} + \cdots$

6. $y'' + y = 0$, when $x = 0, y = A$ and $y' = B$.

    Ans: $y = A[1 - \frac{x^2}{2!} + \cdots] + B[x - \frac{x^3}{3!} + \cdots]$

7. $y'^3 + 3xy'^2 + x - y = 0$, when $x = 0, y = 1$.

    Ans: $y = 1 + x - \frac{1}{2}x^2 + \frac{5}{18}x^3 + \cdots$

8. $3y^2 y' = y^3 - x$, when $x = 0, y = 1$.

    Ans: $y = \frac{1}{81}(81 + 27x - 9x^2 + 5x^3 + \cdots)$

7

1.3. The Method of Undetermined Coefficients

We assume that a series solution of a differential equation takes the form

$$y = c_0 + c_1(x - x_0) + c_2(x - x_0)^2 + \cdots \quad (4)$$

The coefficients c_k, $k = 0, 1, 2, \cdots$ are to be determined so that y satisfies the differential equation. Moreover, the series must be convergent at and near $x = x_0$. In more compact notation the series in Equation (4) can be written as

$$y = \sum_{k=0}^{\infty} c_k (x - x_0)^k \quad (5)$$

The method of undetermined coefficients is based on two facts:

1. If the power series in (5) converges for $|x - x_0| < R$, then the series obtained by differentiating term by term also converges for $|x - x_0| < R$, and represents y'.

2. If the power series in (5) has a sum of zero for $|x - x_0| < R$, then each coefficients c_k, $k = 0, 1, 2, \cdots$ must be zero.

The method of undetermined coefficients is most effective for linear differential equations. We give here some illustrative examples.

***Example* 4**: Find a power series solution for the equation $y'' + xy' + y = 0$ about $x = 0$.

Solution: We start by assuming a solution of the form $y = \sum_{k=0}^{\infty} c_k x^k$.

Then differentiating with respect to x twice, we obtain

$$y' = \sum_{k=1}^{\infty} k c_k x^{k-1} = \sum_{k=0}^{\infty} (k+1) c_{k+1} x^k$$

$$y'' = \sum_{k=2}^{\infty} k(k-1) c_k x^{k-2} = \sum_{k=0}^{\infty} (k+1)(k+2) c_{k+2} x^k$$

Substituting in the differential equation, we obtain

Series Solutions of Differential Equations

$$\sum_{k=0}^{\infty}(k+1)(k+2)c_{k+2}x^k + x\sum_{k=0}^{\infty}(k+1)c_{k+1}x^k + \sum_{k=0}^{\infty}c_k x^k$$

Now, since the sum in the last equation must be identically zero then each coefficient of various powers of x must in turn be zero. In particular, equating the coefficients of x^k to zero, we obtain

$$(k+1)(k+2)c_{k+2} + (k+1)c_k = 0$$

or $c_{k+2} = -\dfrac{c_k}{k+2}, \quad k \geq 0$.

This is called the **recursion formula** or the **recurrence relation** for the coefficients c_k's. If c_0 and c_1 are kept arbitrary, then all the other c_k's will be in terms of these two arbitrary constants; then substituting for $k = 0, 1, 2, \cdots$ in the recursion formula, we obtain

$$c_2 = -\frac{c_0}{2} \qquad\qquad c_3 = -\frac{c_1}{3}$$

$$c_4 = \frac{c_0}{2\cdot 4} \qquad\qquad c_5 = \frac{c_1}{3\cdot 5}$$

$$c_6 = -\frac{c_0}{2\cdot 4\cdot 6} \qquad\qquad c_7 = -\frac{c_1}{3\cdot 5\cdot 7}$$

. . .

$$c_{2n} = \frac{(-1)^n c_0}{2\cdot 4\cdot 6\cdots(2n)} \qquad\qquad c_{2n+1} = \frac{(-1)^n c_1}{3\cdot 5\cdot 7\cdots(2n+1)}$$

And $n = 0, 1, 2, \cdots$. The constants c_0 and c_1 are in fact the initial values of y and y' at $x = 0$. The series solution can be written as

$$y = c_0\left[1 + \sum_{n=1}^{\infty}\frac{(-1)^n x^{2x}}{2\cdot 4\cdot 6\cdots(2n)}\right] + c_1\left[\sum_{n=1}^{\infty}\frac{(-1)^{n+1}x^{2n-1}}{3\cdot 5\cdot 7\cdots(2n-1)}\right]$$

Using the ratio test we can verify that the two series converge for all finite x. □

Special Functions and Orthogonal Polynomials

Example 5: Find a power series solution for the differential equation:
$$(x^2 - 1)y'' + x y' - y = 0 \quad \text{near} \quad x = 0.$$

Solution: Assume a solution of the form $y = \sum_{k=0}^{\infty} c_k x^k$.

Then differentiating with respect to x twice, we obtain

$$y' = \sum_{k=1}^{\infty} k c_k x^{k-1} = \sum_{k=0}^{\infty} (k+1)c_{k+1} x^k$$

$$y'' = \sum_{k=2}^{\infty} k(k-1)c_k x^{k-2} = \sum_{k=0}^{\infty} (k+1)(k+2)c_{k+2} x^k$$

Substituting in the differential equation, we obtain

$$(x^2 - 1)\sum_{k=0}^{\infty} (k+1)(k+2)c_{k+2}x^k + x \sum_{k=0}^{\infty} (k+1)c_{k+1}x^k$$
$$- \sum_{k=0}^{\infty} c_k x^k = 0$$

Equating the coefficients of the general power of x (x^k) to zero, we obtain

$$k(k-1)c_k - (k+1)(k+2)c_{k+2} + k c_k - c_k = 0.$$

Or $c_{k+2} = \dfrac{k-1}{k+2} c_k$.

This is the **recurrence relation** for the coefficients c_k's. If c_0 and c_1 are kept arbitrary, then all the other c_k's will be in terms of these two arbitrary constants; then substituting for $k = 0, 1, 2, \cdots$ in the recurrence relation, we obtain

$k = 0: \quad c_2 = -\dfrac{1}{2}c_0 \qquad\qquad k = 1: \quad c_3 = 0$

$k = 2: \quad c_4 = \dfrac{1}{4}c_2 = -\dfrac{1}{2 \cdot 4}c_0 \qquad k = 3: \quad c_5 = 0$

Series Solutions of Differential Equations

$$k = 4: \quad c_6 = \frac{3}{6}c_4 = -\frac{3 \cdot 1}{2 \cdot 4 \cdot 6}c_0 \quad c_{2n+1} = 0, \quad n \geq 1$$

$$c_{2n} = -\frac{(2n-3)(2n-1)\cdots 3 \cdot 1}{2 \cdot 4 \cdot 6 \cdots 2n}c_0, \quad n \geq 2$$

The general solution is

$$y = A\left\{1 + \frac{1}{2}x^2 + \frac{1}{2 \cdot 4}x^4 + \frac{3 \cdot 1}{2 \cdot 4 \cdot 6}x^6 + \cdots\right\} + Bx, \text{ or}$$

$$y = A\left\{1 + \frac{1}{2}x^2 + \sum_{n=2}^{\infty} \frac{1 \cdot 3 \cdot 5 \cdots (2n-3)}{2 \cdot 4 \cdot 6 \cdots 2n}x^{2n}\right\} + Bx. \quad \square$$

Example 6: Find a power series solution for the differential equation:

$$(x^2 - 1)y'' + 4x\,y' + 2y = 0 \text{ near } x = 0.$$

Solution: Assume a solution of the form $y = \sum_{k=0}^{\infty} c_k x^k$.

Then differentiating with respect to x twice, we obtain

$$y' = \sum_{k=1}^{\infty} k\,c_k\,x^{k-1} = \sum_{k=0}^{\infty} (k+1)c_{k+1}\,x^k$$

$$y'' = \sum_{k=2}^{\infty} k(k-1)c_k\,x^{k-2} = \sum_{k=0}^{\infty} (k+1)(k+2)c_{k+2}\,x^k$$

Substituting in the differential equation, we obtain

$$(x^2 - 1)\sum_{k=0}^{\infty}(k+1)(k+2)c_{k+2}x^k + 4x\sum_{k=0}^{\infty}(k+1)c_{k+1}x^k$$

$$+ 2\sum_{k=0}^{\infty} c_k x^k = 0$$

Equating the coefficients of the general power of x (x^k) to zero, we obtain

$$k(k-1)c_k - (k+1)(k+2)c_{k+2} + 4k\,c_k + 2c_k = 0.$$

Or $c_{k+2} = c_k$.

This is the **recurrence relation** for the coefficients c_k's. If c_0 and c_1 are kept arbitrary, then all the other c_k's will be in terms

11

Special Functions and Orthogonal Polynomials

of these two arbitrary constants; then substituting for $k = 0, 1, 2, \cdots$ in the recurrence relation, we obtain

$k = 0: \quad c_2 = c_0$ $k = 1: \quad c_3 = c_1$

$k = 2: \quad c_4 = c_2 = c_0$ $k = 3: \quad c_5 = c_3 = c_1$

$c_{2n} = c_0, \quad n \geq 1$ $c_{2n+1} = c_1, \quad n \geq 1$

The general solution is

$$y = A\left\{1 + x^2 + x^4 + x^6 + \cdots + x^{2n} + \cdots\right\}$$
$$+ B\left\{x + x^3 + x^5 + x^7 + \cdots + x^{2n+1} + \cdots\right\},$$

or

$$y = A\left\{\sum_{n=0}^{\infty} x^{2n}\right\} + Bx\left\{\sum_{n=0}^{\infty} x^{2n}\right\} = \frac{A + Bx}{1 - x^2}. \qquad \square$$

Note: In the previous examples, the power series solution has been obtained rather in a straightforward manner. This is because the differential equation is "well behaved". We now give some understanding on the behavior of the differential equation.

Definition: A function $g(x)$ is said to be <u>analytic</u> at x_0 if $g(x)$ has a Taylor series expansion which converges to $g(x)$ in some interval about x_0.

For example, $\ln(1+x)$ is analytic at $x = 0$. In fact, it has a Taylor series expansion about $x = 0$ given by

$$\ln(1+x) = x - \frac{x^2}{2} + \frac{x^3}{3} - \frac{x^4}{4} + \cdots$$

and the interval of convergence is $|x| < 1$.

Theorem: If $P(x)$, $Q(x)$ and $f(x)$ are analytic at x_0, then every solution of the linear second order differential equation

$$y'' + P(x)y' + Q(x)y = f(x)$$

is also analytic at x_0, and hence is represented by a power series of the form $\sum_{k=0}^{\infty} c_k (x - x_0)^k$. The interval of convergence of the power series solution is at least as large as the smallest of the interval of convergence of $P(x)$, $Q(x)$ and $f(x)$ about x_0.

12

Series Solutions of Differential Equations

Looking back at the previous examples, we can see that the coefficients of y and its first and second derivatives are all analytic for all finite values of x. Then, we would expect that the two power series in the solution of the equation will converge for all finite values of x. Moreover, the two power series solutions are linearly independent.

Example 7: Find a power series solution for the following differential equation about $x = 0$: $y'' + xy' - y = e^{2x}$.

Solution: The coefficient functions x and -1 and $f(x) = e^{2x}$ are all analytic everywhere. Then, we seek a solution of the form $y = \sum_{k=0}^{\infty} c_k x^k$.

Differentiating with respect to x twice, we obtain

$$y' = \sum_{k=1}^{\infty} k c_k x^{k-1} = \sum_{k=0}^{\infty} (k+1) c_{k+1} x^k$$

$$y'' = \sum_{k=2}^{\infty} k(k-1) c_k x^{k-2} = \sum_{k=0}^{\infty} (k+1)(k+2) c_{k+2} x^k$$

The function e^{2x} has a power series expansion about $x = 0$ given by

$$e^{2x} = \sum_{k=0}^{\infty} \frac{2^k}{k!} x^k .$$

Substituting in the differential equation, we obtain

$$2c_2 - c_0 + \sum_{k=1}^{\infty} [(k+1)(k+2) c_{k+2} + k c_k - c_k] x^k = \sum_{k=0}^{\infty} \frac{2^k}{k!} x^k$$

If we write few terms of the series on both sides of the equation, we get

$$(2c_2 - c_0) + 6c_3 x + (12 c_4 + c_2) x^2 + (20 c_5 + 2 c_3) x^3 + (30 c_6 + 3 c_4) x^4$$
$$+ (42 c_7 + 4 c_5) x^5 = 1 + 2x + 2x^2 + \tfrac{4}{3} x^3 + \tfrac{2}{3} x^4 + \tfrac{5}{15} x^5 + \cdots$$

Equating the coefficients of corresponding powers of x in both sides, we obtain

13

Special Functions and Orthogonal Polynomials

$$2c_2 - c_0 = 1, \qquad 6c_3 = 2, \qquad 12c_4 + c_2 = 2,$$
$$20c_5 + 2c_3 = \tfrac{4}{3}, \quad 30c_6 + 3c_4 = \tfrac{2}{3}, \quad 42c_7 + 4c_5 = \tfrac{4}{15}$$

Hence, the coefficients in terms of c_0 and c_1 are given by

$$c_2 = \tfrac{1}{2}(1+c_0), \qquad c_3 = \tfrac{1}{3}, \quad c_4 = \tfrac{1}{24}(3-c_0),$$
$$c_5 = \tfrac{1}{30}, \quad c_6 = \tfrac{1}{240}(\tfrac{7}{3}+c_0), \qquad c_7 = \tfrac{1}{315}$$

Then the power series solution is

$$y = c_0 + c_1 x + \tfrac{1}{2}(1+c_0)x^2 + \tfrac{1}{3}x^3 + \tfrac{1}{24}(3-c_0)x^4 + \tfrac{1}{30}x^5$$
$$+ \tfrac{1}{240}(\tfrac{7}{3}+c_0)x^6 + \tfrac{1}{315}x^7 + \cdots, \quad \text{or}$$

$$y = c_0\left(1 + \tfrac{1}{2}x^2 - \tfrac{1}{24}x^4 + \tfrac{1}{240}x^6 + \cdots\right) + c_1 x$$
$$+ \left(\tfrac{1}{2}x^2 + \tfrac{1}{3}x^3 + \tfrac{3}{24}x^4 + \tfrac{1}{30}x^5 + \tfrac{7}{720}x^6 + \tfrac{1}{315}x^7 + \cdots\right)$$

where c_0 and c_1 are now the two arbitrary constants. The complementary function is displayed in the first line of solution, whereas the particular integral is shown in the second line. Moreover, since the MacLaurin's series expansion for each coefficient functions is convergent for all finite values of x, then the series solution obtained is also convergent for all finite values of x. □

Example 8: Find the power series solution for the initial value problem

$$y'' + e^x y = 1, \text{ when } x = 0, y = 0 \text{ and } y' = 1.$$

Solution: The coefficient functions e^x and 1 are analytic at $x = 0$, then we assume a series solution of the form

$$y = \sum_{k=0}^{\infty} c_k x^k.$$

Differentiating with respect to x twice, we obtain

$$y' = \sum_{k=1}^{\infty} k c_k x^{k-1} = \sum_{k=0}^{\infty} (k+1) c_{k+1} x^k$$

Series Solutions of Differential Equations

$$y'' = \sum_{k=2}^{\infty} k(k-1)c_k x^{k-2} = \sum_{k=0}^{\infty} (k+1)(k+2)c_{k+2} x^k$$

The function e^{2x} has a power series expansion about $x=0$ given by

$$e^x = \sum_{k=0}^{\infty} \frac{x^k}{k!}.$$

Substituting these values into the differential equation, we get

$$\sum_{k=0}^{\infty} (k+1)(k+2)c_{k+2} x^k + \left(\sum_{k=0}^{\infty} \frac{x^k}{k!}\right)\left(\sum_{k=0}^{\infty} c_k x^k\right) = 1.$$

Writing few terms for each summation, we obtain

$$2c_2 + 6c_3 x + 12c_4 x^2 + 20c_5 x^3 + \cdots$$
$$+ \left(1 + x + \frac{x^2}{2} + \frac{x^3}{6} + \cdots\right)\left(c_0 + c_1 x + c_2 x^2 + c_3 x^3 + \cdots\right) = 1, \text{ or}$$

$$2c_2 + 6c_3 x + 12c_4 x^2 + 20c_5 x^3 + \cdots + c_0 + (c_0+c_1)x$$
$$+ \left(\tfrac{1}{2}c_0 + c_1 + c_2\right)x^2 + \left(\tfrac{1}{6}c_0 + \tfrac{1}{2}c_1 + c_2 + c_3\right)x^3 + \cdots = 1$$

Equating the coefficients of the corresponding powers of x in both sides, we get

$$2c_2 + c_0 = 1, \quad 12c_4 + \tfrac{1}{2}c_0 + c_1 + c_2 = 0$$

$$6c_3 + c_0 + c_1 = 0, \quad 20c_5 + \tfrac{1}{6}c_0 + \tfrac{1}{2}c_1 + c_2 + c_3 = 0$$

$$c_2 = \tfrac{1}{2}(1-c_0), \quad c_4 = -\tfrac{1}{24}(1+2c_1), \text{ thus}$$

$$c_3 = -\tfrac{1}{6}(c_0+c_1), \quad c_5 = -\tfrac{1}{120}(3c_0+2c_1+3).$$

The power series solution is

$$y = c_0 + c_1 x + \tfrac{1}{2}(1-c_0)x^2 - \tfrac{1}{6}(c_0+c_1)x^3 - \tfrac{1}{24}(1+2c_1)x^4$$
$$- \tfrac{1}{120}(3c_0+2c_1+3)x^5 + \cdots$$

For the initial conditions at $x=0$ we have $c_0 = 0$ and $c_1 = 1$, then the solution of the initial value problem is

$$y = x + \tfrac{1}{2}x^2 - \tfrac{1}{6}x^3 - \tfrac{1}{8}x^4 - \tfrac{1}{24}x^5 + \cdots. \qquad \square$$

Special Functions and Orthogonal Polynomials

Example 9: Find a power series solution for the differential equation:
$$y'' + x^2 y = 0 \text{ about } x = 0.$$

Solution: Assume a solution of the form $y = \sum_{k=0}^{\infty} c_k x^k$.

Then differentiating with respect to x twice, we obtain

$$y' = \sum_{k=1}^{\infty} k c_k x^{k-1} = \sum_{k=0}^{\infty} (k+1) c_{k+1} x^k$$

$$y'' = \sum_{k=2}^{\infty} k(k-1) c_k x^{k-2} = \sum_{k=0}^{\infty} (k+1)(k+2) c_{k+2} x^k$$

Substituting in the differential equation, we obtain

$$\sum_{k=0}^{\infty} (k+1)(k+2) c_{k+2} x^k + x^2 \sum_{k=0}^{\infty} c_k x^k = 0.$$

Equating the coefficients of the general power of x (x^k) to zero, we obtain

$$(k+1)(k+1) c_{k+2} + c_{k-2} = 0, \text{ or } c_{k+2} = -\frac{c_{k-2}}{(k+1)(k+2)}.$$

For $k = 0, 1, 2, \cdots$ in this relation, we obtain

$k = 0$: $c_2 = 0$ $k = 1$: $c_3 = 0$

$k = 2$: $c_4 = -\dfrac{c_0}{3 \cdot 4}$ $k = 3$: $c_5 = -\dfrac{c_1}{4 \cdot 5}$

$k = 4$: $c_6 = -\dfrac{c_2}{5 \cdot 6} = 0$ $k = 5$: $c_7 = -\dfrac{c_3}{6 \cdot 7} = 0$

$k = 6$: $c_8 = -\dfrac{c_4}{7 \cdot 8} = \dfrac{c_0}{3 \cdot 4 \cdot 7 \cdot 8}$ $k = 7$: $c_9 = -\dfrac{c_5}{8 \cdot 9} = \dfrac{c_1}{4 \cdot 5 \cdot 8 \cdot 9}$

The general solution is

$$y = a \left\{ 1 - \frac{1}{3 \cdot 4} x^4 + \frac{1}{3 \cdot 4 \cdot 7 \cdot 8} x^8 - \cdots \right\} + b \left\{ x - \frac{1}{4 \cdot 5} x^5 + \frac{1}{4 \cdot 5 \cdot 8 \cdot 9} x^9 - \cdots \right\} \square$$

Note: In all the previous examples, we obtained series solutions about $x = 0$. If it is desired to obtain a series solution about any other point, the following example illustrate will this situation.

16

Series Solutions of Differential Equations

Example 10: Find a power series solution for the differential equation:
$$y'' + (x-1)^2 y' - 4(x-1)y = 0 \text{ about } x = 1.$$

Solution: First, we let $x = t + 1$, the differential Equation becomes
$$\frac{d^2 y}{dt^2} + t^2 \frac{dy}{dt} - 4ty = 0.$$

Assume a solution of the form $y = \sum_{k=0}^{\infty} c_k t^k$.

Then differentiating with respect to t twice, we obtain
$$y' = \sum_{k=1}^{\infty} k c_k t^{k-1} = \sum_{k=0}^{\infty} (k+1) c_{k+1} t^k$$

$$y'' = \sum_{k=2}^{\infty} k(k-1) c_k t^{k-2} = \sum_{k=0}^{\infty} (k+1)(k+2) c_{k+2} t^k$$

Substituting in the differential equation, we obtain
$$\sum_{k=0}^{\infty} (k+1)(k+2) c_{k+2} t^k + t^2 \sum_{k=0}^{\infty} (k+1) c_{k+1} t^k - 4t \sum_{k=0}^{\infty} c_k t^k = 0$$

Equating the coefficients of the general power of t (t^k) to zero, we obtain
$$(k+1)(k+1) c_{k+2} + (k-1) c_{k-1} - 4 c_{k-1} = 0,$$

or $c_{k+2} = -\dfrac{(k-5) c_{k-1}}{(k+1)(k+2)}$, or $c_{k+3} = -\dfrac{(k-4) c_k}{(k+2)(k+3)}$.

For $k = 0, 1, 2, \cdots$, we have

$k = 0$: $c_3 = \dfrac{4}{2 \cdot 3} c_0 = \dfrac{2}{3} c_0$ $k = 1$: $c_4 = \dfrac{3}{3 \cdot 4} c_1 = \dfrac{1}{4} c_1$

$k = 2$: $c_5 = \dfrac{2}{4 \cdot 5} c_2 = 0$ $k = 3$: $c_6 = \dfrac{1}{5 \cdot 6} c_3 = \dfrac{1}{45} c_0$

$k = 4$: $c_7 = 0$ $k = 5$: $c_8 = -\dfrac{1}{7 \cdot 8} c_5 = 0$

The general solution is
$$y = c_0 + c_1 t + \frac{2}{3} c_0 t^3 + \frac{1}{4} c_1 t^4 + \frac{1}{45} c_0 t^6 - \frac{1}{1620} c_0 t^9 + \cdots, \text{ or}$$

Special Functions and Orthogonal Polynomials

$$y = a\left\{1 + \frac{2}{3}t^3 + \frac{1}{45}t^6 - \frac{1}{1620}t^9 + \cdots\right\} + b\left\{t + \frac{1}{4}t^4\right\}.$$

Back-substituting, we obtain

$$y = a\left\{1 + \frac{2}{3}(x-1)^3 + \frac{1}{45}(x-1)^6 - \frac{1}{1620}(x-1)^9 + \cdots\right\}$$

$$+ b\left\{(x-1) + \frac{1}{4}(x-1)^4\right\} \quad \square$$

Exercise 1.2

a. Find the recurrence relation for the coefficients of the power series solution for the following differential equations about $x = 0$.

1. $y'' + xy = 0$
   Ans: $c_{k+2} = -\dfrac{c_k}{(k+1)(k+2)}$

2. $y'' - 2y' + xy = 0$
   Ans: $c_{k+2} = \dfrac{2c_{k+1} - c_{k-1}}{(k+1)(k+2)}$, $k > 0$

3. $y'' - x^3 y = 0$
   Ans: $c_{k+2} = \dfrac{c_{k-3}}{(k+1)(k+2)}$, $k > 2$

4. $y'' + (1-x)y' + 2xy = 0$ Ans: $c_{k+2} = -\dfrac{(k+1)c_{k+1} - kc_k - 2c_{k-1}}{(k+1)(k+2)}$, $k > 0$

5. $y'' + xy' + 2xy = 0$
   Ans: $c_{k+2} = -\dfrac{kc_k + 2c_{k-1}}{(k+1)(k+2)}$, $k > 0$

6. $y'' + y' - x^2 y = 0$
   Ans: $c_{k+2} = -\dfrac{(k+1)c_{k+1} - c_{k-2}}{(k+1)(k+2)}$, $k > 1$

7. $y'' - 8xy' = 1 + 2x$
   Ans: $c_{k+2} = \dfrac{8c_{k-1}}{(k+1)(k+2)}$, $k > 1$

b. Find the first four non-zero terms in the power series solution of the following differential equations about $x = 0$.

1. $2y'' - 4xy' + 8x^2 y = 0$ Ans: $y = c_0 + c_1 x + \frac{1}{3}c_1 x^3 - \frac{1}{3}c_0 x^4 + \cdots$

2. $y'' + 12y' + x^2 y = 0$ Ans: $y = c_0 + c_1 x - 6c_1 x^2 + 24c_1 x^3 + \cdots$

3. $y'' + 2\cos x \cdot y' = x$ Ans: $y = c_0 + c_1 x - c_1 x^2 + \frac{1}{6}(1 + 4c_1)x^3 + \cdots$

4. $y'' - 2\tan x \cdot y' + y = 0$ Ans: $y = c_0 + c_1 x - \frac{1}{2}c_0 x^2 + \frac{1}{6}c_1 x^3 + \cdots$

5. $y'' - e^{-3x} y = 2x^2$ Ans: $y = c_0 + c_1 x + \frac{1}{2}c_0 x^2 + \frac{1}{6}(c_1 - 3c_0)x^3 + \cdots$

c. Find the first five non-zero terms of the solution of the initial value problems.

Series Solutions of Differential Equations

1. $y'' + y' - xy = 0$, when $x = 0, y = -2$ and $y' = 0$

   Ans: $y = -2 - \frac{1}{3}x^3 + \frac{1}{12}x^4 - \frac{1}{60}x^5 - \frac{1}{120}x^6 + \cdots$

2. $y'' + 2xy' + (x-1)y = 0$, when $x = 0, y = 1$ and $y' = 2$

   Ans: $y = 1 + 2x + \frac{1}{2}x^2 - \frac{1}{2}x^3 - \frac{7}{24}x^4 + \cdots$

3. $y'' - xy = 2x$, when $x = 1, y = 3$ and $y' = 0$

   Ans: $y = 3 + \frac{5}{2}(x-1)^2 + \frac{5}{6}(x-1)^3 + \frac{5}{24}(x-1)^4 + \frac{1}{6}(x-1)^5 + \cdots$

4. $y'' + x^2 y = e^x$, when $x = 0, y = -2$ and $y' = 7$

   Ans: $y = -2 + 7x + \frac{1}{2}x^2 + \frac{1}{6}x^3 + \frac{1}{24}x^4 + \cdots$

d. Solve in series the following Differential Equations near $x = 0$:

1. $(1-x^2)y'' + 2x y' - y = 0$

   Ans: $y = a\left\{1 + \frac{1}{2}x^2 - \frac{1}{24}x^4 - \cdots\right\} + b\left\{x - \frac{1}{6}x^3 - \frac{1}{120}x^5 - \cdots\right\}$

2. $(x^2 + 2)y'' + x y' - (1+x)y = 0$

   Ans: $y = a\left\{1 + \frac{1}{4}x^2 + \frac{1}{12}x^3 - \frac{3}{96}x^4 - \cdots\right\} + b\left\{x + \frac{1}{24}x^4 - \cdots\right\}$

3. $(1+x^2)y'' + x y' - y = 0$

   Ans: $y = a\left\{1 + \frac{1}{2}x^2 - \frac{1}{8}x^4 + \frac{1}{15}x^6 - \cdots\right\} + bx$

4. $(1-x^2)y'' - x y' + 4y = 0$

   Ans: $y = a\left\{1 - 2x^2\right\} + b\left\{x - \frac{1}{2}x^3 - \frac{1}{8}x^5 + \frac{1}{16}x^7 + \cdots\right\}$

5. $(2+x^2)y'' + x y' - xy = 1$

   Ans: $y = a\left\{1 + \frac{1}{4}x^2 + \frac{1}{12}x^3 - \frac{1}{32}x^4 + \cdots\right\} + b\left\{x + \frac{1}{24}x^4 + \cdots\right\}$

6. $(1-x^2)y'' + 2x y' + y = 0$

   Ans: $y = a\left\{1 - \frac{1}{2}x^2 + \frac{1}{8}x^4 + \cdots\right\} + b\left\{x - \frac{1}{2}x^3 + \frac{1}{140}x^5 + \cdots\right\}$

7. $y'' + x y' + x^2 y = 0$

   Ans: $y = a\left\{1 - \frac{1}{12}x^4 + \frac{1}{90}x^6 - \cdots\right\} + b\left\{x - \frac{1}{6}x^3 - \frac{1}{40}x^5 - \cdots\right\}$

8. $(1+x^2)y'' + x y' - xy = 0$

   Ans: $y = a\left\{1 + \frac{1}{6}x^3 - \frac{3}{40}x^5 + \cdots\right\} + b\left\{x - \frac{1}{6}x^3 + \frac{1}{12}x^4 + \frac{3}{45}x^5 + \cdots\right\}$

1.4. The Method of Frobenius

In the previous discussion, we dealt with differential equations which are "*well behaved*" at and near x_0. We found that, for linear differential equations, if the coefficient functions are analytic at x_0, then we have no problem finding a power series solution about x_0 by the method of undetermined coefficients or by the use of Taylor's series. Now, what about if one or more of the coefficient functions is not analytic at x_0? Look at this example:

Breakdown example: Find a power series solution for the following differential equation about $x = 0$: $4x^2 y'' + y = 0$.

Solution: Assume as before a series solution of the form $y = \sum_{k=0}^{\infty} c_k x^k$.

Differentiating with respect to x twice, we obtain

$$y' = \sum_{k=1}^{\infty} k c_k x^{k-1} = \sum_{k=0}^{\infty} (k+1) c_{k+1} x^k$$

$$y'' = \sum_{k=2}^{\infty} k(k-1) c_k x^{k-2} = \sum_{k=0}^{\infty} (k+1)(k+2) c_{k+2} x^k$$

Substituting in the differential equation, we obtain

$$4 \sum_{k=0}^{\infty} k(k-1) c_k x^k + \sum_{k=0}^{\infty} c_k x^k = 0$$

Then equating the coefficient of x^k to zero, we get

$[4k(k-1)+1] c_k = 0$ or $(2k-1)^2 c_k = 0$.

Hence, $c_k = 0$, $k = 0, 1, 2, \cdots$. Thus, we obtain only the trivial solution $y(x) = 0$!!. Were is the power series solution? Something is wrong. In fact, our assumption does not give us the required solution.

In this section, we treat this situation in some details. First, we state some definitions that might help.

Consider the linear second order homogeneous differential equation

$$y'' + P(x) y' + Q(x) y = 0 \tag{6}$$

Definition 1: *The point x_0 is an ordinary point of the differential Equation (6) if $P(x)$ and $Q(x)$ are analytic at x_0.*

Series Solutions of Differential Equations

With this definition, if x_0 is an ordinary point of the differential Equation (6), then we can use either the Taylor's series method or the method of undetermined coefficients to find a power series solution about x_0.

Definition 2: If x_0 is not an ordinary point of the differential equation (6), then it is a singular point, look back at the previous example.

Definition 3: The point x_0 is a regular singular point of the differential equation (6) if it is not an ordinary point and if $(x-x_0)P(x)$ and $(x-x_0)^2 Q(x)$ are analytic at x_0. Regular singular points are simply called regular points.

Definition 4: If x_0 is neither ordinary nor regular singular point, it is called an irregular singular point.

Example 11: Find all of the singular points of the following two differential equations and classify each of them as either regular or irregular.

i) $x^2(x-3)^2 y'' + 4x(x^2 - x - 6)y' + (x^2 - x - 2)y = 0$

ii) $x^{5/2}(x-2)y'' - x^{5/2} y' + (x-2)y = 0$.

Solution: i) We have $P(x) = \dfrac{4(x+2)}{x(x-3)}$, $Q(x) = \dfrac{(x-2)(x+1)}{x^2(x-3)^2}$.

Then $x = 0$ and $x = 3$ are singular points for the equation.

For $x = 0$, we have $xP(x) = \dfrac{4(x+2)}{x-3}$ and

$x^2 Q(x) = \dfrac{(x-2)(x+1)}{(x-3)^2}$, both are analytic at $x = 0$. Then

$x = 0$ is a regular singular point.

For $x = 3$, we have

$(x-3)P(x) = \dfrac{4(x+2)}{x}$ and $(x-3)^2 Q(x) = \dfrac{(x-2)(x+1)}{x^2}$,

both are analytic at $x = 3$. Then $x = 3$ is a regular singular point.

ii) We have $P(x) = -\dfrac{1}{x-2}$, $Q(x) = \dfrac{x-2}{x^{5/2}}$, Then $x = 0$ and

$x = 2$ are singular points for the differential equation.

For $x = 0$, we have $xP(x) = -\dfrac{x}{x-2}$ and $x^2 Q(x) = \dfrac{x-2}{x^{1/2}}$,

21

the second function is not analytic at $x = 0$. Then $x = 0$ is an irregular singular point.

For $x = 2$, we have

$$(x-2)P(x) = -1 \text{ and } (x-2)^2 Q(x) = \frac{(x-2)^3}{x^{5/2}}, \quad \text{both are}$$

analytic at $x = 2$. Then $x = 2$ is a regular singular point. □

Solutions about irregular singular points are beyond the scope of this book, so they will not be discussed here. On the other hand, the method of Frobenius deals with solutions of differential equations about regular singular points.

Theorem: *If x_0 is a regular point of the differential equation,*

$$y'' + P(x)y' + Q(x)y = 0, \text{ then there is at least one solution of}$$

the form $\sum_{k=0}^{\infty} c_k (x-x_0)^{k+\alpha}$. *The interval of convergence of the series solution is not less than the distance from x_0 to the nearest of the other singular points of the equation.*

The index α need not be an integer. If we substitute y and its derivatives into the differential equation as we did for the method of undetermined coefficients, we hope to determine the coefficients c's as well as the index α. It is not straightforward as we might expect. The values of α can be either a positive or negative integer, zero or a fraction. Different values of the index α, as we will see, dictate some precautions to be taken.

For the sake of simplification, whenever we wish to obtain solutions about a point other than $x = 0$, we first translate the origin to that point and then proceed with the solution procedure. Hence, we concentrate our efforts on solutions valid about $x = 0$.

If $x = 0$ is a regular singular point of the differential equation

$$y'' + P(x)y' + Q(x)y = 0 \tag{7}$$

then, we assume a Frobenius series solution of the form

$$y = \sum_{k=0}^{\infty} c_k (x-x_0)^{k+\alpha}, \quad c_0 \neq 0 \tag{8}$$

The index α is obtained, as we shall see, from a quadratic equation in α, called the ***indicial equation***. In general, for each of the two roots of the quadratic equation α_1 and α_2 corresponds a series solution of the differential equation. In some cases, only one series solution is obtained using the Frobenius assumption, and we must find another way to obtain the second solution. Moreover, even is the Frobenius assumption gives us two solutions, these two solutions <u>may not always be linearly independent</u>!

Using the method of Frobenius, the following cases, depending on the values of the index α, will be studied.

Case I: **The roots of the indicial equation are distinct and the difference between them is not an integer.**

***Example* 12**: Find a series solution for the following differential equation about the origin: $2xy'' + (1+x)y' - 2y = 0$.

Solution: The point $x = 0$ is clearly a regular singular point of the differential equation. Assume a solution of the form

$$y = \sum_{k=0}^{\infty} c_k x^{k+\alpha}, \quad c_0 \neq 0, \text{ then } y' = \sum_{k=0}^{\infty} (k+\alpha) c_k x^{k+\alpha-1}, \text{ and}$$

$$y'' = \sum_{k=0}^{\infty} (k+\alpha)(k+\alpha-1) c_k x^{k+\alpha-2}$$

Substituting in the differential equation, we obtain

$$2x \sum_{k=0}^{\infty} (k+\alpha)(k+\alpha-1) c_k x^{k+\alpha-2} + (1+x) \sum_{k=0}^{\infty} (k+\alpha) c_k x^{k+\alpha-1} - 2 \sum_{k=0}^{\infty} c_k x^{k+\alpha} = 0$$

Now, to get the indicial equation, <u>we equate the coefficient of the lowest power of x, ($x^{\alpha-1}$), to zero</u>, we obtain $\alpha(2\alpha - 1) c_0 = 0$.

Since $c_0 \neq 0$, the ***indicial equation*** is $\boxed{\alpha(2\alpha - 1) = 0}$. The roots are $\alpha = 0$ or $\frac{1}{2}$. The difference between the two roots is <u>not an integer</u>. Then, for each value of α corresponds a solution.

Equating the coefficients of $x^{k+\alpha}$ to zero, we obtain

23

Special Functions and Orthogonal Polynomials

$(k+\alpha+1)(2k+2\alpha+1)c_{k+1} + (k+\alpha-2)c_k = 0$ or

$$c_{k+1} = -\frac{(k+\alpha-2)}{(k+\alpha+1)(2k+2\alpha+1)}c_k, \quad k = 0, 1, 2, \cdots$$

This is the **recurrence relation** for the coefficients.

For $\alpha = 0$: The recurrence relation becomes

$$c_{k+1} = -\frac{(k-2)}{(k+1)(2k+1)}c_k, \quad k = 0, 1, 2, \cdots$$

For various values of k, we have

$k = 0$: $c_1 = 2c_0$

$k = 1$: $c_2 = \frac{1}{2 \cdot 3}c_1 = \frac{1}{3}c_0$

$k = 2$: $c_3 = 0$

Then for $k \geq 2$: $c_k = 0$.

The solution corresponding to the root $\alpha = 0$ of the indicial equation is $y_1 = c_0(1 + 2x + \frac{1}{3}x^2)$, where c_0 is an arbitrary constant.

For $\alpha = 1/2$: The recurrence relation becomes

$$c_{k+1} = -\frac{(2k-3)}{2(2k+3)(k+1)}c_k, \quad k = 0, 1, 2, \cdots$$

For different values of k, we have

$k = 0$: $c_1 = -\frac{(-3)}{2 \cdot 3}c_0$

$k = 1$: $c_2 = -\frac{(-1)}{4 \cdot 5}c_1 = \frac{(-3)(-1)}{2 \cdot 3 \cdot 4 \cdot 5}c_0$

$k = 2$: $c_3 = -\frac{(1)}{6 \cdot}c_2 = -\frac{(-3)(-1)(1)}{2 \cdot 3 \cdot 4 \cdot 5 \cdot 6 \cdot 7}c_0$

and in general for $k = n-1$, we have

$$c_n = (-1)^n \frac{(-3)(-1)(1)\cdots(2n-5)}{(2 \cdot 4 \cdot 6 \cdots (2n))[3 \cdot 5 \cdot 7 \cdots (2n+1)]}c_0, \quad n < 0,$$

Series Solutions of Differential Equations

and the solution corresponding to the root $\alpha = \frac{1}{2}$ of the indicial equation is

$$y_2 = c_0 \sqrt{x} \sum_{n=0}^{\infty} \frac{(-1)^n 3x^n}{2^n n!(2n-3)(2n-1)(2n+1)}.$$

Finally, the general solution is

$$y = A\left(1 + 2x + \frac{1}{3}x^2\right) + B\sqrt{x} \sum_{n=0}^{\infty} \frac{(-1)^n 3x^n}{2^n n!(2n-3)(2n-1)(2n+1)}$$

We can use the ratio test to verify that the series appearing in the solution converges for all finite values of x. □

Example 13: Find a series solution for the following differential equation about the origin: $2x^2 y'' + x y' - (x+1)y = 0$.

Solution: Here $P(x) = \frac{1}{2x}$ and $Q(x) = -\frac{x+1}{2x^2}$. Both $x P(x)$ and $x^2 Q(x)$ are analytic at the origin, then the point $x = 0$ is a regular singular point of the differential equation. We assume a solution of the form

$$y = \sum_{k=0}^{\infty} c_k x^{k+\alpha}, \quad c_0 \neq 0, \text{ then } y' = \sum_{k=0}^{\infty} (k+\alpha) c_k x^{k+\alpha-1}, \text{ and}$$

$$y'' = \sum_{k=0}^{\infty} (k+\alpha)(k+\alpha-1) c_k x^{k+\alpha-2}$$

Substituting in the differential equation, we obtain

$$2x^2 \sum_{k=0}^{\infty} (k+\alpha)(k+\alpha-1) c_k x^{k+\alpha-2} + x \sum_{k=0}^{\infty} (k+\alpha) c_k x^{k+\alpha-1}$$

$$-(x+1) \sum_{k=0}^{\infty} c_k x^{k+\alpha} = 0$$

We equate the coefficient of the lowest power of x, (x^{α}), to zero, to obtain $2\alpha(\alpha-1)c_0 + \alpha c_0 - c_0 = 0$. Since $c_0 \neq 0$, the **indicial**

25

equation is $2\alpha^2 - \alpha - 1 = 0$. The roots are $\alpha = 1, -\frac{1}{2}$. The difference between the two roots is <u>not an integer</u>. Then, for each value of α corresponds a solution.

Equating the coefficients of $x^{k+\alpha}$ to zero, we obtain

$$2(k+\alpha)(k+\alpha-1)c_k + (k+\alpha)c_k - c_{k-1} - c_k = 0 \text{ or}$$

$$\boxed{c_k = \frac{c_{k-1}}{(k+\alpha-1)(2k+2\alpha+1)}}.$$

This is the **recurrence relation** for the coefficients.

For $\alpha = 1$ The recurrence relation becomes $c_k = \frac{c_{k-1}}{k(2k+3)}$. Then for various values of k, we have

$k = 1: \quad c_1 = \frac{c_0}{1 \cdot 5} \qquad\qquad k = 2: \quad c_2 = \frac{c_1}{2 \cdot 7} = \frac{c_0}{1 \cdot 2 \cdot 5 \cdot 7}$

$k = 3: \quad c_3 = \frac{c_2}{3 \cdot 9} = \frac{c_0}{1 \cdot 2 \cdot 3 \cdot 5 \cdot 7 \cdot 9} \quad \cdots$

The first solution is

$$y_1 = ax\left\{1 + \frac{x}{1 \cdot 5} + \frac{x^2}{1 \cdot 2 \cdot 5 \cdot 7} + \frac{x^3}{1 \cdot 2 \cdot 3 \cdot 5 \cdot 7 \cdot 9} + \cdots\right\}.$$

For $\alpha = -1/2$ The recurrence relation becomes $c_k = \frac{c_{k-1}}{k(2k-3)}$. Then for various values of k, we have

$k = 1: \quad c_1 = \frac{c_0}{1 \cdot (-1)} = -c_0 \qquad\qquad k = 2: \quad c_2 = \frac{c_1}{2 \cdot 1} = -\frac{c_0}{2}$

$k = 3: \quad c_3 = \frac{c_2}{3 \cdot 3} = -\frac{c_0}{9} \quad \cdots$

The second solution is

$$y_2 = \frac{b}{\sqrt{x}}\left\{1 - x - \frac{x^2}{2} - \frac{x^3}{9} - \cdots\right\}.$$

Series Solutions of Differential Equations

Finally, the general solution will be

$$y = ax\left\{1+\frac{x}{1!5}+\frac{x^2}{2!5\cdot 7}+\frac{x^3}{3!5\cdot 7\cdot 9}+\cdots\right\}$$

$$+\frac{b}{\sqrt{x}}\left\{1-x-\frac{x^2}{2}-\frac{x^3}{9}-\cdots\right\} \square$$

Example 14: Find a series solution for the following differential equation about the origin: $2x(1-x)y''+(1-x)y'+3y = 0$.

Solution: Here $P(x) = \dfrac{1}{2x}$ and $Q(x) = \dfrac{3}{2x(1-x)}$. Both $xP(x)$ and $x^2 Q(x)$ are analytic at $x = 0$, then the point $x = 0$ is a regular singular point of the differential equation. We assume a solution of the form

$$y = \sum_{k=0}^{\infty} c_k x^{k+\alpha}, \quad c_0 \neq 0, \text{ then } y' = \sum_{k=0}^{\infty}(k+\alpha)c_k x^{k+\alpha-1}, \text{ and}$$

$$y'' = \sum_{k=0}^{\infty}(k+\alpha)(k+\alpha-1)c_k x^{k+\alpha-2}$$

Substituting in the differential equation, we obtain

$$(2x-2x^2)\sum_{k=0}^{\infty}(k+\alpha)(k+\alpha-1)c_k x^{k+\alpha-2}$$

$$+(1-x)\sum_{k=0}^{\infty}(k+\alpha)c_k x^{k+\alpha-1}+3\sum_{k=0}^{\infty}c_k x^{k+\alpha}=0$$

We equate the coefficient of the lowest power of x, ($x^{\alpha-1}$), to zero, to obtain $2\alpha(\alpha-1)c_0+\alpha c_0 = 0$. Since $c_0 \neq 0$, the **indicial equation** is $2\alpha^2 - \alpha = 0$. The roots are $\alpha = 0, \dfrac{1}{2}$. The difference between the two roots is <u>not an integer</u>. Then, for each value of α corresponds a solution.

Equating the coefficients of $x^{k+\alpha}$ to zero, we obtain
$2(k+\alpha+1)(k+\alpha)c_{k+1}-2(k+\alpha)(k+\alpha-1)c_k$

$+(k+\alpha+1)c_{k+1}-(k+\alpha)c_k+3c_k = 0$

27

Special Functions and Orthogonal Polynomials

or

$$\boxed{c_{k+1} = \frac{2k+2\alpha-3}{2k+2\alpha+1}c_k}.$$

This is the **recurrence relation** for the coefficients.

For $\alpha = 0$ The recurrence relation becomes

$$c_{k+1} = \frac{(2k-3)c_k}{2k+1}.$$

Then for various values of k, we have

$k = 0:$ $c_1 = (-3)c_0$ $\qquad k = 1:$ $c_2 = \frac{-c_1}{3} = \frac{(-3)(-1)c_0}{1 \cdot 3}$

$k = 2:$ $c_3 = \frac{c_2}{5} = \frac{(-3)(-1)(1)\ldots c_0}{1 \cdot 3 \cdot 5}$ $\ldots,$ in general

$$c_n = \frac{(-3)(-1) \cdot 1 \cdot 3 \cdot 5 \cdots (2n-5)c_0}{1 \cdot 3 \cdot 5 \cdots (2n-1)} = \frac{3c_0}{(2n-1)(2n-3)}, \quad n \geq 1.$$

The first solution is

$$y_1 = a\left\{1 + \sum_{n=1}^{\infty} \frac{3x^n}{(2n-1)(2n-3)}\right\}.$$

For $\alpha = 1/2$ The recurrence relation becomes

$$c_{k+1} = \frac{(k-1)c_k}{k+1}.$$

Then for various values of k, we have

$k = 0:$ $c_1 = -c_0$ $\qquad k = 1:$ $c_2 = 0$, $c_3 = c_4 = c_5 = \cdots = 0$.

The second solution is

$$y_2 = b\sqrt{x}\{1-x\}.$$

Finally, the general solution will be

$$y = a\left\{1 + \sum_{n=1}^{\infty} \frac{3x^n}{(2n-1)(2n-3)}\right\} + b\sqrt{x}\{1-x\}. \qquad \square$$

Series Solutions of Differential Equations

Exercise 1.3

Obtain the series solution of the following differential equations about $x = 0$

1. $2x(x-1)y'' + 3(x-1)y' - y = 0$

$$\text{Ans: } y = A\left(1 - \sum_{n=1}^{\infty} \frac{x^n}{4n^2 - 1}\right) + \left(x^{-1/2} - x^{1/2}\right)$$

2. $2x^2(x+1)y'' + x(7x-1)y' + y = 0$ \quad Ans:

$$y = A\sum_{n=0}^{\infty} (-1)^n (2n+3)(2n+5) x^{n+1} + B\sum_{n=0}^{\infty} (-1)^n (n+1)(n+2) x^{n+1/2}$$

3. $3xy'' + (2-x)y' - 2y = 0$

$$\text{Ans: } y = A\sum_{n=0}^{\infty} \frac{(3n+4)}{4 \cdot 3^n n!} x^{n+1/3} + B\left(1 + \sum_{n=1}^{\infty} \frac{(n+1)x^n}{2 \cdot 5 \cdot 8 \cdots (3n-1)}\right)$$

4. $2x^2 y'' - x(2x+1)y' + (1-5x)y = 0$ \quad Ans:

$$y = A\sum_{n=0}^{\infty} \frac{(2n+3)(2n+5)}{n!} x^{n+1} + B\left(x^{1/2} + \sum_{n=1}^{\infty} \frac{2^{n-1}(n+1)(n+2)}{1 \cdot 3 \cdot 5 \cdots (2n-1)} x^{n+1/2}\right)$$

5. $2xy'' + (1-2x^2)y' - 4xy = 0$

$$\text{Ans: } y = A\, x^{1/2} e^{x^2/2} + B\left(1 + \sum_{n=1}^{\infty} \frac{2^n x^{2n}}{3 \cdot 7 \cdot 11 \cdots (4n-1)}\right)$$

6. $2xy'' + (1+2x)y' - 5y = 0$

$$\text{Ans: } y = A\, x^{1/2}\left(1 + \tfrac{4}{3}x + \tfrac{4}{15}x^2\right) + B\sum_{n=0}^{\infty} \frac{15(-1)^{n+1} x^n}{n!(2n-5)(2n-3)(2n-1)}$$

7. $9x^2 y'' + 3x(x+3)y' - (1+4x)y = 0$

$$\text{Ans: } y = A\, x^{1/3}(1 + \tfrac{1}{5}x) + B\sum_{n=0}^{\infty} \frac{10(-1)^n x^{n-1/3}}{3^n n!(3n-5)(3n-2)}$$

29

Special Functions and Orthogonal Polynomials

8. $2x^2 y'' + 5xy' - 2y = 0$ **Ans:** $y = A x^{1/2} + B x^{-2}$

9. $4x y'' + 2y' + y = 0$

    Ans: $y = a\left\{1 - \frac{1}{2!}x + \frac{1}{4!}x^2 - \frac{1}{6!}x^3 + \cdots\right\} + b\sqrt{x}\left\{1 - \frac{1}{3!}x + \frac{1}{5!}x^2 + \cdots\right\}$
    $= a\cos\sqrt{x} + b\sin\sqrt{x}$

10. $8x^2 y'' - 2x y' + y = 0$ **Ans:** $y = a\sqrt{x} + b\sqrt[4]{x}$

11. $9x(1-x)y'' - 12y' + 4y = 0$

    Ans: $y = a\left\{1 + \frac{1}{3}x + \frac{1\cdot 4}{3\cdot 6}x^2 + \cdots\right\} + b x^{7/3}\left\{1 + \frac{8}{10}x + \frac{8\cdot 11}{10\cdot 13}x^2 + \cdots\right\}$

12. $2x^2 y'' - x y' + (1 - x^2)y = 0$

    Ans: $y = ax\left\{1 + \frac{x^2}{2\cdot 5} + \frac{x^4}{2\cdot 4\cdot 5\cdot 9} + \cdots\right\} + b\sqrt{x}\left\{1 + \frac{x^2}{2\cdot 3} + \frac{x^4}{2\cdot 3\cdot 4\cdot 7} + \cdots\right\}$

13. $4x y'' + 2(1-x)y' - y = 0$

    Ans: $y = a\left\{1 + \frac{x}{2\cdot 1!} + \frac{x^2}{2^2\cdot 2!} + \frac{x^3}{2^3\cdot 3!} + \cdots\right\} + b\sqrt{x}\left\{1 + \frac{x}{1\cdot 3} + \frac{x^2}{1\cdot 3\cdot 5} + \frac{x^3}{1\cdot 3\cdot 5\cdot 7} + \cdots\right\}$

14. $3x y'' + (1-x)y' - y = 0$

    Ans: $y = a\left\{1 + x + \frac{x^2}{4} + \frac{x^3}{4\cdot 7} + \cdots\right\} + b x^{2/3}\left\{1 + \frac{x}{3} + \frac{x^2}{3\cdot 6} + \frac{x^3}{3\cdot 6\cdot 9} + \cdots\right\}$

15. $2x(1-x)y'' + (1-x)y' + 3y = 0$

    Ans: $y = a\left\{1 - 3x + \frac{3x^2}{1\cdot 3} + \frac{3x^3}{3\cdot 5} + \frac{3x^4}{5\cdot 7}\cdots\right\} + b(1-x)\sqrt{x}$

Series Solutions of Differential Equations

Case II: The two roots of the indicial equation are equal

In this case, $\alpha_1 = \alpha_2 = a$, and we obtain only one Frobenius series solution. Suppose that this solution in term of the double root α is $y_1 = g(x,\alpha)$, then if we substitute this function in the differential equation we get

$$\frac{\partial^2 g}{\partial x^2} + P(x)\frac{\partial g}{\partial x} + Q(x)g = F(x,\alpha). \tag{9}$$

We have used partial derivatives since g is a function of two variables x and α. You may notice the function $F(x,\alpha)$ to the right of Equation (9). This function vanishes when $\alpha = a$, since $g(x,a)$ is a solution of the differential equation. In fact, because $\alpha = a$ is a double root, we will also have $\left.\frac{\partial}{\partial \alpha}F(x,\alpha)\right|_{\alpha=a} = 0$.

Now, differentiating Equation (9) with respect to α, we obtain

$$\frac{\partial}{\partial \alpha}\left(\frac{\partial^2 g}{\partial x^2}\right) + P(x)\frac{\partial}{\partial \alpha}\left(\frac{\partial g}{\partial x}\right) + Q(x)\frac{\partial}{\partial \alpha}(g) = \frac{\partial}{\partial \alpha}F(x,\alpha)$$

Interchanging the order of differentiation, we obtain

$$\frac{\partial^2}{\partial x^2}\left(\frac{\partial g}{\partial \alpha}\right) + P(x)\frac{\partial^2}{\partial x^2}\left(\frac{\partial g}{\partial \alpha}\right) + Q(x)\left(\frac{\partial g}{\partial \alpha}\right) = \frac{\partial}{\partial \alpha}F(x,\alpha) \tag{10}$$

Clearly, $\frac{\partial g}{\partial \alpha}$ satisfies the differential equation when $\alpha = a$, hence

$$y_1 = g(x,a) \quad \text{and} \quad y_2 = \left.\frac{\partial}{\partial \alpha}g(x,\alpha)\right|_{\alpha=a}$$

are the two solutions of the equation. They are, in fact, linearly independent.

Another approach is as follows. Since, in the case of a double root, Frobenius method produces only one series solution, say y_1, given by

$$y_1 = x^a \sum_{k=0}^{\infty} c_k x^k,$$

We can use the method of variation of parameters to obtain the second linearly independent solution by assuming a solution of the form

$$y(x) = y_1 \cdot v(x).$$

In this case, we will end up with

$$v(x) = \ln x + \sum_{k=1}^{\infty} \beta_k x^{k+a}.$$

Special Functions and Orthogonal Polynomials

Example 15: Find the series solution for the following equation about $x = 0$:
$$(x^2 - x)y'' + (2x + 1)y' - \frac{1}{x}y = 0.$$

Solution: The coefficient functions are
$$P(x) = \frac{2x+1}{x(x-1)} \text{ and } Q(x) = -\frac{1}{x^2(x-1)}.$$

Then $xP(x)$ and $x^2 Q(x)$ are analytic at $x = 0$. Then $x = 0$ is a regular point. Assume a Frobenius series solution of the form

$$y = \sum_{k=0}^{\infty} c_k x^{k+\alpha}, \quad c_0 \neq 0, \text{ then } y' = \sum_{k=0}^{\infty} (k+\alpha) c_k x^{k+\alpha-1}, \text{ and}$$

$$y'' = \sum_{k=0}^{\infty} (k+\alpha)(k+\alpha-1) c_k x^{k+\alpha-2}$$

Substituting for y and its derivatives into the differential equation, we get

$$(x^2 - x) \sum_{k=0}^{\infty} (k+\alpha)(k+\alpha-1) c_k x^{k+\alpha-2}$$
$$+ (2x+1) \sum_{k=0}^{\infty} (k+\alpha) c_k x^{k+\alpha-1} - \frac{1}{x} \sum_{k=0}^{\infty} c_k x^{k+\alpha} = 0$$

To get the indicial equation, we equate the coefficient of the lowest power of x, which is $x^{\alpha-1}$, to zero, we obtain

$(\alpha-1)^2 c_0 = 0$, and since $c_0 \neq 0$, we have $(\alpha-1)^2 = 0$. Then $\alpha = 1, 1$. We have a double root. Then, one solution is $g(x,1)$ corresponding to $\alpha = 1$, and the other is $y_2 = \left.\frac{\partial}{\partial \alpha} g(x, \alpha)\right|_{\alpha=1}$.

Now, equating the coefficients of $x^{k+\alpha}$ to zero, we obtain

Series Solutions of Differential Equations

$$[(k+\alpha)(k+\alpha-1)+2(k+\alpha)]c_k$$
$$+[-(k+\alpha+1)(k+\alpha)+(k+\alpha+1)-1]c_{k+1}=0.$$

Rearranging, we get the recurrence relation

$$c_{k+1} = \frac{k+\alpha+1}{k+\alpha}c_k, \quad k \geq 0, \ \alpha = 1 \text{ and } k+\alpha \neq 0$$

For various values of k, we have

$k = 0:$ $\quad c_1 = \dfrac{\alpha+1}{\alpha}c_0$

$k = 1:$ $\quad c_2 = \dfrac{\alpha+2}{\alpha}c_0$

$k = 2:$ $\quad c_3 = \dfrac{\alpha+3}{\alpha}c_0$

And for $k = n-1:$ $\quad c_n = \dfrac{\alpha+n}{\alpha}c_0$

Then, the solution in terms of α is $y = c_0 \displaystyle\sum_{n=0}^{\infty} \dfrac{\alpha+n}{\alpha} x^{n+\alpha}$.

Differentiating with respect to α, we obtain

$$\frac{\partial y}{\partial \alpha} = c_0 \ln x \sum_{n=0}^{\infty} \frac{n+\alpha}{\alpha} x^{n+\alpha} - c_0 \sum_{n=0}^{\infty} \frac{n}{\alpha^2} x^{n+\alpha}.$$

Then the two solutions are

$$y_1 = y(x,1) = c_0 \sum_{n=0}^{\infty} (n+1)x^{n+1}, \text{ and}$$

$$y_2 = \left.\frac{\partial y}{\partial \alpha}\right|_{\alpha=1} = c_0\left(\ln x \sum_{n=0}^{\infty}(n+1)x^{n+1} - \sum_{n=0}^{\infty} n x^{n+1}\right).$$

The general solution is

$$y = A\sum_{n=0}^{\infty}(n+1)x^{n+1} + B\left(\ln x \sum_{n=0}^{\infty}(n+1)x^{n+1} - \sum_{n=0}^{\infty}n x^{n+1}\right) \quad \square$$

Special Functions and Orthogonal Polynomials

Example 16: Find the series solution for the following equation about $x = 0$:

$$xy'' + y' + xy = 0.$$

Solution: The coefficient functions are $P(x) = 1/x$ and $Q(x) = 1$. Then $xP(x)$ and $x^2Q(x)$ are analytic at $x = 0$. Then $x = 0$ is a regular point. Assume a Frobenius series solution of the form

$$y = \sum_{k=0}^{\infty} c_k x^{k+\alpha}, \quad c_0 \neq 0, \text{ then } y' = \sum_{k=0}^{\infty} (k+\alpha) c_k x^{k+\alpha-1}, \text{ and}$$

$$y'' = \sum_{k=0}^{\infty} (k+\alpha)(k+\alpha-1) c_k x^{k+\alpha-2}$$

Substituting for y and its derivatives into the differential equation, we get

$$x \sum_{k=0}^{\infty} (k+\alpha)(k+\alpha-1) c_k x^{k+\alpha-2}$$

$$+ \sum_{k=0}^{\infty} (k+\alpha) c_k x^{k+\alpha-1} + x \sum_{k=0}^{\infty} c_k x^{k+\alpha} = 0$$

Equating the coefficient of the lowest power of x, ($x^{\alpha-1}$), to zero, we obtain $\alpha(\alpha-1)c_0 + \alpha c_0 = 0$. Since $c_0 \neq 0$, the **indicial equation** is $\alpha^2 = 0$. The roots are $\boxed{\alpha = 0, 0}$. We have a double root.

Equating the coefficient of the next lowest power of x, (x^{α}), to zero, we obtain $\alpha(\alpha+1)c_1 + (\alpha+1)c_1 = 0$, or $(\alpha+1)^2 c_1 = 0$. This implies that $\boxed{c_1 = 0}$.

Equating the coefficients of $x^{k+\alpha}$ to zero, we obtain

$(k+\alpha+1)(k+\alpha) c_{k+1} + (k+\alpha+1) c_k + c_{k-1} = 0$ or

$$\boxed{c_{k+1} = -\frac{c_{k-1}}{(k+\alpha+1)^2}}.$$

This is the **recurrence relation** for the coefficients.

For various values of k, we have

Series Solutions of Differential Equations

$$k = 1: \quad c_2 = -\frac{c_0}{(\alpha+2)^2} \qquad k = 2: \quad c_3 = -\frac{c_1}{(\alpha+3)^2} = 0$$

$$k = 3: \quad c_4 = \frac{c_0}{(\alpha+2)^2(\alpha+4)^2} \qquad \boxed{c_{2n+1} = 0, \quad n = 0,1,2,3,\cdots}$$

$$k = 5: \quad c_6 = -\frac{c_0}{(\alpha+2)^2(\alpha+4)^2(\alpha+6)^2} \quad , \text{ and}$$

$$\boxed{c_{2n} = \frac{(-1)^n c_0}{(\alpha+2)^2(\alpha+4)^2\cdots(\alpha+2n)^2}, \quad n = 1, 2, 3, \cdots}$$

The solution takes the form

$$\boxed{y = x^\alpha \left\{ 1 - \sum_{n=1}^{\infty} \frac{(-1)^n x^{2n}}{(\alpha+2)^2(\alpha+4)^2\cdots(\alpha+2n)^2} \right\}}$$

The first solution is obtained by letting $\alpha = 0$

$$y_1 = \left\{ 1 - \sum_{n=1}^{\infty} \frac{(-1)^n x^{2n}}{2^2 \cdot 4^2 \cdots (2n)^2} \right\}.$$

Now, we know that the second solution is $y_2 = \left.\frac{\partial y}{\partial \alpha}\right|_{\alpha=0}$, then differentiation the expression of y with respect to α, we obtain

$$\frac{\partial y}{\partial \alpha} = x^\alpha \ln x \left\{ 1 - \sum_{n=1}^{\infty} \frac{(-1)^n x^{2n}}{(\alpha+2)^2(\alpha+4)^2\cdots(\alpha+2n)^2} \right\}$$

$$- x^\alpha \sum_{n=1}^{\infty} (-1)^n x^{2n} \frac{d}{d\alpha}\left[\frac{1}{(\alpha+2)^2(\alpha+4)^2\cdots(\alpha+2n)^2} \right]$$

To obtain $\dfrac{d}{d\alpha}\left[\dfrac{1}{(\alpha+2)^2(\alpha+4)^2\cdots(\alpha+2n)^2}\right] = \dfrac{dw}{d\alpha}$, we use logarithmic differentiation to get

$$\frac{dw}{d\alpha} = \frac{-2}{(\alpha+2)^2(\alpha+4)^2\cdots(\alpha+2n)^2}\left[\frac{1}{\alpha+2} + \frac{1}{\alpha+4} + \cdots + \frac{1}{\alpha+2n} \right].$$

Finally,

35

$$y_2 = \frac{\partial y}{\partial \alpha}\bigg|_{\alpha=0} = \ln x \left\{1 - \sum_{n=1}^{\infty} \frac{(-1)^n x^{2n}}{2^2 4^2 \cdots (2n)^2}\right\} + \sum_{n=1}^{\infty} (-1)^n x^{2n} \frac{dw}{d\alpha}\bigg|_{\alpha=0}$$

$$y_2 = y_1 \ln x + 2 \sum_{n=1}^{\infty} \frac{(-1)^n x^{2n}}{2^2 \cdot 4^2 \cdots (2n)^2} \left[\frac{1}{2} + \frac{1}{4} + \cdots + \frac{1}{2n}\right].$$

The general solution is

$$y = a\left\{1 - \sum_{n=1}^{\infty} \frac{(-1)^n x^{2n}}{2^2 \cdot 4^2 \cdots (2n)^2}\right\}$$

$$+ b\, y_1 \ln x + 2 \sum_{n=1}^{\infty} \frac{(-1)^n x^{2n}}{2^2 \cdot 4^2 \cdots (2n)^2} \left[\frac{1}{2} + \frac{1}{4} + \cdots + \frac{1}{2n}\right].\square$$

The last two examples reveal that the second solution takes the form

$$y_2 = y_1 \ln x + \sum_{n=1}^{\infty} \beta_n x^{n+\alpha}.$$

In fact, this is always the case when the indicial equation has a double root. We state here a theorem for this particular case. The proof is omitted.

Theorem: If $x = 0$ is a regular singular point of the differential equation:

$y'' + P(x)y' + Q(x)y = 0$, and if the roots of the corresponding indicial equation are equal ($\alpha_1 = \alpha_2$), then there is a Frobenius series solution

$$y_1 = \sum_{n=0}^{\infty} c_n x^{n+\alpha_1}, \quad c_0 \neq 0 \tag{11}$$

Also, the second linearly independent solution is given by

$$y_2 = y_1 \ln x + \sum_{n=1}^{\infty} \beta_n x^{n+\alpha_1}. \tag{12}$$

The coefficients β_n are obtained by substituting y_2 and its derivatives into the differential equation. By doing so, we obtain a recurrence relation for β_n. The following example illustrates the use of this theorem.

Series Solutions of Differential Equations

***Example* 17**: Find the series solution for the following equation about $x = 0$:
$$xy'' + (x+1)y' + 2y = 0$$

Solution: It is routine to check that $x = 0$ is a regular singular point of this differential equation. Assume a Frobenius series solution of the form

$$y = \sum_{k=0}^{\infty} c_k x^{k+\alpha}, \quad c_0 \neq 0, \text{ then } y' = \sum_{k=0}^{\infty} (k+\alpha) c_k x^{k+\alpha-1}, \text{ and}$$

$$y'' = \sum_{k=0}^{\infty} (k+\alpha)(k+\alpha-1) c_k x^{k+\alpha-2}$$

Substituting for y and its derivatives in the differential equation, we get

$$x \sum_{k=0}^{\infty} (k+\alpha)(k+\alpha-1) c_k x^{k+\alpha-2}$$
$$+ (x+1) \sum_{k=0}^{\infty} (k+\alpha) c_k x^{k+\alpha-1} + 2 \sum_{k=0}^{\infty} c_k x^{k+\alpha} = 0$$

To get the indicial equation, we equate the coefficient of the lowest power of x, which is $x^{\alpha-1}$, to zero, we obtain $\alpha^2 c_0 = 0$, and since $c_0 \neq 0$, we have $\alpha^2 = 0$. Then $\alpha = 0, 0$. We have a double root. Then, one solution is $g(x, 0)$ corresponding to $\alpha = 0$, and the other is

$$y_2 = \left.\frac{\partial}{\partial \alpha} g(x, \alpha)\right|_{\alpha=0}.$$

Now, equating the coefficients of $x^{k+\alpha}$ to zero, we obtain

$$(k+\alpha-1)^2 c_{k+1} + (k+\alpha+2) c_k = 0.$$

Rearranging, we get the recurrence relation

$$c_{k+1} = -\frac{k+\alpha+2}{(k+\alpha+1)^2} c_k, \quad k \geq 0.$$

<u>For $\alpha = 0$</u>, we have $c_{k+1} = -\frac{k+2}{(k+1)^2} c_k, \quad k \geq 0.$

37

Special Functions and Orthogonal Polynomials

For various values of k, we have

$k = 0:$ $\quad c_1 = -\dfrac{2}{1^2} c_0$

$k = 1:$ $\quad c_2 = -\dfrac{3}{2^2} c_1 = \dfrac{2 \cdot 3}{1^2 \cdot 2^2} c_0$

$k = 2:$ $\quad c_3 = -\dfrac{4}{3^2} c_2 = -\dfrac{2 \cdot 3 \cdot 4}{1^2 \cdot 2^2 \cdot 3^2} c_0$

And for $k = n-1:$ $\quad c_n = (-1)^n \dfrac{(n+1)}{n!} c_0$

The first solution is $y_1 = c_0 \displaystyle\sum_{n=0}^{\infty} (-1)^n \dfrac{(n+1)}{n!} x^n$.

Now, according to the previous theorem, the second solution will be of the form

$$y_2 = y_1 \ln x + \sum_{n=1}^{\infty} \beta_n x^n.$$

Differentiating twice with respect to x, we obtain

$$y_2' = \dfrac{1}{x} y_1 + y_1' \ln x + \sum_{n=1}^{\infty} n \beta_n x^{n-1} \quad \text{and}$$

$$y_2'' = -\dfrac{1}{x^2} y_1 + \dfrac{2}{x} y_1' + y_1'' \ln x + \sum_{n=2}^{\infty} n(n-1) \beta_n x^{n-2}.$$

Substituting these values into the differential equation, we obtain

$$-\dfrac{1}{x} y_1 + 2 y_1' + x y_1'' \ln x + \sum_{n=2}^{\infty} n(n-1) \beta_n x^{n-1} + y_1 + x y_1' \ln x$$

$$+ \sum_{n=1}^{\infty} n \beta_n x^n + \dfrac{1}{x} y_1 + y_1' \ln x + \sum_{n=1}^{\infty} n \beta_n x^{n-1} + 2 y_1 \ln x + 2 \sum_{n=1}^{\infty} \beta_n x^n$$

38

Series Solutions of Differential Equations

We can see that $\ln x$ is multiplied by $[xy_1'' + (x+1)y_1' + 2y_1]$ and since y_1 is a solution of the differential equation then

$$xy_1'' + (x+1)y_1' + 2y_1 = 0,$$

and the term containing $\ln x$ vanishes. Rearranging and simplifying, we get

$$2y_1' + y_1 + \sum_{n=1}^{\infty}[n(n+1)\beta_{n+1} + (n+2)\beta_n]x^n + \sum_{n=1}^{\infty} n\beta_n x^{n-1} = 0$$

Now, substituting for y_1 (let $c_0 = 1$ for convenience), we obtain

$$2\sum_{n=1}^{\infty}(-1)^n \frac{n(n+1)}{n!} x^{n-1} + \sum_{n=0}^{\infty}(-1)^n \frac{(n+1)}{n!} x^n$$

$$+ \sum_{n=1}^{\infty}[n(n+1)\beta_{n+1} + (n+2)\beta_n]x^n + \sum_{n=1}^{\infty} n\beta_n x^{n-1} = 0$$

Equating the coefficient of x^0 to zero, we get $\beta_1 = 3$.

Equating the coefficient of x^n to zero, we get the recurrence relation

$$\beta_{n+1} = (-1)^n \frac{n+3}{(n+1)^2 n!} - \frac{n+2}{(n+1)^2}\beta_n, \ n \geq 1.$$

From which

$$\beta_2 = -\frac{13}{4}, \ \beta_3 = \frac{31}{18}, \ \beta_4 = -\frac{163}{228}, \cdots,$$

and the second solution is

$$y_2 = y_1 \ln x + \left(3x - \frac{13}{4}x^2 + \frac{31}{18}x^3 - \frac{163}{228}x^4 + \cdots\right)$$

and the general solution is

$$y = A\sum_{n=0}^{\infty}(-1)^n \frac{(n+1)}{n!}x^n + B\left(\begin{array}{c}\ln x \sum_{n=0}^{\infty}(-1)^n \frac{(n+1)}{n!}x^n \\ +\left(3x - \frac{13}{4}x^2 + \frac{31}{18}x^3 + \cdots\right)\end{array}\right)$$

□

Special Functions and Orthogonal Polynomials

There is another approach to obtain the second solution. The solution in terms of α is given by

$$y = x^\alpha \left\{ 1 - \frac{\alpha+2}{(\alpha+1)^2} x + \frac{\alpha+3}{(\alpha+1)^2(\alpha+2)} x^2 - \frac{\alpha+4}{(\alpha+1)^2(\alpha+2)(\alpha+3)} x^3 + \cdots \right\}.$$

Differentiating this expression with respect to α, we obtain

$$\frac{\partial y}{\partial \alpha} = x^\alpha \ln x \left\{ 1 - \frac{\alpha+2}{(\alpha+1)^2} x + \frac{\alpha+3}{(\alpha+1)^2(\alpha+2)} x^2 - \frac{\alpha+4}{(\alpha+1)^2(\alpha+2)(\alpha+3)} x^3 + \cdots \right\}$$

$$+ x^\alpha \left\{ -\frac{\alpha+2}{(\alpha+1)^2} \left[\frac{1}{\alpha+2} - \frac{2}{\alpha+1} \right] x + \frac{\alpha+3}{(\alpha+1)^2(\alpha+2)} \left[\frac{1}{\alpha+3} - \frac{2}{\alpha+1} - \frac{1}{\alpha+2} \right] x^2 + \cdots \right\}$$

Letting $\alpha = 0$ in this last expression, we obtain the second solution as

$$y_2(x) = \ln x \left\{ \sum_{n=0}^{\infty} (-1)^n \frac{n+1}{n!} x^n \right\}$$

$$+ \left\{ 3x - \frac{3}{2!} \left[2 + \frac{1}{2} - \frac{1}{3} \right] x^2 + \frac{4}{3!} \left[2 + \frac{1}{2} + \frac{1}{3} - \frac{1}{4} \right] x^3 + \cdots \right\}.$$

The general solution will be

$$y = A \sum_{n=0}^{\infty} (-1)^n \frac{(n+1)}{n!} x^n + B \left(\begin{array}{c} \ln x \sum_{n=0}^{\infty} (-1)^n \frac{(n+1)}{n!} x^n \\ + \left(3x - \frac{13}{4} x^2 + \frac{31}{18} x^3 + \cdots \right) \end{array} \right) \quad \square$$

40

Exercise 1.4

Obtain the two linearly independent solutions of the following equation using the method of Frobenius:

1. $x(x-1)y'' + (3x-1)y' + y = 0$ Ans: $y_1 = \sum_{n=0}^{\infty} x^n$, $y_2 = y_1 \ln x$

2. $xy'' - y' - xy = 0$

   Ans: $y_1 = 1 + \sum_{n=1}^{\infty} \frac{x^{2n}}{2^2 \cdot 4^2 \cdot 6^2 \cdots (2n)^2}$, $y_2 = y_1 \ln x - \frac{x^2}{4} - \frac{3x^4}{8 \cdot 16} - \cdots$

3. $x^2 y'' + 3xy' + (1-2x)y = 0$

   Ans: $y_1 = \frac{1}{x} + \sum_{n=1}^{\infty} \frac{2^n x^{n-1}}{(n!)^2}$, $y_2 = y_1 \ln x - \sum_{n=1}^{\infty} \frac{2^{n+1}\left(1+\frac{1}{2}+\frac{1}{3}+\cdots+\frac{1}{n}\right)}{(n!)^2} x^{n-1}$

4. $x^2 y'' + x(x-1)y' + (1-x)y = 0$

   Ans: $y_1 = x$, $y_2 = y_1 \ln x + \sum_{n=1}^{\infty} \frac{(-1)^n x^{n+1}}{n \cdot n!}$

5. $x(1+x)y'' + (1+5x)y' + 3y = 0$

   Ans: $y_1 = 1 + \frac{1}{2}\sum_{n=1}^{\infty}(-1)^n (n+1)(n+2) x^n$

   $y_2 = y_1 \ln x - \frac{3}{2}(y_1 - 1) + \frac{1}{2}\sum_{n=1}^{\infty}(-1)^n (2n+3) x^n$

6. $x(x-2)y'' + 2(x-1)y' - 2y = 0$

   Ans: $y_1 = 1 - x$, $y_2 = y_1 \ln x + \frac{5}{2}x - \sum_{n=2}^{\infty} \frac{(n+1)x^n}{2^n n(n-1)}$

7. Show that the series solution of the equation $xy'' + y' + y = 0$ is of the form $y = (A + B\ln x)\sum_{n=0}^{\infty} c_n x^n + B\sum_{n=0}^{\infty} \beta_n x^n$, and find the values of the coefficients c_n and β_n for $n = 0, 1, 2, 3$.

41

Special Functions and Orthogonal Polynomials

Case III: The roots of the indicial equation are distinct differ by an integer

This case is best described by the following theorem.

Theorem*: *If $x = 0$ is a regular singular point of the differential equation:*
$y'' + P(x)y' + Q(x)y = 0$, and if α_1 and α_2 are the two roots of the indicial equation, and if $(\alpha_1 - \alpha_2)$ is a positive integer, then there is a Frobenius series solution of the form

$$y_1 = \sum_{n=0}^{\infty} c_n x^{n+\alpha_1}, \quad c_0 \neq 0 \tag{13}$$

Also, the second linearly independent solution is given by

$$y_2 = m y_1 \ln x + \sum_{n=0}^{\infty} \beta_n x^{n+\alpha_2}. \tag{14}$$

The constants m and β_n is obtained by substituting expression (14) into the differential equation and equating the coefficients of various powers of x to zero.

The number m may or may not be zero. If $m = 0$, then the solution y_2 will not contain a logarithmic term and we have the **Nonlogarithmic case** (see *Example* 1 below). On the other hand, if $m \neq 0$, the solution y_2 will contain a logarithmic term, and we have the **Logarithmic case** (see *Example* 2 below).

The solution procedure is then:

Step 1: Obtain the first solution y_1 by assuming a Frobenius series solution of the form (13) then evaluating the coefficients c_n.

Step 2: Substitute y_2 from expression (14) into the differential equation to obtain the constant m and the coefficients β_n.

Step 3: The general solution is $y = A y_1 + B y_2$.

Example 18: **Nonlogarithmic case**: Find the series solution for the following equation about $x = 0$: $xy'' - (4+x)y' + 2y = 0$

Solution: The point $x = 0$ is a regular singular point of this differential equation. Assume a Frobenius series solution of the form

---

[*] For the proof of this theorem see for example G. Birkoff and G.-C. Rota, *Ordinary differential equations*, p. 261, Blaisdell, Waltham, Mass., 1969.

Series Solutions of Differential Equations

$$y = \sum_{k=0}^{\infty} c_k x^{k+\alpha}, \quad c_0 \neq 0, \text{ then } y' = \sum_{k=0}^{\infty} (k+\alpha) c_k x^{k+\alpha-1}, \text{ and}$$

$$y'' = \sum_{k=0}^{\infty} (k+\alpha)(k+\alpha-1) c_k x^{k+\alpha-2}$$

Substituting for y and its derivatives in the differential equation, we get

$$\sum_{k=0}^{\infty} [(k+\alpha)(k+\alpha-1) - 4(k+\alpha)] c_k x^{k+\alpha-1}$$
$$- \sum_{k=0}^{\infty} (k+\alpha-2) c_k x^{k+\alpha} \equiv 0$$

To get the indicial equation, we equate the coefficient of the lowest power of x, which is $x^{\alpha-1}$, to zero, we obtain $\alpha(\alpha-5)c_0 = 0$, and since $c_0 \neq 0$, we have $\alpha(\alpha-5) = 0$. Then $\alpha_1 = 5$ and $\alpha_2 = 0$. The roots are distinct and the difference between them is an integer (5). Now, equating the coefficients of $x^{k+\alpha}$ to zero, we obtain

$$(k+\alpha+1)(k+\alpha-4) c_{k+1} = (k+\alpha-2) c_k .$$

Rearranging, we get the recurrence relation

$$c_{k+1} = \frac{k+\alpha-2}{(k+\alpha+1)(k+\alpha-4)} c_k, \quad k \geq 0.$$

For $\underline{\alpha = 5}$, we have $c_{k+1} = \dfrac{k+5}{(k+6)(k+1)} c_k, \quad k \geq 0.$

For various values of k, we have

$k = 0:$ $\qquad c_1 = \dfrac{3}{6 \cdot 1} c_0$

$k = 1:$ $\qquad c_2 = \dfrac{4}{7 \cdot 2} c_1 = \dfrac{3 \cdot 4}{6 \cdot 7 \cdot 1 \cdot 2} c_0$

$k = 2:$ $\qquad c_3 = \dfrac{5}{8 \cdot 3} c_2 = \dfrac{3 \cdot 4 \cdot 5}{6 \cdot 7 \cdot 8 \cdot 1 \cdot 2 \cdot 3} c_0$

And for $k = n-1$: $c_n = \dfrac{3\cdot 4\cdot 5}{(n+5)(n+4)(n+3)n!}c_0$

The first solution is $y_1 = c_0 \displaystyle\sum_{n=0}^{\infty} \dfrac{3\cdot 4\cdot 5}{(n+5)(n+4)(n+3)n!} x^{n+5}$.

Now, according to the previous theorem, the second solution will be of the form $y_2 = my_1 \ln x + \displaystyle\sum_{n=0}^{\infty} \beta_n x^n$.

Differentiating twice with respect to x, we obtain

$$y_2' = \dfrac{m}{x} y_1 + my_1' \ln x + \sum_{n=1}^{\infty} n \beta_n x^{n-1} \text{ and}$$

$$y_2'' = -\dfrac{m}{x^2} y_1 + \dfrac{2m}{x} y_1' + my_1'' \ln x + \sum_{n=2}^{\infty} n(n-1)\beta_n x^{n-2}.$$

Substituting these values into the differential equation, we obtain

$$m \ln x \, [xy_1'' - (4+x)y_1' + 2y_1] - my_1\left(1+\dfrac{5}{x}\right) + 2my_1'$$

$$+ \sum_{n=0}^{\infty} [(n+1)(n+2)\beta_{n+2} - (n+1)\beta_{n+1}]x^{n+1}$$

$$- \sum_{n=0}^{\infty} [4(n+1)\beta_{n+1} - 2\beta_n]x^n = 0$$

We can see that $\ln x$ is multiplied by $[xy_1'' - (4+x)y_1' + 2y_1]$ and since y_1 is a solution of the differential equation then $xy_1'' - (4+x)y_1' + 2y_1 = 0$, and the term containing $\ln x$ vanishes.

$$-my_1\left(1+\dfrac{5}{x}\right) + 2my_1' + \sum_{n=0}^{\infty} [(n+1)(n+2)\beta_{n+2} - (n+1)\beta_{n+1}]x^{n+1}$$

$$- \sum_{n=0}^{\infty} [4(n+1)\beta_{n+1} - 2\beta_n]x^n = 0$$

Now, substituting for y_1 (let $c_0 = 1$ for convenience), we obtain

Series Solutions of Differential Equations

$$-m\left(1+\frac{5}{x}\right)\sum_{n=0}^{\infty}\frac{3\cdot 4\cdot 5}{(n+5)(n+4)(n+3)n!}x^{n+5}+2m\sum_{n=0}^{\infty}\frac{3\cdot 4\cdot 5}{(n+4)(n+3)n!}x^{n+4}$$

$$+\sum_{n=0}^{\infty}[(n+1)(n+2)\beta_{n+2}-(n+1)\beta_{n+1}]x^{n+1}-\sum_{n=0}^{\infty}[4(n+1)\beta_{n+1}-2\beta_n]x^n=0$$

Equating the coefficients of various powers of x to zero, we get

$x^0: -4\beta_1 + 2\beta_0 = 0$ \qquad\qquad $\beta_1 = \frac{1}{2}\beta_0$

$x^1: -6\beta_2 + \beta_1 = 0$ \qquad\qquad $\beta_2 = \frac{1}{12}\beta_0$

$x^2: -12\beta_3 + 2\beta_2 + 6\beta_3 - 2\beta_2 = 0$ \qquad\qquad $\beta_3 = 0$

$x^3: -16\beta_4 + 2\beta_3 + 12\beta_4 - 3\beta_3 = 0$ \qquad\qquad $\beta_4 = 0$

$x^4: -20\beta_5 + 20\beta_5 + 2\beta_4 - 4\beta_4 + 10m = 0$ \qquad β_5 arbitrary, $m = 0$

In fact, β_5 represents the arbitrary constant for the first series solution and there is no need to proceed any further. The second solution is now $y_2 = \beta_0\left(1 + \frac{1}{2}x + \frac{1}{12}x^2\right)$, and the general solution is

$$y = A\sum_{n=0}^{\infty}\frac{3\cdot 4\cdot 5}{(n+5)(n+4)(n+3)n!}x^{n+5} + B\left(1 + \frac{1}{2}x + \frac{1}{12}x^2\right) \quad \square$$

Example 19: *Logarithmic case*: Find the series solution for the following equation about $x = 0$: $xy'' - y = 0$.

Solution: The point $x = 0$ is a regular singular point of this differential equation. Assume a Frobenius series solution of the form

$$y = \sum_{k=0}^{\infty} c_k x^{k+\alpha}, \quad c_0 \neq 0, \text{ then } y' = \sum_{k=0}^{\infty}(k+\alpha)c_k x^{k+\alpha-1}, \text{ and}$$

$$y'' = \sum_{k=0}^{\infty}(k+\alpha)(k+\alpha-1)c_k x^{k+\alpha-2}$$

Substituting for y and its derivatives in the differential equation, we get

$$\sum_{k=0}^{\infty}(k+\alpha)(k+\alpha-1)c_k x^{k+\alpha-1} - \sum_{k=0}^{\infty}c_k x^{k+\alpha} \equiv 0$$

45

Special Functions and Orthogonal Polynomials

To get the indicial equation, we equate the coefficient of the lowest power of x, which is $x^{\alpha-1}$, to zero, we obtain $\alpha(\alpha-1)c_0 = 0$, and since $c_0 \neq 0$, we have $\alpha(\alpha-1) = 0$. Then $\alpha_1 = 1$ and $\alpha_2 = 0$. The roots are distinct and the difference between them is an integer ($\alpha_1 - \alpha_2 = 1$). Now, equating the coefficients of $x^{k+\alpha}$ to zero, we obtain $(k+\alpha+1)(k+\alpha)c_{k+1} = c_k$.

Rearranging, we get the recurrence relation

$$c_{k+1} = \frac{1}{(k+\alpha+1)(k+\alpha)} c_k, \quad k \geq 0.$$

<u>For $\alpha = 1$</u>, we have $c_{k+1} = \frac{1}{(k+1)(k+2)} c_k, \quad k \geq 0.$

For various values of k, we have

$k = 0:$ $\quad c_1 = \frac{1}{1 \cdot 2} c_0$

$k = 1:$ $\quad c_2 = \frac{1}{2 \cdot 3} c_1 = \frac{1}{1 \cdot 2 \cdot 3 \cdot 2} c_0$

$k = 2:$ $\quad c_3 = \frac{1}{3 \cdot 4} c_2 = \frac{1}{1 \cdot 2 \cdot 3 \cdot 4 \cdot 2 \cdot 3} c_0$

And for $k = n-1:$ $\quad c_n = \frac{1}{n!(n+1)!} c_0, \quad n \geq 1$

The first solution is $y_1 = c_0 \sum_{n=0}^{\infty} \frac{1}{n!(n+1)!} x^{n+1}$.

The second solution will be of the form $y_2 = m y_1 \ln x + \sum_{n=0}^{\infty} \beta_n x^n$.

Differentiating twice with respect to x, we obtain

$$y_2' = \frac{m}{x} y_1 + m y_1' \ln x + \sum_{n=1}^{\infty} n \beta_n x^{n-1} \quad \text{and}$$

$$y_2'' = -\frac{m}{x^2} y_1 + \frac{2m}{x} y_1' + m y_1'' \ln x + \sum_{n=2}^{\infty} n(n-1) \beta_n x^{n-2}.$$

Series Solutions of Differential Equations

Substituting these values into the differential equation, we obtain

$$m \ln x \, [xy_1'' - y_1] - \frac{m}{x} y_1 + 2m y_1'$$
$$+ \sum_{n=0}^{\infty} (n+1)(n+2) \beta_{n+2} x^{n+1} - \sum_{n=0}^{\infty} \beta_n x^n = 0$$

We can see that $\ln x$ is multiplied by $[xy_1'' - y_1]$ and since y_1 is a solution of the differential equation then $xy_1'' - y_1 = 0$, and the term containing $\ln x$ vanishes.

$$-\frac{m}{x} y_1 + 2m y_1' + \sum_{n=0}^{\infty} (n+1)(n+2) \beta_{n+2} x^{n+1} - \sum_{n=0}^{\infty} \beta_n x^n = 0$$

Now, substituting for y_1 (let $c_0 = 1$ for convenience), we obtain

$$-m \sum_{n=0}^{\infty} \frac{1}{n!(n+1)!} x^n + 2m \sum_{n=0}^{\infty} \frac{n+1}{n!(n+1)!} x^n$$
$$+ \sum_{n=0}^{\infty} (n+1)(n+2) \beta_{n+2} x^{n+1} - \sum_{n=0}^{\infty} \beta_n x^n = 0$$

Equating the coefficient of x^0 to zero, we obtain

$$-m + 2m - \beta_0 = 0 \quad \text{or} \quad m = \beta_0$$

Equating the coefficients of x^n to zero, we get

$$\frac{2m(n+1)}{n!(n+1)!} - \frac{m}{n!(n+1)!} + n(n+1) \beta_{n+1} - \beta_n = 0$$

Rearranging, the recurrence relation for the coefficients β_n is

$$\beta_{n+1} = \frac{1}{n(n+1)} \left[\beta_n - \frac{\beta_0 (2n+1)}{n!(n+1)!} \right], \quad n \geq 1,$$

and for various values of k, we have

$n = 1$: $\quad \beta_2 = \frac{1}{2} \beta_1 - \frac{3}{4} \beta_0$

$n = 2$: $\quad \beta_3 = \frac{1}{12} \beta_1 - \frac{7}{36} \beta_0$

$n = 3$: $\quad \beta_4 = \frac{1}{144} \beta_1 - \frac{35}{1728} \beta_0$

The second solution is

47

Special Functions and Orthogonal Polynomials

$$y_2 = \beta_0 y_1 \ln x + \beta_0 + \beta_1 x + \left(\frac{1}{2}\beta_1 - \frac{3}{4}\beta_0\right)x^2$$
$$+ \left(\frac{1}{12}\beta_1 - \frac{7}{36}\beta_0\right)x^3 + \left(\frac{1}{144}\beta_1 - \frac{35}{1728}\beta_0\right)x^4 + \cdots$$

or

$$y_2 = \beta_0 y_1 \ln x + \beta_0 \left(1 - \frac{3}{4}x^2 - \frac{7}{36}x^3 - \frac{35}{1728}x^4 - \cdots\right)$$
$$+ \beta_1 \left(x + \frac{1}{2}x^2 + \frac{1}{12}x^3 + \frac{1}{144}x^4 + \cdots\right)$$

We can recognize that the expression multiplied by β_1 is nothing but the first solution. Then we can drop it from the second solution to obtain

$$y_2 = \beta_0 \left[y_1 \ln x + \left(1 - \frac{3}{4}x^2 - \frac{7}{36}x^3 - \frac{35}{1728}x^4 - \cdots\right)\right],$$

and the general solution is

$$y = A \sum_{n=0}^{\infty} \frac{x^{n+1}}{n!(n+1)!} + B \left[\ln x \sum_{n=0}^{\infty} \frac{x^{n+1}}{n!(n+1)!} + \left(1 - \frac{3}{4}x^2 - \frac{7}{36}x^3 - \frac{35}{1728}x^4 - \cdots\right)\right] \quad \square$$

Exercise 1.5

Obtain series solutions for the following equation about $x = 0$.

1. $xy'' + 2x(x-2)y' + 2(2-3x)y = 0$

$$\text{Ans: } y = c_0\left(x - 2x^2 + 2x^3\right) + c_3 \sum_{n=3}^{\infty} \frac{6(-2)^{n-3} x^{n+1}}{n!}$$

2. $xy'' - (3+x)y' + 2y = 0$

$$\text{Ans: } y = c_0\left(1 + \frac{2}{3}x + \frac{1}{6}x^2\right) + c_4 \sum_{n=4}^{\infty} \frac{24(n-3)x^n}{n!}$$

3. $x(1-x)y'' - 3y' + 2y = 0$

$$\text{Ans: } y = c_0\left(1 + \frac{2}{3}x + \frac{1}{3}x^2\right) + c_4 \sum_{n=4}^{\infty} (n-3)x^n$$

Series Solutions of Differential Equations

4. $xy'' + (4+3x)y' + 3y = 0$

$$\text{Ans: } y = c_0\left(x^{-3} - 3x^{-2} + \frac{9}{2}x^{-1}\right) + 6c_3 \sum_{n=3}^{\infty} \frac{(-3)^{n-3} x^{n-3}}{n!}$$

5. $x(x+3)y'' - 9y' - 6y = 0$

$$\text{Ans: } y = c_0\left(1 - \frac{2}{3}x + \frac{1}{3}x^2 - \frac{4}{27}x^3\right) + c_4 \sum_{n=4}^{\infty} \frac{(-1)^n (n+1) x^n}{5 \cdot 3^{n-4}}$$

6. $x^2 y'' + x^2 y' - 2y = 0$

$$\text{Ans: } y = c_0\left(x^{-1} - \frac{1}{2}\right) + 6c_3 \sum_{n=3}^{\infty} \frac{(-1)^{n+1}(n-2) x^{n-1}}{n!}$$

7. $xy'' + (x^3 - 1)y' + x^2 y = 0$

$$\text{Ans: } y = c_0 \sum_{n=0}^{\infty} \frac{(-1)^n x^{3n}}{3^n n!} + c_2\left[x^2 + \sum_{n=1}^{\infty} \frac{(-1)^n x^{3n+2}}{5 \cdot 8 \cdot 11 \cdots (3n+2)}\right]$$

8. $x^2 y'' + x(x-2)y' + (x^2 + 2)y = 0$

$$\text{Ans: } y_1 = x^2 - x^3 + \frac{1}{3}x^4 - \frac{1}{36}x^5 - \frac{7}{720}x^6 + \cdots$$

$$y_2 = y_1 \ln x - x + \frac{3}{2}x^3 - \frac{31}{36}x^4 + \frac{65}{432}x^5 + \frac{61}{4320}x^6 + \cdots$$

9. $xy'' + (x-1)y' - 2y = 0$

$$\text{Ans: } y_1 = x^2, \quad y_2 = -y_1 \ln x + 1 - 2x + 2x^2 - 2\sum_{n=3}^{\infty} \frac{(-1)^n x^n}{n!(n-2)}$$

10. $x^2 y'' - 3xy' + (x^3 - 5)y = 0$ $\quad$ Ans: $y_1 = 2 \sum_{n=0}^{\infty} \dfrac{(-1)^n x^n}{3^{2n} n!(n+1)!}$

$$y_2 = -\frac{1}{54} y_1 \ln x + x^{-1} + \frac{1}{9}x^2 + \frac{1}{324}x^5$$

$$- \sum_{n=3}^{\infty} \frac{(-1)^n x^{3n-1}}{3^{2n} n!(n-2)!}\left[1 + \frac{1}{n-1} + \frac{1}{n} - 2\left(1 + \frac{1}{2} + \frac{1}{3} + \cdots + \frac{1}{n}\right)\right]$$

49

Special Functions and Orthogonal Polynomials

11. $x(1-x)y'' - 4xy' - 2y = 0$

$$\text{Ans: } y_1 = \sum_{n=0}^{\infty} n(n+1)x^n, \quad y_2 = y_1 \ln x + \sum_{n=0}^{\infty} (1+n-n^2)x^n$$

12. $x(1-x)y'' + 2(1-x)y' + 2y = 0$

$$\text{Ans: } y_1 = 2x - 2, \quad y_2 = y_1 \ln x + x^{-1} + 1 - 5x + \sum_{n=3}^{\infty} \frac{2x^{n-1}}{(n-1)(n-2)}$$

13. $xy'' + (1-x)y' + 3y = 0$

$$\text{Ans: } y_2 = y_1 \ln x + 7x - \frac{23}{4}x^2 + \frac{11}{12}x^3 - 6 \sum_{n=4}^{\infty} \frac{x^n}{n!n(n-1)(n-2)(n-3)}$$

14. $x^2 y'' + x(1-x)y' + 2xy = 0$ **Ans:**

$y_1 = 2 - 4x + x^2, \quad y_2 = y_1 \ln x + 10x - \frac{9}{2}x^2 + \frac{1}{9}x^3 + \frac{1}{144}x^4 + \frac{1}{1800}x^5 + \cdots$

15. $x(1-x)y'' + 4y' + 2y = 0$

$$\text{Ans: } y = a\left\{1 - \frac{x}{2} + \frac{x^2}{10} + \cdots\right\} + \frac{b}{x^3}\left\{1 - 5x + 10x^2 + \cdots\right\}$$

16. $xy'' + y' + x^2 y = 0$

$$\text{Ans: } y = (a + b \ln x)\left\{1 - \frac{x^3}{3^2} + \frac{x^6}{3^4 (2!)^2} + \cdots\right\} + 2b\left\{\frac{x^3}{3^3} - \frac{\left(1+\frac{1}{2}\right)}{3^5 (2!)^2} x^6 + \cdots\right\}$$

1.5. Solutions for Large Values of x

Sometimes it is desirable to obtain series solutions for differential equations as x, the independent variable, approaches infinity. This case is often called the asymptotic behavior of the differential equation. In this case, we use the transformation $x = 1/u$.

The behavior at the point of infinity is obtained by studying the behavior of the transformed equation at $u = 0$. If $u = 0$ is an ordinary point of the transformed equation, then the point at infinity for the original equation is also ordinary. The same apply for regular and irregular singular points.

Example 20: Determine the nature of the point at infinity for the differential equation $2x(x+1)\dfrac{d^2 y}{dx^2} + (5x-3)\dfrac{dy}{dx} + y = 0$, then obtain a series solution for large values of x.

Solution: Let $x = \dfrac{1}{u}$, then we have $\dfrac{dy}{dx} = \dfrac{dy}{du} \cdot \dfrac{du}{dx} = -u^2 \dfrac{dy}{du}$

and $\dfrac{d^2 y}{dx^2} = \dfrac{d}{dx}\left(-u^2 \dfrac{dy}{du}\right) = u^4 \dfrac{d^2 y}{du^2} + 2u^3 \dfrac{dy}{du}$

Substituting these expressions into the differential equation and rearranging, we obtain

$$2u^2(u+1)\dfrac{d^2 y}{du^2} + u(7u-1)\dfrac{dy}{du} + y = 0$$

Clearly $u = 0$ is a regular singular of this point equation. Then, the point at infinity for the original equation is also a regular singular point.

The solution of the transformed equation is found to be

$$y = A\sum_{n=0}^{\infty}(-1)^n (2n+3)(2n+5) u^{n+1} + B\sum_{n=0}^{\infty}(-1)^n (n+1)(n+2) u^{n+1/2}$$

Then the solution of the original equation for large values of x is

$$y = A\sum_{n=0}^{\infty}(-1)^n (2n+3)(2n+5) x^{-n-1} + B\sum_{n=0}^{\infty}(-1)^n (n+1)(n+2) x^{-n-\frac{1}{2}}$$

Example 21: Determine the nature of the point at infinity for the differential equation $(1-x^2)y'' - 2x y' + 6y = 0$, then obtain a series solution for large values of x.

Solution: Let $x = \dfrac{1}{u}$, then we have $\dfrac{dy}{dx} = \dfrac{dy}{du} \cdot \dfrac{du}{dx} = -u^2 \dfrac{dy}{du}$

and $\dfrac{d^2y}{dx^2} = \dfrac{d}{dx}\left(-u^2 \dfrac{dy}{du}\right) = u^4 \dfrac{d^2y}{du^2} + 2u^3 \dfrac{dy}{du}$

Substituting these expressions in the differential equation and rearranging, we obtain

$$u^2(u^2-1)\dfrac{d^2y}{du^2} + 2u^3 \dfrac{dy}{du} + 6y = 0.$$

Here, $P(u) = \dfrac{2u}{u^2-1}$ and $Q(u) = \dfrac{6}{u^2(u^2-1)}$. Clearly $u = 0$ is a regular point of the differential equation, and the point at infinity is also a regular point for the original differential equation. Assume that the solution is of the form

$$u = \sum_{k=0}^{\infty} c_k u^{k+\alpha}, \text{ then } u' = \sum_{k=0}^{\infty} (k+\alpha) c_k u^{k+\alpha-1} \text{ and}$$

$$u'' = \sum_{k=0}^{\infty} (k+\alpha)(k+\alpha-1) c_k u^{k+\alpha-2}.$$

Substituting these vales in the differential equation, we get

$$u^2(u^2-1)\sum_{k=0}^{\infty}(k+\alpha)(k+\alpha-1)c_k u^{k+\alpha-2}$$
$$+ 2u^3 \sum_{k=0}^{\infty}(k+\alpha)c_k u^{k+\alpha-1} + 6 \sum_{k=0}^{\infty} c_k u^{k+\alpha} = 0$$

Equating the coefficient of the lowest power of u, (u^α), to zero, to obtain $-\alpha(\alpha-1)c_0 + 6c_0 = 0$. Since $c_0 \neq 0$, the **indicial equation** is $\alpha^2 - \alpha - 6 = 0$. The roots are $\alpha = 3, -2$.

Equating the coefficients of $u^{\alpha+1}$ to zero, we obtain
$(\alpha+1)\alpha c_1 + 6c_1 = 0$. Since $\alpha = 3, -2$, then c_1 must be zero.

Equating the coefficients of $u^{k+\alpha}$ to zero, we obtain

Series Solutions of Differential Equations

$$(k+\alpha-2)(k+\alpha-3)c_{k-2} - (k+\alpha)(k+\alpha-1)c_k$$
$$+2(k+\alpha-2)c_{k-2} + 6c_k = 0$$

Rearranging, we obtain the recurrence relation for the coefficients as

$$\boxed{c_k = \frac{(k+\alpha-2)(k+\alpha-1)}{(k+\alpha-3)(k+\alpha+2)} c_{k-2}, \quad k \geq 2}.$$

For various values of k, we have

$$k=2: \quad c_2 = \frac{\alpha(\alpha+1)}{(\alpha-1)(\alpha+4)} c_0$$

$$k=4: \quad c_4 = \frac{\alpha(\alpha+2)(\alpha+3)}{(\alpha-1)(\alpha+4)(\alpha+6)} c_0$$

The solution in terms of α is

$$y = c_0 u^\alpha \left\{ 1 + \frac{\alpha(\alpha+1)}{(\alpha-1)(\alpha+4)} u^2 + \frac{\alpha(\alpha+2)(\alpha+3)}{(\alpha-1)(\alpha+4)(\alpha+6)} u^4 + \cdots \right\}$$

For $\alpha = 3$, the first solution is

$$y_1 = au^3 \left\{ 1 + \frac{3 \cdot 4 \cdot u^2}{2 \cdot 7} + \frac{3 \cdot 5 \cdot 6 \cdot u^4}{2 \cdot 7 \cdot 9} + \cdots \right\}.$$

For $\alpha = -2$, The second solution is $y_2 = \dfrac{b}{u^2} \left\{ 1 - \dfrac{u^2}{3} \right\}$.

The general solution will be

$$y = au^3 \left\{ 1 + \frac{3 \cdot 4 \cdot u^2}{2 \cdot 7} + \frac{3 \cdot 5 \cdot 6 \cdot u^4}{2 \cdot 7 \cdot 9} + \cdots \right\} + \frac{b}{u^2} \left\{ 1 - \frac{u^2}{3} \right\}.$$

Replacing u by $1/x$, the general solution of the original differential equation is

$$y = \frac{a}{x^3} \left\{ 1 + \frac{3 \cdot 4}{2 \cdot 7 \cdot x^2} + \frac{3 \cdot 5 \cdot 6}{2 \cdot 7 \cdot 9 \cdot x^4} + \cdots \right\} + b \left\{ x^2 - \frac{1}{3} \right\}. \quad □$$

Special Functions and Orthogonal Polynomials

Exercise **1.6**

Determine the nature of the point at infinity for the following differential equations, then find the series solutions for large values of x:

1. $2x^3 y'' + x(3x-1)y' - 2y = 0$

$$\text{Ans: } y = a\left(1 + \frac{2}{x} + \frac{1}{3x^2}\right) + b\, x^{-1/2} \sum_{n=0}^{\infty} \frac{(-1)^n 3\, x^{-n}}{2^n\, n!\,(2n-3)(2n-1)(2n+1)}$$

2. $3x^3 y'' + x(4x+1)y' - 2y = 0$

$$\text{Ans: } y = a \sum_{n=0}^{\infty} \frac{(3n+4)}{4 \cdot 3^n\, n!} x^{-n-1/3} + b\left[1 + \sum_{n=1}^{\infty} \frac{(n+1)x^{-n}}{2 \cdot 5 \cdot 8 \cdots (3n-1)}\right]$$

3. $(1-x^2)y'' - 2xy' + 2y = 0$

$$\text{Ans: } y = a\, x^{-1} + b\, x^{-2}\left(1 + \frac{3}{5x^2} + \frac{3}{7x^4} + \cdots\right)$$

Chapter Two

Gamma and Beta Functions
and Others

Gamma Function $\Gamma(n)$

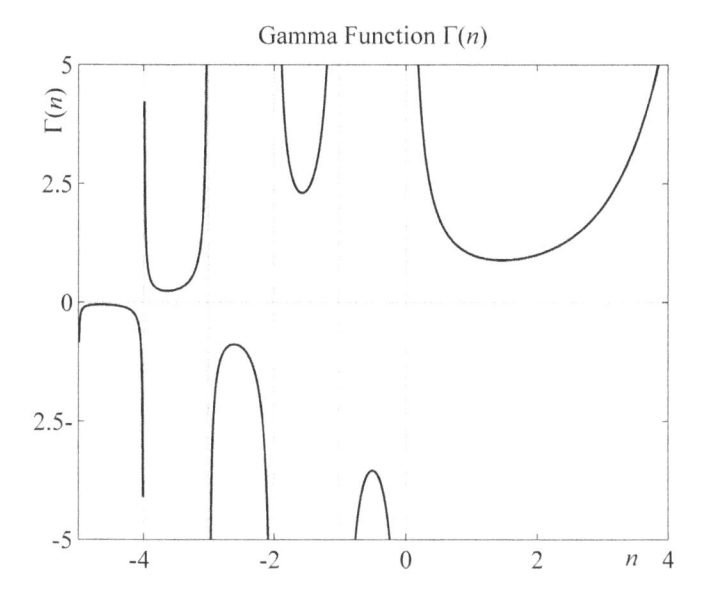

$$\Gamma(n) = \frac{1}{n} \prod_{k=1}^{\infty} \left[\left(1 + \frac{1}{k}\right)^n \left(1 + \frac{n}{k}\right)^{-1} \right]$$

Special Functions and Orthogonal Polynomials

Chapter 2
Gamma and Beta Functions and Others

2.1. Gamma Function

The Gamma function, usually denoted by $\Gamma(n)$, is seen as a generalization of the factorial. It was Euler[1], a Swiss mathematician, who first worked on the curve of the function in 1729. The name and the notation $\Gamma(n)$ of Gamma function has been set by Legendre[2] in 1809. This function is also called *Euler Gamma Function* or *the Eulerian Integral of the Second Kind*. Gamma Function is defined by the improper integral (*Euler Integral Form*)

$$\Gamma(n) = \int_0^\infty e^{-x} x^{n-1} dx \qquad (1)$$

The variable n may be complex. The integral is absolutely convergent for $\mathrm{Re}(n) > 0$.

Gamma function shows up in various applications and areas such as definite integrals, asymptotic series, zeta function, number theory, Bessel function, etc.... It has numerous properties that we will study them in some details.

Gauss[3] gave an alternate notation called the Pi-function as $\Pi(n) = \Gamma(n+1)$.

Property 1: $\boxed{\Gamma(1) = 1}$

Proof: From the definition, we have

$$\Gamma(1) = \int_0^\infty e^{-x} x^{1-1} dx = \int_0^\infty e^{-x} dx = \left[-e^{-x}\right]_0^\infty = 1 \qquad \square$$

---

1. Leonhard Euler (1707-1783) studied Beta and Gamma functions, which he introduced first in 1729.
2. Adrien-Marie Legendre (1752-1833).
3. Carl Friedrich Gauss (1777-1855).

Special Functions and Orthogonal Polynomials

Note: We took the liberty of integrating directly without resorting to the procedure of limits for the improper integral. This is because we know that the integral is convergent for $n > 0$.

Property 2: $\boxed{\Gamma(n+1) = n\Gamma(n),\ n > 0}$

Proof: Again from the definition, we have

$$\Gamma(n+1) = \int_0^\infty e^{-x} x^n dx = -\int_0^\infty x^n de^{-x}$$

Integrating by parts once, we obtain

$$\Gamma(n+1) = -\left[e^{-x} x^n\right]_0^\infty + n \int_0^\infty e^{-x} x^{n-1} dx$$

The quantity in the square brackets vanishes at both limits (use l'Hôpital's rule $(n-1)$ times for the upper limit), then

$$\Gamma(n+1) = n\Gamma(n),\ n > 0. \qquad \square$$

Note: For positive values of n, we need only to know the values of $\Gamma(n)$ for $1 \leq n \leq 2$. We can then determine the values of $\Gamma(n)$ for $n > 2$ in terms of $\Gamma(n)$ for $1 \leq n \leq 2$ using property 2. We give here a table for some values of $\Gamma(n)$, $0 < n \leq 2$. The minimum value for $\Gamma(n)$ is $0.8856031944...$ at $n = 1.46163...$

n	$\Gamma(n)$	n	$\Gamma(n)$	n	$\Gamma(n)$
1.00	1.00000	1.55	0.88887	1/5	4.5909
1.05	0.97350	1.60	0.89352	1/4	3.6256
1.10	0.95135	1.65	0.90012	1/3	2.6789
1.15	0.93304	1.70	0.90864	2/5	2.2182
1.20	0.91817	1.75	0.91906	3/5	1.4892
1.25	0.90640	1.80	0.93138	2/3	1.3541
1.30	0.89747	1.85	0.9618	3/4	1.2254
1.35	0.89115	1.90	0.96177	4/5	1.1642
1.40	0.88726	1.95	0.97988		
1.45	0.88566	2.00	1.00000		
1.50	0.88623				

Gamma and Beta Functions and Others

Property 3: If n is a positive integer, then $\boxed{\Gamma(n+1) = n!}$

Proof: From **Property** 2, we have

$$\Gamma(n+1) = n\Gamma(n) = n(n-1)\Gamma(n-1) = \cdots = n(n-1)(n-2)\cdots 3\cdot 2\cdot 1\cdot \Gamma(1),$$

we obtain

$$\Gamma(n+1) = n!, \text{ if n is a positive integer.} \qquad \square$$

If $\alpha \geq 0$, but α is not necessarily an integer, we have
$$\Gamma(n+\alpha+1) = (n+\alpha)\Gamma(n+\alpha) = (n+\alpha)(n+\alpha-1)\Gamma(n+\alpha-1)$$
$$= \cdots = (n+\alpha)(n+\alpha-1)\cdots(1+\alpha)\Gamma(1+\alpha)$$

then,

$$\Gamma(n+\alpha+1) = (n+\alpha)(n+\alpha-1)\cdots(1+\alpha)\Gamma(1+\alpha) \qquad (2)$$

Equation (2) demonstrates the **Factorial Property of Gamma Function**.

Property 4: $\boxed{\Gamma(n) = 2\int_0^\infty e^{-t^2} t^{2n-1} dt}$

Proof: Letting $x = t^2$ in the definition of Gamma function, we get $dx = 2t\, dt$,

and $\Gamma(n) = \int_0^\infty e^{-t^2}(t^2)^{n-1}\cdot 2t\, dt = 2\int_0^\infty e^{-t^2} t^{2n-1} dt$. $\qquad \square$

Property 5: $\boxed{\Gamma\left(\frac{1}{2}\right) = \sqrt{\pi}}$

Proof: From **Property** 4, we have $\Gamma\left(\frac{1}{2}\right) = 2\int_0^\infty e^{-t^2} dt$.

To evaluate the integral to the right, we proceed as follows. Letting

$$I = \int_0^\infty e^{-x^2} dx = \int_0^\infty e^{-y^2} dy \text{ , then}$$

$$I^2 = \int_0^\infty e^{-x^2} dx \cdot \int_0^\infty e^{-y^2} dy = \int_0^\infty\int_0^\infty e^{-(x^2+y^2)} dx\, dy \text{ .}$$

This is a double integral where the area of integration is the whole of the

59

first quadrant. To evaluate this integral, it is clearly easier to change to polar coordinates where $x = r\cos\theta$ and $y = r\sin\theta$, and the unit area is $rd\theta dr$. The values of r and θ that cover the first quadrant are r from 0 to ∞, and θ from 0 to $\pi/2$. Therefore

$$I^2 = \int_0^{\pi/2}\int_0^\infty e^{-r^2}\cdot rd\theta dr = \frac{1}{2}\int_0^{\pi/2} d\theta \cdot \int_0^\infty e^{-r^2} dr^2 = \frac{1}{2}\cdot\frac{\pi}{2}\cdot 1 = \frac{\pi}{4}.$$

Thus, $I = \int_0^\infty e^{-t^2} dt = \frac{\sqrt{\pi}}{2}$, and $\Gamma\left(\frac{1}{2}\right) = \sqrt{\pi}$. □

We have seen that by letting $x = t^2$ in the integral of Equation (1) we have

$$\Gamma(n) = 2\int_0^\infty e^{-t^2} t^{2n-1} dt$$

This is another integral form of Gamma function. Other substitutions will result in different integral forms of Gamma function.

Property 6: $\quad \Gamma(n) = \int_0^1 \left(\ln\frac{1}{y}\right)^{n-1} dy$

Proof: This form of Gamma function was proposed by Euler in early in 1730.

In Equation (1), letting $y = e^{-x}$, then $x = \ln\frac{1}{y}$ and $dx = -\frac{dy}{y}$.

The integral equals $\Gamma(n) = -\int_1^0 y\left(\ln\frac{1}{y}\right)^{n-1}\frac{dy}{y} = \int_0^1 \left(\ln\frac{1}{y}\right)^{n-1} dy$. □

Property 7: $\quad \Gamma(n) = \alpha^n \int_0^\infty e^{-\alpha x} x^{n-1} dx$

Proof: Replacing x with αx in Equation (1), we get

$$\Gamma(n) = \int_0^\infty e^{-\alpha x}(\alpha x)^{n-1}\alpha dx = \alpha^n \int_0^\infty e^{-\alpha x} x^{n-1} dx. \quad □$$

Gamma and Beta Functions and Others

Property 8: $$\Gamma(n+1) = \int_0^\infty e^{-y^{1/n}} dy$$

Proof: In Equation (1), letting $y = x^n$, $x = y^{1/n}$ then $dx = \frac{1}{n} y^{-1+1/n} dy$,

and the integral becomes

$$\Gamma(n) = \int_0^\infty e^{-y^{1/n}} \left(y^{1/n}\right)^{n-1} \cdot \frac{1}{n} y^{-1+1/n} dy = \frac{1}{n} \int_0^\infty e^{-y^{1/n}} dy,$$

therefore, $\Gamma(n+1) = \int_0^\infty e^{-y^{1/n}} dy$. □

For negative values of n, Gamma function behaves differently, and we have the following property.

Property 9: If n is zero or a negative integer, then $\Gamma(n) = \infty$.

Proof: From **Property 2**, we have $\Gamma(n) = \dfrac{\Gamma(n+1)}{n}$, then for:

$n = 0$: $\Gamma(0) = \dfrac{\Gamma(1)}{0} = \infty$ $\qquad$ $n = -1$: $\Gamma(-1) = \dfrac{\Gamma(0)}{-1} = \infty$

$n = -2$: $\Gamma(-2) = \dfrac{\Gamma(-1)}{-2} = \infty$ $\qquad$ $n = -3$: $\Gamma(-3) = \dfrac{\Gamma(-2)}{-3} = \infty$.

And so on. In general, $\Gamma(n) = \infty$, n negative integer or zero. □

Property 10: Gauss[4] Formula for Gamma Function:

$$\Gamma(n) = \frac{1}{n} \prod_{k=1}^\infty \left[\left(1+\frac{1}{k}\right)^n \left(1+\frac{n}{k}\right)^{-1}\right]$$

Proof: We know that: $e^{-x} = \lim\limits_{k \to \infty} \left(1 - \dfrac{x}{k}\right)^k$, then substituting this expression

in Euler integral form, we obtain

$$\Gamma(n) = \int_0^\infty e^{-x} x^{n-1} dx = \lim_{k \to \infty} \int_0^k \left(1-\frac{x}{k}\right)^k x^{n-1} dx.$$

---

[4] Carl Friedrich Gauss (1777-1855)

61

Special Functions and Orthogonal Polynomials

Let $t = \dfrac{x}{k}$, we obtain

$$\Gamma(n) = \lim_{k \to \infty} \int_0^1 (1-t)^k \, k^{n-1} t^{n-1} k \, dt = \lim_{k \to \infty} k^n \int_0^1 (1-t)^k \, t^{n-1} dt \,.$$

Now, if k is an integer, then the integral to the right of this last equation becomes

$$\int_0^1 (1-t)^k t^{n-1} dt = \frac{1}{n} \int_0^1 (1-t)^k \, dt^n = \frac{1}{n} \Big[(1-t)^k t^n \Big]_0^1 - \frac{1}{n} \int_0^1 t^n d(1-t)^k$$

$$= \frac{k}{n} \int_0^1 (1-t)^{k-1} t^n dt = \frac{k(k-1)}{n(n+1)} \int_0^1 (1-t)^{k-2} t^{n+1} dt$$

$$= \frac{k(k-1)\cdots 2\cdot 1}{n(n+1)\cdots(n+k-1)} \int_0^1 t^{n+k-1} dt = \frac{k(k-1)\cdots 2\cdot 1}{n(n+1)\cdots(n+k-1)} \left[\frac{t^{n+k}}{n+k} \right]_0^1$$

$$= \frac{k!}{n(n+1)\cdots(n+k)} \,.$$

Then, we have

$$\Gamma(n) = \lim_{k \to \infty} \frac{k^n k!}{n(n+1)\cdots(n+k)} = \frac{1}{n} \lim_{k \to \infty} \frac{1\cdot 2\cdot 3 \cdots\cdot k \cdot k^n}{(n+1)(n+2)\cdots(n+k)}$$

$$= \frac{1}{n} \lim_{k \to \infty} \frac{1}{(1+n)(1+n/2)(1+n/3)\cdots(1+n/k)} \cdot k^n$$

$$= \frac{1}{n} \lim_{k \to \infty} \frac{1}{(1+n)(1+n/2)(1+n/3)\cdots(1+n/k)} \cdot \frac{2^n 3^n \cdots k^n}{1^n 2^n \cdots (k-1)^n}$$

We can multiply the expression inside the limit by $\dfrac{(k+1)^n}{k^n}$, since

$\lim\limits_{k \to \infty} \dfrac{(k+1)^n}{k^n} = 1$, then we get

$$\Gamma(n) = \frac{1}{n} \lim_{k \to \infty} \frac{1}{(1+n)(1+n/2)(1+n/3)\cdots(1+n/k)} \cdot \frac{2^n 3^n \cdots (k+1)^n}{1^n 2^n \cdots k^n}$$

$$= \frac{1}{n} \prod_{k=1}^{\infty} \frac{1}{1+\dfrac{n}{k}} \cdot \frac{(k+1)^n}{k^n} = \frac{1}{n} \prod_{k=1}^{\infty} \left[\left(1 + \frac{1}{k}\right)^n \left(1 + \frac{n}{k}\right)^{-1} \right].$$

This product formula is valid for all values of n except $n = 0, -1, -2, \cdots$ □

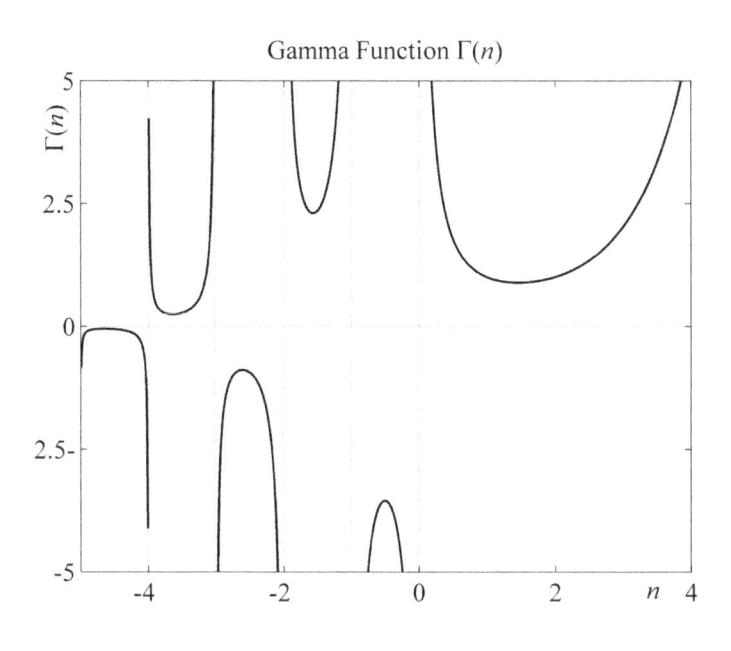

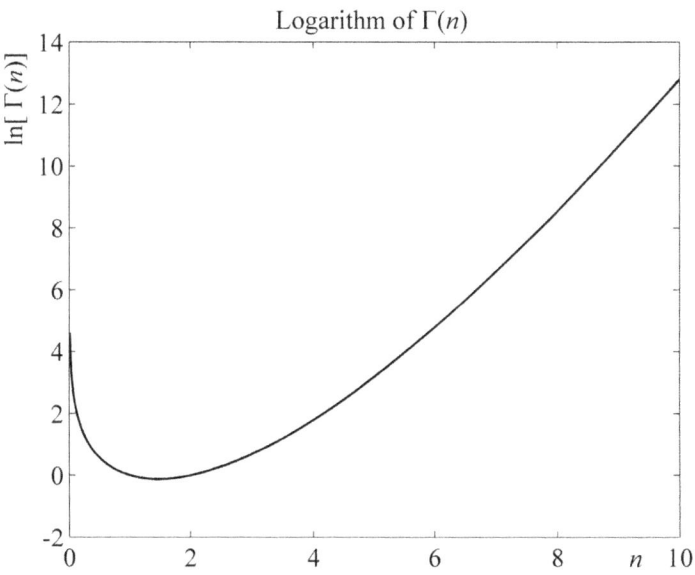

Gamma Function $\Gamma(n)$

Logarithm of $\Gamma(n)$

Property 11: **Weierstrass[5] Form of Gamma function**:

If n is a real number except $n = 0, -1, -2, \cdots$, then

$$\Gamma(n) = \left[n e^{\gamma n} \prod_{k=1}^{\infty} \left(1 + \frac{n}{k}\right) e^{-n/k} \right]^{-1}$$

Where γ is Euler's (or Euler-Mascheroni) constant given by

$$\gamma = \lim_{n \to \infty} \left[\sum_{k=1}^{n} \frac{1}{k} - \ln n \right] = 0.577215665.$$

Proof: In the derivation of Gauss' Formula we had

$$\Gamma(n) = \lim_{k \to \infty} \frac{k^n \cdot k!}{n(n+1)\cdots(n+k)}$$

$$= \lim_{k \to \infty} \left[n^{-1} \left(1 + \frac{n}{1}\right)^{-1} \left(1 + \frac{n}{2}\right)^{-1} \cdots \left(1 + \frac{n}{k}\right)^{-1} k^n \right], \text{ then}$$

$$\frac{1}{\Gamma(n)} = \lim_{k \to \infty} \left[n \left(1 + \frac{n}{1}\right) \left(1 + \frac{n}{2}\right) \cdots \left(1 + \frac{n}{k}\right) \cdot e^{-n \ln k} \right]$$

$$= \lim_{k \to \infty} \left[\frac{n \left(1 + \frac{n}{1}\right) \cdot e^{-n} \left(1 + \frac{n}{2}\right) \cdot e^{-n/2} \cdots \left(1 + \frac{n}{k}\right) \cdot e^{-n/k}}{e^{\left(1 + \frac{1}{2} + \frac{1}{3} + \cdots + \frac{1}{k} + \ln k\right) \cdot n}} \times \right]$$

Therefore, $\dfrac{1}{\Gamma(n)} = n e^{\gamma n} \prod_{k=1}^{\infty} \left(1 + \dfrac{n}{k}\right) e^{-n/k}$.[6] □

Property 12: **Hankel's Formula for Gamma Function**:

$$\Gamma(n) = \frac{1}{2i \sin(n\pi)} \int_C e^t t^{n-1} dt$$

, where C is the contour starting at $-\infty$ below the real axis, enclosing the origin and returning to $-\infty$ above the real axis.

---

5. Karl Weierstrass (1815-1895)

[6] We have used the relation

$$k^n = e^{n \ln k} = e^{n(\ln k - 1 - 1/2 - \cdots - 1/k)} \cdot e^{n + n/2 + \cdots + n/k}$$

Gamma and Beta Functions and Others

***Property* 13: Legendre Duplication Formula:**

$$\boxed{\Gamma(2n) = \frac{2^{2n-1}}{\sqrt{\pi}} \Gamma(n)\Gamma(n+1/2)}$$

If n is a positive integer, then $\Gamma(n+1/2) = \dfrac{(2n)!}{2^{2n}n!}\sqrt{\pi}$.

***Property* 14: Triplication Formula:**

$$\boxed{\Gamma(3n) = \frac{3^{3n-1/2}}{2\pi} \Gamma(n)\Gamma(n+1/3)\Gamma(n+2/3)}$$

***Property* 15: Gauss Multiplication Formula:**

$$\boxed{\Gamma(n)\Gamma\left(n+\frac{1}{k}\right)\cdots\Gamma\left(n+\frac{k-1}{k}\right) = (2\pi)^{(k-1)/2} k^{-kn+1/2}\Gamma(kn)}$$

***Property* 16: Reflection Property of Gamma Function:**

$$\boxed{\Gamma(n)\Gamma(1-n) = \frac{\pi}{\sin \pi n}}$$

Proof: From Weierstrass' Formula, we have:

$$\frac{1}{\Gamma(n)}\cdot\frac{1}{\Gamma(-n)} = -n^2 e^{\gamma n} e^{-\gamma n} \prod_{k=1}^{\infty}\left(1+\frac{n}{k}\right)e^{-n/k}\left(1-\frac{n}{k}\right)e^{n/k}$$

$$= -n^2 \prod_{k=1}^{\infty}\left(1-\frac{n^2}{k^2}\right)$$

But, we have $\Gamma(-n) = -\dfrac{\Gamma(1-n)}{n}$. Substituting this value, we get

$$\frac{1}{\Gamma(n)}\cdot\frac{1}{\Gamma(1-n)} = n\prod_{k=1}^{\infty}\left(1-\frac{n^2}{k^2}\right).$$

Also, from the well known infinite product $\sin(n\pi) = \pi n \prod_{k=1}^{\infty}\left(1-\dfrac{n^2}{k^2}\right)$,

we obtain $\Gamma(n)\cdot\Gamma(1-n) = \dfrac{\pi}{\sin n\pi}$, where n and $1-n$ are not negative integers or zero. □

Property 17: **Stirling's Formula**

The behavior of Gamma function as n becomes large was given by Stirling[7] and De Moivre[8]. Depending on whether n is an integer or not, we have two formulas.

If the integer n tends to infinity, we have the asymptotic formula

$$\Gamma(n+1) = n! \sim \sqrt{2\pi n}\; n^n e^{-n}.$$

As $x \to \infty$, we have the Stirling's asymptotic formula with higher order correction terms. This is given as

$$\Gamma(x+1) = \sqrt{2\pi x}\; x^x e^{-x}\left(1 + \frac{1}{12x} + \frac{1}{288x^2} - \frac{139}{51840x^3} - \cdots\right)$$

Proof: We will prove the first formula only. From Euler integral formula, if n is a positive integer, we have

$$\Gamma(n+1) = n! = \int_0^\infty e^{-x} x^n dx$$

Let us put the integrand in the form $e^{-x} x^n = e^{n \ln x - x} = e^{g(x)}$.
If we expand $g(x) = n \ln x - x$ in Taylor's expansion about $x = n$ (the maximum of $g(x)$), we get $g(x) = -n + n \ln n - \frac{(x-n)^2}{2n} + \cdots$.

Substituting these in the expression for $n!$, we obtain for large n,

$$n! = \int_0^\infty e^{-n + n \ln n - \frac{(x-n)^2}{2n} + \cdots} dx \approx e^{-n} n^n \int_{-\infty}^\infty e^{-\frac{(x-n)^2}{2n}} dx.$$

Note that we made the integral on the right of this equation run from ∞ to $-\infty$. This is because n is large and the error in doing so will be negligible. Letting $t = (x - n)$, then

$$n! \approx e^{-n} n^n \int_{-\infty}^\infty e^{-t^2/2n} dt = e^{-n} n^n \sqrt{2n\pi}.\text{[9]} \qquad \square$$

---

[7] James Stirling (1692-1730)

[8] Abraham de Moivre (1667-1754)

[9] We have used the fact that $\int_{-\infty}^\infty e^{-t^2} dt = \sqrt{\pi}$

Gamma and Beta Functions and Others

Property 18: Digamma Function $\Psi(n)$:

The Digamma function is defined for any non-negative integer or zero as the logarithmic derivative of $\Gamma(n)$, i.e.,

$\Psi(n) = \dfrac{d}{dn}\{\ln[\Gamma(n)]\}$. We will show that

$$\boxed{\Psi(n) = -\gamma + \sum_{k=1}^{\infty}\left(\dfrac{x-1}{k(n+k-1)}\right), \quad n \neq 0, -1, -2, \cdots}$$

Proof: From Weierstass' Formula $\dfrac{1}{\Gamma(n)} = ne^{\gamma n}\prod_{k=1}^{\infty}\left(1+\dfrac{n}{k}\right)e^{-n/k}$, taking the logarithm of both sides, we get

$$-\ln[\Gamma(n)] = \ln n + \gamma n + \sum_{k=1}^{\infty}\left[\ln\left(1+\dfrac{n}{k}\right) - \dfrac{n}{k}\right].$$

Differentiating this expression with respect to n, we obtain

$$\Psi(n) = \dfrac{d}{dn}\{\ln[\Gamma(n)]\} = \dfrac{\Gamma'(n)}{\Gamma(n)} = -\gamma - \dfrac{1}{n} + \underbrace{\sum_{k=1}^{\infty}\left(\dfrac{1}{k} - \dfrac{1}{n+k}\right)}_{\sum_{k=1}^{\infty}\left(\dfrac{1}{k} - \dfrac{1}{n+k-1}\right)}$$

$$= -\gamma + \sum_{k=1}^{\infty}\left(\dfrac{n-1}{k(n+k-1)}\right), \quad n \neq 0, -1, -2, \cdots \qquad \square$$

Property 19: Polygamma Functions:

$$\Psi_m(n) = \dfrac{d^m \Psi(n)}{dn^m} = \sum_{k=1}^{\infty}\left(\dfrac{(-1)^{m+1}m!}{(n+k-1)^{m+1}}\right)$$

Proof: From the definition of the Digamma function $\Psi(n) = \dfrac{\Gamma'(n)}{\Gamma(n)}$, differentiation several times, we obtain

$$\Psi'(n) = \dfrac{\Gamma(n)\Gamma''(n) - \Gamma'^2(n)}{\Gamma^2(n)} = \sum_{k=1}^{\infty}\dfrac{1}{(n+k-1)^2}$$

$$\Psi''(n) = -\sum_{k=1}^{\infty} \frac{2!}{(n+k-1)^3},$$

$$\Psi'''(n) = \sum_{k=1}^{\infty} \frac{3!}{(n+k-1)^4},$$

. . .

$$\Psi^{(m)}(n) = \sum_{k=1}^{\infty} \frac{(-1)^{m+1} m!}{(n+k-1)^{m+1}} = \frac{d^m \Psi(n)}{dn^m} = \Psi_m(n) \qquad \square$$

The Digamma and Polygamma functions satisfy the following relations:

$$\Psi(n+1) = \Psi(n) + \frac{1}{n}$$

$$\Psi(n+k) = \Psi(n) + \frac{1}{n} + \frac{1}{n+1} + \frac{1}{n+2} + \cdots + \frac{1}{n+k-1}, \quad n \geq 1$$

$$\Psi_m(n+1) = \Psi_m(n) + \frac{(-1)^m m!}{n^{m+1}}$$

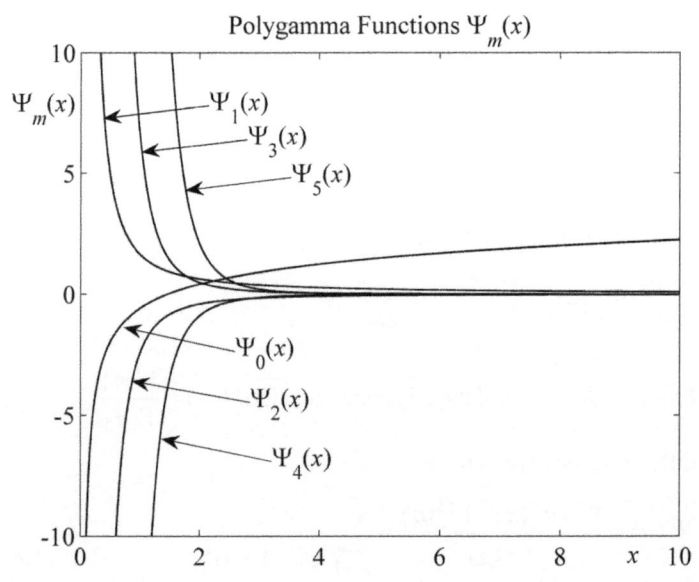

$$\Psi(1-n) = \Psi(n) + \pi \cot n\pi, \quad \Psi(2n) = \frac{1}{2}\Psi(n) + \frac{1}{2}\Psi\left(n + \frac{1}{2}\right) + \ln 2.$$

Polygamma Functions $\Psi_m(x)$

[Graph showing polygamma functions $\Psi_0(x)$ through $\Psi_5(x)$ on axes with x from 0 to 10 and y from -10 to 10]

Gamma and Beta Functions and Others

From the previous properties of Gamma function, it is possible to evaluate some integrals that are otherwise difficult to evaluate.

Example 1: Evaluate the following integrals:

i) $\int_0^\infty x^4 e^{-x}\, dx$ ii) $\int_0^\infty x^3 e^{-2x}\, dx$ iii) $\int_0^\infty \sqrt{x}\, e^{-x^3}\, dx$

iv) $\int_0^\infty 2^{-3x^2}\, dx$ v) $\int_0^1 \frac{dx}{\sqrt{-\ln x}}$

Solution: i) From the definition of Gamma function and **Property 3**, we obtain

$$\int_0^\infty x^4 e^{-x}\, dx = \Gamma(5) = 4! = 24$$

ii) From **Property 7**, we have

$$\int_0^\infty x^3 e^{-2x}\, dx = \frac{1}{2^4}\Gamma(4) = \frac{3!}{16} = \frac{3}{8}$$

iii) Letting $x^3 = t$, then $dx = \frac{1}{3} t^{-2/3} dt$ and the integral becomes

$$\int_0^\infty \sqrt{x}\, e^{-x^3}\, dx = \int_0^\infty t^{1/6} e^{-t} \cdot \tfrac{1}{3} t^{-2/3} dt = \tfrac{1}{3}\int_0^\infty t^{-1/2} e^{-t} dt = \tfrac{1}{3}\Gamma\!\left(\tfrac{1}{2}\right) = \frac{\sqrt{\pi}}{3}$$

iv) $\int_0^\infty 2^{-3x^2} dx = \int_0^\infty \left(e^{\ln x}\right)^{-3x^2} dx = \int_0^\infty e^{-(3\ln 2)x^2} dx$

Letting $3\ln 2 = a$ and $ax^2 = t$, then the integral becomes

$$\frac{1}{2\sqrt{a}}\int_0^\infty t^{-1/2} e^{-t} dt = \frac{\Gamma\!\left(\tfrac{1}{2}\right)}{2\sqrt{a}} = \frac{1}{2}\sqrt{\frac{\pi}{a}}$$

Thus $\int_0^\infty 2^{-3x^2} dx = \frac{1}{2}\sqrt{\frac{\pi}{3\ln 2}}$.

v) Letting $-\ln x = t$, then $x = e^{-t}$ and when $x = 1$, $t = 0$ and when $x = 0$, $y = \infty$, then the integral becomes

$$\int_0^1 \frac{dx}{\sqrt{-\ln x}} = \int_0^\infty t^{-1/2} e^{-t} dt = \sqrt{\pi}.\qquad\square$$

Example 2: Evaluate the integral $\int_0^\infty e^{-\alpha x^2} \cos \beta x \, dx$, hence deduce that:

$$\int_0^\infty e^{-x^2} \cos 2x \, dx = \frac{\sqrt{\pi}}{2e}.$$

Solution: This integral contains two parameters α and β, then let

$$I(\alpha, \beta) = \int_0^\infty e^{-\alpha x^2} \cos \beta x \, dx.$$

Differentiating with respect to β, we get

$$\frac{\partial I}{\partial \beta} = \int_0^\infty -x \, e^{-\alpha x^2} \sin \beta x \, dx = \frac{1}{2\alpha} \int_0^\infty \sin \beta x \, d e^{-\alpha x^2}.$$

Integrating by parts, we obtain

$$\frac{\partial I}{\partial \beta} = \left[\frac{e^{-\alpha x^2}}{2\alpha} \sin \beta x \right]_0^\infty - \frac{\beta}{2\alpha} \int_0^\infty e^{-\alpha x^2} \cos \beta x \, dx.$$

The term in the brackets vanishes at both limits, then

$$\frac{\partial I}{\partial \beta} = -\frac{\beta}{2\alpha} I.$$

This is a first order differential equation that can be solved by separation of variables. The solution is

$$I = c e^{-\beta^2 /(4\alpha)},$$

where c is the constant of integration.

To evaluate c, we have

$$I(\alpha, 0) = \int_0^\infty e^{-\alpha x^2} dx \quad \text{(here is where Gamma function will appear!).}$$

Let $t = \alpha x^2$, then $dx = \frac{t^{1/2} dt}{2\sqrt{\alpha}}$, and

Gamma and Beta Functions and Others

$$I(\alpha,0) = \frac{1}{2\sqrt{\alpha}} \int_0^\infty e^{-t} t^{1/2} dt = \frac{1}{2\sqrt{\alpha}} \Gamma\left(\frac{1}{2}\right) = \frac{1}{2}\sqrt{\frac{\pi}{\alpha}}.$$

From this we have $c = \frac{1}{2}\sqrt{\frac{\pi}{\alpha}}$, and the integral becomes

$$\int_0^\infty e^{-\alpha x^2} \cos\beta x\, dx = \frac{1}{2}\sqrt{\frac{\pi}{\alpha}} e^{-\beta^2/(2\alpha)}$$

Now, if $\alpha = 1$ and $\beta = 2$, we get $\int_0^\infty e^{-x^2} \cos 2x\, dx = \frac{\sqrt{\pi}}{2e}$. □

Example 3: Evaluate the following integrals:

i) $\int_0^1 \ln\left(\frac{1}{x}\right)^3 dx$ ii) $\int_0^\infty e^{-\sqrt[3]{y}}\, dy$

Solution: i) From **Property 6**, we have $\int_0^1 \ln\left(\frac{1}{x}\right)^3 dx = \Gamma(4) = 3! = 6$.

ii) From **Property 8**, we have $\int_0^\infty e^{-\sqrt[3]{y}}\, dy = 3\Gamma(3) = 6$. □

Example 4: Show that $\Gamma'(n) = -\Gamma(n)\left\{\frac{1}{n} + \gamma + \sum_{k=1}^\infty \left[\frac{1}{k+n} - \frac{1}{k}\right]\right\}$, and hence deduce that $\Gamma'(1) = -\gamma$ and $\Gamma'\left(\frac{1}{2}\right) = -(\gamma + \ln 2)\sqrt{\pi}$. Also, if n is a positive integer, show that

$$\Gamma'(n) = -(n-1)!\left\{\frac{1}{n} + \gamma - \sum_{k=1}^n \frac{1}{k}\right\}.$$

Solution: From Weierstrass' form of Gamma Function (**Property 11**):

$$\Gamma(n) = \left[ne^{\gamma n} \prod_{k=1}^\infty \left(1 + \frac{n}{k}\right) e^{-n/k}\right]^{-1}.$$

71

Special Functions and Orthogonal Polynomials

Taking the logarithm of both sides, we get

$$-\ln[\Gamma(n)] = \ln n + \gamma n + \sum_{k=1}^{\infty}\left[\ln\left(1+\frac{n}{k}\right)-\frac{n}{k}\right].$$

Differentiating with respect to n, we obtain

$$-\frac{\Gamma'(n)}{\Gamma(n)} = \frac{1}{n} + \gamma + \sum_{k=1}^{\infty}\left[\frac{1}{k+n}-\frac{1}{k}\right], \text{ or }$$

$$\Gamma'(n) = -\Gamma(n)\left\{\frac{1}{n}+\gamma+\sum_{k=1}^{\infty}\left[\frac{1}{k+n}-\frac{1}{k}\right]\right\}.$$

Letting $n = 1$, we get

$$\Gamma'(1) = -\Gamma(1)\left\{1+\gamma+\left[\left(\tfrac{1}{2}-1\right)+\left(\tfrac{1}{3}-\tfrac{1}{2}\right)+\cdots+\left(\tfrac{1}{k+1}-\tfrac{1}{k}\right)+\cdots\right]\right\}$$

$$= -(1+\gamma-1) = -\gamma.$$

Letting $n = \frac{1}{2}$, we get

$$\Gamma'\left(\tfrac{1}{2}\right) = -\Gamma\left(\tfrac{1}{2}\right)\left\{2+\gamma-\sum_{k=1}^{\infty}\frac{1}{k(2k+1)}\right\}$$

$$= \sqrt{\pi}\left\{-2-\gamma+\sum_{k=1}^{\infty}\left[\frac{1}{k}-\frac{2}{2k+1}\right]\right\}$$

$$= \sqrt{\pi}\left\{-2-\gamma+\left[1+\tfrac{1}{2}+\tfrac{1}{3}+\cdots-2\left(\tfrac{1}{3}+\tfrac{1}{5}+\tfrac{1}{7}+\cdots\right)\right]\right\}$$

$$= \sqrt{\pi}\left\{-2-\gamma+\left[2-\left(1-\tfrac{1}{2}+\tfrac{1}{3}-\tfrac{1}{4}+\cdots\right)\right]\right\}$$

$$= \sqrt{\pi}\{-2-\gamma+2-\ln 2\} = -\sqrt{\pi}\{\gamma+\ln 2\}$$

Finally, if n is a positive integer, we obtain

$$\Gamma'(n) = -\Gamma(n)\left\{\tfrac{1}{n}+\gamma+\left[\left(\tfrac{1}{1+n}-1\right)+\left(\tfrac{1}{2+n}-\tfrac{1}{2}\right)+\left(\tfrac{1}{3+n}-\tfrac{1}{3}\right)+\cdots\right]\right\}$$

$$= -(n-1)!\left\{\tfrac{1}{n}+\gamma-\sum_{k=1}^{n}\tfrac{1}{k}\right\}. \qquad \square$$

Gamma and Beta Functions and Others

Example 5: Express $2 \cdot 5 \cdot 8 \cdots (3n-1)$ in terms of Gamma function.

Solution: From **Property 2**, $\Gamma(n+1) = n\Gamma(n)$, we have

$$\Gamma\left(n+\frac{2}{3}\right) = \left(n-\frac{1}{3}\right)\Gamma\left(n-\frac{1}{3}\right) = \left(n-\frac{1}{3}\right)\left(n-\frac{4}{3}\right)\Gamma\left(n-\frac{4}{3}\right) = \cdots$$

$$= \left(n-\frac{1}{3}\right)\left(n-\frac{4}{3}\right)\left(n-\frac{7}{3}\right)\cdots\frac{8}{3}\cdot\frac{5}{3}\cdot\frac{2}{3}\Gamma\left(\frac{2}{3}\right)$$

Therefore

$$2 \cdot 5 \cdot 8 \cdots (3n-1) = 3^n \frac{\Gamma\left(n+\frac{2}{3}\right)}{\Gamma\left(\frac{2}{3}\right)}. \qquad \square$$

Example 6: If n is a positive integer, show that: $\int_0^1 x^m \ln^n x \, dx = \frac{(-1)^n n!}{(m+1)^{n+1}}$.

Solution: Let $\ln x = -t$ or $x = e^{-t}$ and $dx = -e^{-t} dt$. Also, when $x = 0$, $t = \infty$; and when $x = 1$, $t = 0$. The integral becomes

$$I = \int_0^1 x^m \ln^n x \, dx = \int_{\infty}^0 e^{-mt}(-t)^n(-e^{-t})dt = (-1)^n \int_0^{\infty} e^{-(m+1)t} t^n dt$$

Let $(m+1)t = y$ then $dt = \frac{dy}{m+1}$; and, when $t = 0$, $y = 0$; and when $t = \infty$, $y = \infty$. The integral now becomes

$$I = \int_0^1 x^m \ln^n x \, dx = (-1)^n \int_0^{\infty} e^{-y} \frac{y^n}{(m+1)^n} \cdot \frac{dy}{m+1}$$

$$= \frac{(-1)^n}{(m+1)^{n+1}} \int_0^{\infty} e^{-y} y^n dy = \frac{(-1)^n \Gamma(n+1)}{(m+1)^{n+1}} = \frac{(-1)^n n!}{(m+1)^{n+1}} \qquad \square$$

Example 7: If $\alpha, \beta, m, n > 0$, show that:

$$\int_0^{\infty}\int_0^{\infty} e^{-(\alpha x^2 + \beta y^2)} x^{2m-1} y^{2n-1} dx \, dy = \frac{\Gamma(m)\Gamma(n)}{4\alpha^m \beta^n}.$$

Solution: In this double integral we can see that the two dummy variables x and y can be easily separated, then we can write the integral as the product of two separate definite integrals as

73

Special Functions and Orthogonal Polynomials

$$I = \int_0^\infty \int_0^\infty e^{-(\alpha x^2 + \beta y^2)} x^{2m-1} y^{2n-1} dx\, dy$$

$$= \underbrace{\int_0^\infty e^{-\alpha x^2} x^{2m-1} dx}_{I_1} \cdot \underbrace{\int_0^\infty e^{-\beta y^2} y^{2n-1} dy}_{I_2}$$

For I_1, let $\alpha x^2 = t$, then $dt = 2\alpha x\, dx$, and the integral becomes,

$$I_1 = \int_0^\infty e^{-\alpha x^2} x^{2m-1} dx = \frac{1}{2\alpha} \int_0^\infty e^{-t} t^{m-1} dt = \frac{\Gamma(m)}{2\alpha^m}, \quad m, \alpha > 0.$$

Similarly $I_2 = \int_0^\infty e^{-\beta y^2} y^{2n-1} dy = \frac{\Gamma(n)}{2\beta^n}, \quad n, \beta > 0.$

Therefore, $\int_0^\infty \int_0^\infty e^{-(\alpha x^2 + \beta y^2)} x^{2m-1} y^{2n-1} dx\, dy = \frac{\Gamma(m)\Gamma(n)}{4\alpha^m \beta^n}$ □

Example 8: Show that: $\Gamma\left(n + \frac{1}{2}\right) = \frac{1 \cdot 3 \cdot 5 \cdots (2n-1)}{2^n} \sqrt{\pi}$.

Solution: From **Property 2**, $\Gamma(n+1) = n\Gamma(n)$, we have

$$\Gamma\left(n + \frac{1}{2}\right) = \left(n - \frac{1}{2}\right)\Gamma\left(n - \frac{3}{2}\right)\left(n - \frac{5}{2}\right) \cdots \frac{5}{2} \cdot \frac{3}{2} \cdot \frac{1}{2} \Gamma\left(\frac{1}{2}\right)$$

$$= \frac{(2n-1)(2n-3)(2n-5) \cdots 5 \cdot 3 \cdot 1}{2^n} \sqrt{\pi} \quad \square$$

Example 9: Using Hankel's formula for Gamma function, show that:

$\Gamma(n+1) = n\Gamma(n), \ n > 0$.

Solution: Hankel's formula is given by $\Gamma(n) = \dfrac{1}{2i \sin(n\pi)} \int_C e^t t^{n-1} dt$, where C is the contour starting at $-\infty$ below the real axis, enclosing the origin and returning to $-\infty$ above the real axis. Then

$$\Gamma(n+1) = \frac{1}{2i\sin[(n+1)\pi]} \int_C e^t t^n \, dt$$

$$= -\frac{1}{2i\sin(n\pi)} \left\{ \left[e^t t^n \right]_{-\infty-i0}^{-\infty+i0} - n \int_C e^t t^{n-1} \, dt \right\}.$$

The expression in the square brackets vanishes at both limits. Thus

$$\Gamma(n+1) = n \int_C e^t t^{n-1} \, dt = n\Gamma(n). \qquad \square$$

***Example* 10:** The Riemann[10]-Zeta function $\zeta(n)$ is defined by:

$$\zeta(n) = \sum_{k=1}^{\infty} \frac{1}{k^n}, \quad \mathrm{Re}(n) > 1.$$

Show that it can be expressed in terms of Gamma function as

$$\zeta(n) = \frac{1}{\Gamma(n)} \int_0^{\infty} \frac{t^{n-1}}{e^t - 1} \, dt.$$

Solution: From Euler integral form $\Gamma(n) = \int_0^{\infty} e^{-x} x^{n-1} \, dx$, let $x = kt$, where k is a positive integer, we obtain

$$\Gamma(n) = \int_0^{\infty} e^{-kt} k^{n-1} t^{n-1} \cdot k \, dt = k^n \int_0^{\infty} e^{-kt} t^{n-1} \, dt.$$

From this, we have $\dfrac{1}{k^n} = \dfrac{1}{\Gamma(n)} \int_0^{\infty} e^{-kt} t^{n-1} \, dt.$

Summing both sides from $k = 1 \to \infty$, we get

---

10. Georg Friedrich Bernhard Riemann (1826-1866)

$$\sum_{k=1}^{\infty}\frac{1}{k^n} = \frac{1}{\Gamma(n)}\int_0^{\infty} t^{n-1}\left(\sum_{k=1}^{\infty} e^{-kt}\right)dt = \frac{1}{\Gamma(n)}\int_0^{\infty} t^{n-1}\left(\frac{1}{1-e^{-t}}-1\right)dt$$

$$= \frac{1}{\Gamma(n)}\int_0^{\infty} t^{n-1}\left(\frac{e^{-t}}{1-e^{-t}}\right)dt = \frac{1}{\Gamma(n)}\int_0^{\infty} \frac{t^{n-1}}{e^t-1}dt.$$

Thus $\zeta(n) = \dfrac{1}{\Gamma(n)}\displaystyle\int_0^{\infty} \dfrac{t^{n-1}}{e^t-1}dt$. $\square$

Note: It can be shown that $\zeta(n)$ is infinite for $n=1$ and that $\zeta(2) = \dfrac{\pi^2}{6}$ and $\zeta(4) = \dfrac{\pi^4}{90}$. Riemann-Zeta function is related to the prime numbers. Euler gave the Product formula

$$\zeta(n) = \prod_{p\ prime}\left(1-\frac{1}{p^n}\right)^{-1}.$$

$\zeta(n)$ satisfies the functional equation

$$\zeta(n) = \pi^{n-1}2^n \sin\left(\frac{\pi n}{2}\right)\Gamma(1-n)\zeta(1-n).$$

Riemann Zeta Function $\zeta(n)$

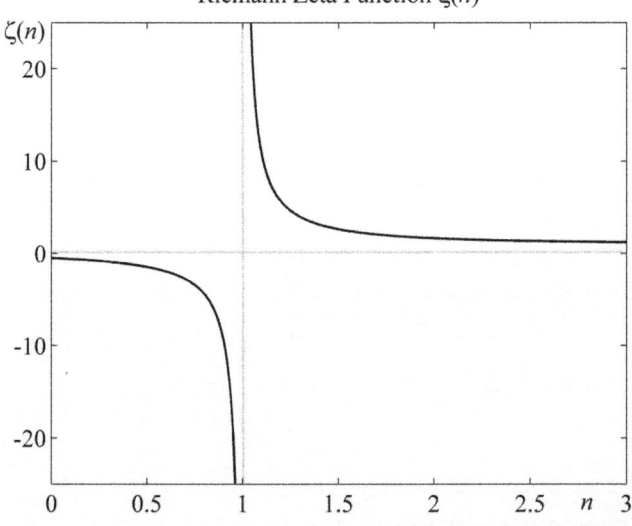

2.2. Beta Function

The Beta function, denoted by $\beta(p,q)$, can be expressed as a definite integral with 0 and 1 as limits, and is given by

$$\beta(p,q) = \int_0^1 x^{p-1}(1-x)^{q-1} dx \tag{4}$$

The integral is convergent for $p,q > 0$. The properties of Beta function are given below.

***Property* 20: Symmetry** : $\boxed{\beta(p,q) = \beta(q,p)}$

Proof: Let $1-x = y$ in the integral of Equation (4), then when $x = 0, y = 1$ and when $x = 1, y = 0$, and the integral becomes

$$\beta(p,q) = -\int_1^0 (1-y)^{p-1} y^{q-1} dy = \int_0^1 y^{q-1}(1-y)^{p-1} dy = \beta(q,p) \quad \square$$

***Property* 21**: $\beta(p,q) = 2 \int_0^{\pi/2} \sin^{2p-1}\theta \cos^{2q-1}\theta \, d\theta$

Proof: Let $x = \sin^2 \theta$ in the integral of Equation (4), then $dx = 2\sin\theta\cos\theta \, d\theta$ and when $x = 0$, $\theta = 0$ and when $x = 1$, $\theta = \pi/2$, then

$$\beta(p,q) = \int_0^{\pi/2} (\sin^2\theta)^{p-1}(\cos^2\theta)^{q-1} \cdot 2\sin\theta\cos\theta \, d\theta, \text{ or}$$

$$\beta(p,q) = 2\int_0^{\pi/2} \sin^{2p-1}\theta \cos^{2q-1}\theta \, d\theta. \quad \square$$

***Property* 22**: $\boxed{\beta(p,q) = \dfrac{\Gamma(p)\Gamma(q)}{\Gamma(p+q)}}$

Proof: From *Property* 4 of Gamma function, we have

$$\Gamma(p) = 2\int_0^\infty e^{-x^2} x^{2p-1} dx \text{ , and } \Gamma(q) = 2\int_0^\infty e^{-y^2} y^{2q-1} dy \text{ , then}$$

$$\Gamma(p)\cdot\Gamma(q) = 2\int_0^\infty e^{-x^2} x^{2p-1} dx \cdot 2\int_0^\infty e^{-y^2} y^{2q-1} dy$$

$$= 4\int_0^\infty\int_0^\infty e^{-(x^2+y^2)} x^{2p-1} y^{2q-1} dx\, dy$$

This is a double integral where the area of integration is the first quadrant. To evaluate this integral, we change to polar coordinates where $x = r\cos\theta$ and $y = r\sin\theta$, and the unit area is $r\, d\theta\, dr$. The values of r and θ that cover the first quadrant are r from 0 to ∞, and θ from 0 to $\pi/2$. Therefore

$$\Gamma(p)\cdot\Gamma(q) = 4\int_0^{\pi/2}\int_0^\infty e^{-r^2} (r\cos\theta)^{2p-1} (r\sin\theta)^{2q-1} r\, d\theta\, dr$$

$$= 2\underbrace{\int_0^{\pi/2} \cos^{2p-1}\theta \sin^{2q-1}\theta\, d\theta}_{\beta(p,q)} \cdot 2\underbrace{\int_0^\infty e^{-r^2} r^{2p+2q-1} dr}_{\Gamma(p+q)}$$

Thus, $\beta(p,q) = \dfrac{\Gamma(p)\Gamma(q)}{\Gamma(p+q)}$. □

Note: From the last property, which gives a relation between Beta and Gamma functions, we can evaluate the values of $\beta(p,q)$ in terms of the known values of Gamma function. Putting $p = q = 1/2$ in

Property 22, we obtain $\beta\left(\dfrac{1}{2},\dfrac{1}{2}\right) = \dfrac{\Gamma\left(\dfrac{1}{2}\right)\Gamma\left(\dfrac{1}{2}\right)}{\Gamma(1)} = \pi$.

Property 23: $\boxed{\beta(p,q) = \int_0^\infty \dfrac{y^{q-1} dy}{(1+y)^{p+q}}}$

Proof: Let $x = \dfrac{1}{1+y}$ in Equation (4), then $y = \dfrac{1-x}{x}$, $dx = -\dfrac{dy}{(1+y)^2}$,

$1-x = \dfrac{y}{1+y}$, and the integral becomes

Gamma and Beta Functions and Others

$$\beta(p,q) = -\int_{\infty}^{0} \frac{1}{(1+y)^{p-1}} \cdot \left(\frac{y}{1+y}\right)^{q-1} \frac{dy}{(1+y)^2} = \int_{0}^{\infty} \frac{y^{q-1}dy}{(1+y)^{p+q}}.$$

Also, because of the symmetric property,

$$\beta(p,q) = \int_{0}^{\infty} \frac{y^{q-1}dy}{(1+y)^{p+q}}. \qquad \square$$

Property 24: $\boxed{\dfrac{\beta(p,q+1)}{q} = \dfrac{\beta(p+1,q)}{p} = \dfrac{\beta(p,q)}{p+q}}$

Proof: From **Property 22**, we have

$$\frac{\beta(p+1,q)}{p} = \frac{\Gamma(p+1)\Gamma(q)}{p\Gamma(p+q+1)} = \frac{p\Gamma(p)\Gamma(q)}{p(p+q)\Gamma(p,q)} = \frac{\beta(p,q)}{p+q}$$

Similarly, we have

$$\frac{\beta(p,q+1)}{q} = \frac{\Gamma(p)\Gamma(q+1)}{q\Gamma(p+q+1)} = \frac{q\Gamma(p)\Gamma(q)}{q(p+q)\Gamma(p,q)} = \frac{\beta(p,q)}{p+q}. \qquad \square$$

Example 11: Evaluate the following integrals:

i) $\int_{0}^{1} x^4(1-x)^3 dx$ ii) $\int_{0}^{\pi/2} \sin^6\theta \cos^7\theta\, d\theta$

iii) $\int_{0}^{\pi/2} \cos^8\theta\, d\theta$ iv) $\int_{0}^{\pi/2} \sqrt{\tan\theta}\, d\theta$

Solution: i) $\int_{0}^{1} x^4(1-x)^3 dx = \beta(5,4) = \dfrac{\Gamma(5)\Gamma(4)}{\Gamma(9)} = \dfrac{4!\,3!}{8!} = \dfrac{1}{280}.$ $\square$

ii) $\int_{0}^{\pi/2} \sin^6\theta \cos^7\theta\, d\theta = \dfrac{1}{2}\beta\!\left(\tfrac{7}{2},4\right) = \dfrac{1}{2}\dfrac{\Gamma\!\left(\tfrac{7}{2}\right)\cdot\Gamma(4)}{\Gamma\!\left(\tfrac{15}{2}\right)} = \dfrac{16}{3003}.$ $\square$

iii) $\int_{0}^{\pi/2} \cos^8\theta\, d\theta = \dfrac{1}{2}\beta\!\left(\tfrac{1}{2},\tfrac{9}{2}\right) = \dfrac{1}{2}\dfrac{\Gamma\!\left(\tfrac{1}{2}\right)\Gamma\!\left(\tfrac{9}{2}\right)}{\Gamma(5)} = \dfrac{105\pi}{768}.$ $\square$

iv) $\int_0^{\pi/2} \sqrt{\tan\theta}\,d\theta = \int_0^{\pi/2} \sin^{1/2}\theta \cos^{1/2}\theta\, d\theta = \frac{1}{2}\beta\left(\frac{3}{4},\frac{1}{4}\right) = \frac{1}{2}\frac{\Gamma\left(\frac{3}{4}\right)\Gamma\left(\frac{1}{4}\right)}{\Gamma(1)}$

From the property $\Gamma(n)\Gamma(1-n) = \dfrac{\pi}{\sin n\pi}$, we get

$\Gamma\left(\frac{3}{4}\right)\Gamma\left(\frac{1}{4}\right) = \dfrac{\pi}{\sin 3\pi/4} = \sqrt{2}\pi$, thus

$\int_0^{\pi/2} \sqrt{\tan\theta}\,d\theta = \frac{1}{2}\sqrt{2}\pi = \dfrac{\pi}{\sqrt{2}}$. □

Example 12: Prove the Legendre Duplication Formula:

$$\sqrt{\pi}\,\Gamma(2n) = 2^{2n-1}\Gamma(n)\,\Gamma\left(n+\frac{1}{2}\right).$$

Solution: From the relation between Gamma and Beta functions, we have

$$\frac{\Gamma(n)\,\Gamma(m)}{\Gamma(n+m)} = \beta(n,m) = 2\int_0^{\pi/2} \sin^{2n-1}\theta \cos^{2m-1}\theta\, d\theta.$$

Letting $m = 1/2$, we obtain $\dfrac{\Gamma(n)\sqrt{\pi}}{\Gamma\left(n+\frac{1}{2}\right)} = 2\int_0^{\pi/2} \sin^{2n-1}\theta\, d\theta.$

Also, from integral calculus, we have

$$\int_0^{\pi/2} \sin^{2n-1}\theta\,d\theta = \frac{(2n-1)(2n-4)\cdots 4\cdot 2}{(2n-1)(2n-3)\cdots 3\cdot 1}\cdot 1$$

$$= \frac{[(2n-2)(2n-4)\cdots 4\cdot 2]^2}{(2n-1)(2n-2)(2n-3)\cdots 3\cdot 2\cdot 1}$$

$$= \frac{2^{2n-2}[(n-1)!]^2}{(2n-1)!} = \frac{2^{2n-2}[\Gamma(n)]^2}{\Gamma(2n)}$$

Therefore $\dfrac{\Gamma(n)\sqrt{\pi}}{\Gamma\left(n+\frac{1}{2}\right)} = \dfrac{2^{2n-1}[\Gamma(n)]^2}{\Gamma(2n)}$, or

$\sqrt{\pi}\,\Gamma(2n) = 2^{2n-1}\Gamma(n)\,\Gamma\left(n+\frac{1}{2}\right).$ □

Gamma and Beta Functions and Others

Note: The integral $\int_0^{\pi/2} \sin^{2n-1}\theta \, d\theta$, in this last example is one of what is called Walli's Integrals. These integrals are defined as

$$W_n = \int_0^{\pi/2} \sin^n \theta \, d\theta = \int_0^{\pi/2} \cos^n \theta \, d\theta$$

In terms of Beta and Gamma functions, they take the form

$$W_n = \frac{1}{2}\beta\left(\frac{n+1}{2}, \frac{1}{2}\right) = \frac{1}{2} \cdot \frac{\Gamma\left(\frac{n+1}{2}\right)\Gamma\left(\frac{1}{2}\right)}{\Gamma\left(\frac{n+2}{2}\right)} = \frac{\sqrt{\pi}}{2} \cdot \frac{\Gamma\left(\frac{n+1}{2}\right)}{\Gamma\left(\frac{n+2}{2}\right)}.$$

Replacing n by $2n+1$, we get

$$W_{2n+1} = \frac{\sqrt{\pi}}{2} \cdot \frac{\Gamma\left(\frac{2n+2}{2}\right)}{\Gamma\left(\frac{2n+3}{2}\right)} = \frac{\sqrt{\pi}}{2} \cdot \frac{\Gamma(n+1)}{\Gamma(n+3/2)} = \frac{\sqrt{\pi} \cdot n!}{(2n+1)\Gamma(n+1/2)}.$$

Also Walli's integrals satisfy the following relations

$$W_{2n+1} = \frac{4^n \cdot (n!)^2}{(2n+1)!},$$

$$W_{2n} = \frac{1}{2}\beta\left(n+\frac{1}{2}, \frac{1}{2}\right) = \frac{\Gamma\left(n+\frac{1}{2}\right)\Gamma\left(\frac{1}{2}\right)}{2 \cdot \Gamma(n+1)} \quad \text{and}$$

$$W_{2n} = \frac{1 \cdot 3 \cdot 5 \cdots (2n-1)}{2^{n+1} n!} \cdot \pi = \frac{\pi}{2} \cdot \frac{(2n)!}{4^n \cdot (n!)^2}.$$

Example 13: If $\int_0^\infty \frac{x^{n-1} dx}{1+x} = \frac{\pi}{\sin n\pi}$, show that $\Gamma(n)\Gamma(1-n) = \frac{\pi}{\sin n\pi}$.

Solution: From **Property 22**, we have

$$\beta(n, 1-n) = \frac{\Gamma(n)\Gamma(1-n)}{\Gamma(1)} = \Gamma(n)\Gamma(1-n).$$

And from **Property 23**, we obtain

81

$$\beta(n,1-n) = \int_0^\infty \frac{x^{n-1} dx}{(1+x)^{n+1-n}} = \int_0^\infty \frac{x^{n-1} dx}{(1+x)} = \frac{\pi}{\sin n\pi}.$$

Thus $\Gamma(n)\Gamma(1-n) = \dfrac{\pi}{\sin n\pi}$. □

Example 14: If n is a positive integer, show that

$$\beta(p,n+1) = \frac{n!}{p(p+1)(p+2)\cdots(p+n)}$$

Solution: From the relation between Beta and Gamma functions, we have

$$\beta(p,n+1) = \frac{\Gamma(p)\Gamma(n+1)}{\Gamma(p+n+1)} = \frac{\Gamma(p)\cdot n!}{(p+n)(n+p-1)\cdots(p+1)p\cdots 3\cdot 2\cdot 1}$$

$$= \frac{\Gamma(p)\cdot n!}{p(p+1)\cdots(p+n)}. \quad \square$$

Example 15: Show that: $\displaystyle\int_0^1 \frac{x^n\, dx}{\sqrt{1-x^2}} = \begin{cases} \dfrac{1\cdot 3\cdot 5\cdots(n-1)}{2\cdot 4\cdot 6\cdots n}\cdot\dfrac{\pi}{2} & n \text{ even} \\[2mm] \dfrac{2\cdot 4\cdot 6\cdots(n-1)}{1\cdot 3\cdot 5\cdots n} & n \text{ odd} \end{cases}$

Solution: Let $x = \sin\theta$, then $dx = \cos\theta\, d\theta$, and the integral becomes

$$I = \int_0^1 \frac{x^n\, dx}{\sqrt{1-x^2}} = \int_0^{\pi/2} \frac{\sin^n\theta \cos\theta}{\cos\theta}\, d\theta = \int_0^{\pi/2} \sin^n\theta\, d\theta$$

$$= \frac{1}{2}\beta\left(\frac{n+1}{2},\frac{1}{2}\right) = \frac{1}{2}\frac{\Gamma\left(\dfrac{n+1}{2}\right)\Gamma\left(\dfrac{1}{2}\right)}{\Gamma\left(\dfrac{n+2}{2}\right)}$$

If n is even, then

$$I = \frac{1}{2}\cdot\frac{1\cdot 3\cdot 5\cdots(n-1)\cdot\pi}{\dfrac{n}{2}\cdot\dfrac{n-2}{2}\cdot\dfrac{n-4}{2}\cdots 3\cdot 2\cdot 1\cdot 2^{n/2}} = \frac{1\cdot 3\cdot 5\cdots(n-1)}{2^{n/2}\cdot 1\cdot 2\cdot 3\cdots n/2}\cdot\frac{\pi}{2}$$

$$= \frac{1\cdot 3\cdot 5\cdots(n-1)}{2\cdot 4\cdot 6\cdots n}\cdot\frac{\pi}{2}$$

Gamma and Beta Functions and Others

If n is odd, then

$$I = \frac{1}{2} \cdot \frac{\frac{n-1}{2} \cdot \frac{n-3}{2} \cdot \frac{n-5}{2} \cdots 3 \cdot 2 \cdot 1}{1 \cdot 3 \cdot 5 \cdots n}$$

$$= \frac{1}{2} \cdot \frac{1 \cdot 2 \cdot 3 \cdots \left(\frac{n-1}{2}\right) \cdot 2^{(n-1)/2}}{1 \cdot 3 \cdot 5 \cdots n} = \frac{2 \cdot 4 \cdot 6 \cdots (n-1)}{1 \cdot 3 \cdot 5 \cdots n} \quad \square$$

Example 16: Using Beta function, show that:

$$\int_a^b \frac{dx}{(b-x)^{1-\alpha}(x-a)^\alpha} = \frac{\pi}{\sin n\alpha}, \quad 0 < \alpha < 1$$

Solution: From the definition of Beta function $\beta(p,q) = \int_0^1 y^{p-1}(1-y)^{q-1} dy$,

let $y = \frac{x-a}{b-a}$, then $dy = \frac{dx}{b-a}$, and the integral becomes

$$\beta(p,q) = \int_a^b \frac{(x-a)^{p-1}}{(b-a)^{p-1}} \cdot \frac{(b-x)^{q-1}}{(b-a)^{q-1}} \cdot \frac{dx}{b-a}$$

$$= \frac{1}{(b-a)^{p+q-1}} \int_a^b (x-a)^{p-1} \cdot (b-x)^{q-1} dx$$

Let $q = \alpha$ and $p = 1-\alpha$, we get

$$\beta(\alpha, 1-\alpha) = \int_a^b (x-a)^{-\alpha} \cdot (b-x)^{\alpha-1} dx = \int_a^b \frac{(b-x)^{\alpha-1}}{(x-a)^\alpha} dx$$

$$= \frac{\Gamma(\alpha)\Gamma(1-\alpha)}{\Gamma(1)} = \frac{\pi}{\sin n\alpha}, \quad 0 < \alpha < 1. \quad \square$$

Example 17: Show that: $\int_0^{\pi/2} \tan^n x \, dx = \frac{\pi}{2} \sec \frac{n\pi}{2}, |n| < 1$, then evaluate the

integral $\int_0^{\pi/2} \sqrt{\tan x} \, dx$.

Solution: We have,

$$I = \int_0^{\pi/2} \tan^n x \, dx = \int_0^{\pi/2} \sin^n x \cos^{-n} x \, dx = \frac{1}{2}\beta\left(\frac{n+1}{2}, \frac{-n+1}{2}\right)$$

$$= \frac{1}{2} \cdot \frac{\Gamma\left(\frac{n+1}{2}\right)\Gamma\left(\frac{-n+1}{2}\right)}{\Gamma(1)} = \frac{1}{2} \cdot \Gamma\left(\frac{n+1}{2}\right)\Gamma\left(1-\frac{n+1}{2}\right)$$

$$= \frac{1}{2} \cdot \frac{\pi}{\sin\left(\frac{n+1}{2}\cdot\pi\right)} = \frac{1}{2} \cdot \frac{\pi}{\sin\left(\frac{n\pi}{2}+\frac{\pi}{2}\right)}$$

$$= \frac{1}{2} \cdot \frac{\pi}{\cos\left(\frac{n\pi}{2}\right)} = \frac{\pi}{2} \cdot \sec\left(\frac{n\pi}{2}\right)$$

To evaluate $\int_0^{\pi/2} \sqrt{\tan x} \, dx$, put $n = 1/2$, we get

$$\int_0^{\pi/2} \sqrt{\tan x} \, dx = \frac{\pi}{2} \cdot \sec\frac{\pi}{4} = \frac{\pi}{2}\cdot\sqrt{2} = \frac{\pi}{\sqrt{2}}. \qquad \square$$

2.3. Other Special Functions Defined by Integrals

There are many functions that are defined by integrals. These functions appear frequently in many engineering applications. We state here some of them.

2.3.1. Incomplete Gamma Functions:

Two incomplete Gamma functions are given by

$$\Gamma(n,\alpha) = \int_\alpha^\infty e^{-x} x^{n-1} dx \quad \text{and} \quad \gamma(n,\alpha) = \int_0^\alpha e^{-x} x^{n-1} dx,$$

and we have
$$\Gamma(n) = \Gamma(n,\alpha) + \gamma(n,\alpha).$$

These two functions are regularized to give the two functions $P(n,\alpha)$ and $Q(n,\alpha)$ as

$$P(n,\alpha) = \frac{\gamma(n,\alpha)}{\Gamma(n)} = \frac{1}{\Gamma(n)}\int_0^\alpha e^{-x} x^{n-1} dx, \text{ and}$$

$$Q(n,\alpha) = \frac{\Gamma(n,\alpha)}{\Gamma(n)} = \frac{1}{\Gamma(n)} \int_\alpha^\infty e^{-x} x^{n-1} dx .$$

It is to be noticed that $P(n,0)=0$ and $P(n,\infty)=1$. Also the complement of $P(n,\alpha)$ is $Q(n,\alpha)$, i.e. $Q(n,\alpha)=1-P(n,\alpha)$.

$P(n,\alpha)$ and $Q(n,\alpha)$ satisfy the following relations

$$\frac{d}{dn}P(n,\alpha) = \frac{e^{-\alpha}\alpha^{n-1}}{\Gamma(n)}, \qquad \frac{d}{dn}Q(n,\alpha) = -\frac{e^{-\alpha}\alpha^{n-1}}{\Gamma(n)}$$

$$\frac{d^2}{dn^2}P(n,\alpha) = \frac{e^{-\alpha}(n-\alpha-1)\alpha^{n-2}}{\Gamma(n)}, \qquad \frac{d}{dn}Q(n,\alpha) = \frac{e^{-\alpha}(1+\alpha-n)\alpha^{n-1}}{\Gamma(n)}$$

$$\int P(n,\alpha)\,d\alpha = \frac{\alpha\Gamma(n)-\alpha\Gamma(n,\alpha)+\Gamma(n+1,\alpha)}{\Gamma(n)},$$

$$\int Q(n,\alpha)\,d\alpha = \frac{\alpha\Gamma(n,\alpha)-\Gamma(n+1,\alpha)}{\Gamma(n)}.$$

2.3.2. The Error Functions:

The error function is given by the integral

$$\operatorname{erf}(x) = \frac{2}{\sqrt{\pi}} \int_0^x e^{-t^2} dt .$$

The complementary error function is

$$\operatorname{erfc}(x) = \frac{2}{\sqrt{\pi}} \int_x^\infty e^{-t^2} dt ,$$

and we have $\operatorname{erf}(x) + \operatorname{erfc}(x) = 1$.

Also the error function is generalized as

$$\operatorname{erf}_n(x) = \frac{1}{\Gamma\{(n+1)/n\}} \int_0^x e^{-t^n} dt ,$$

and we have $\operatorname{erf}(x) = \operatorname{erf}_2(x)$.

2.3.3. The Exponential Integral Functions:

$$\operatorname{Ei}(x) = \int_{-\infty}^x \frac{e^t}{t} dt \quad \text{and} \quad \operatorname{E}_1(x) = \int_x^\infty \frac{e^{-t}}{t} dt$$

The Error Functions erf(x) and erfc(x)

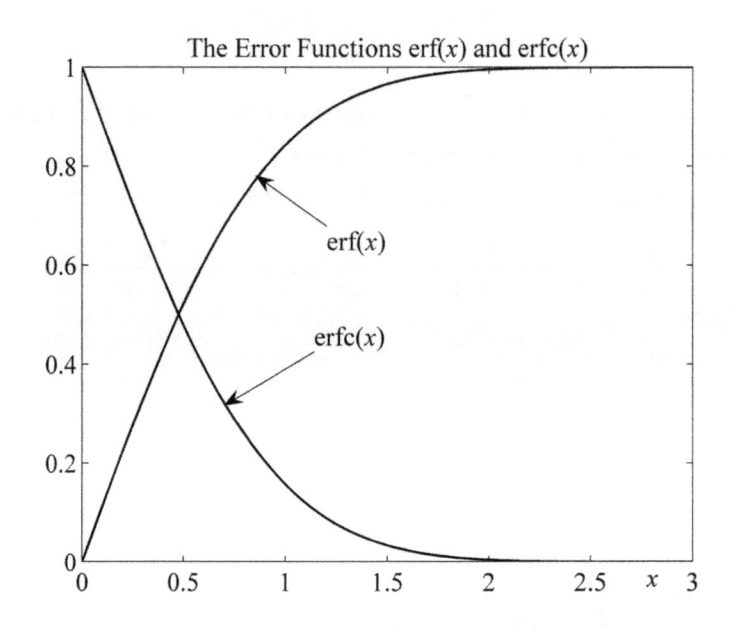

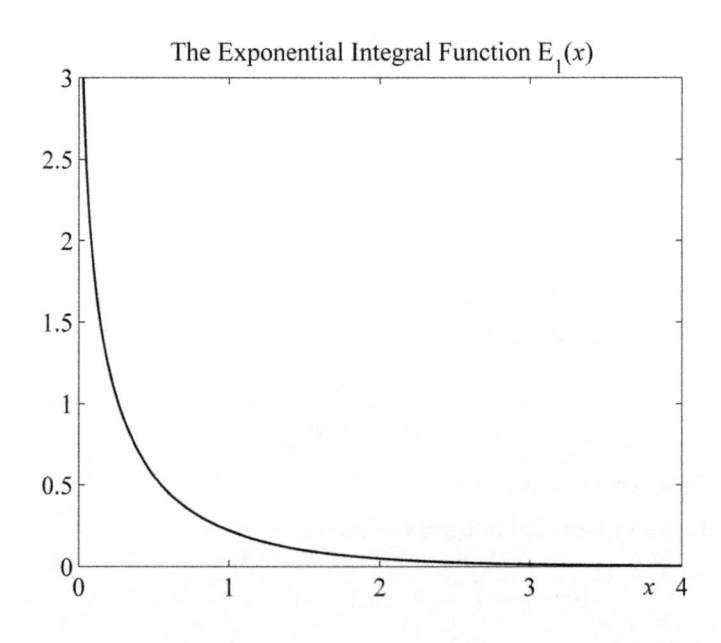

The Exponential Integral Function $E_1(x)$

Gamma and Beta Functions and Others

2.3.4. The Sine and Cosine Integral Functions:

Sine Integral function: $\operatorname{Si}(x) = \int_0^x \frac{\sin\theta}{\theta} d\theta$

Cosine integral function: $\operatorname{Ci}(x) = \int_x^\infty \frac{\cos\theta}{\theta} d\theta$

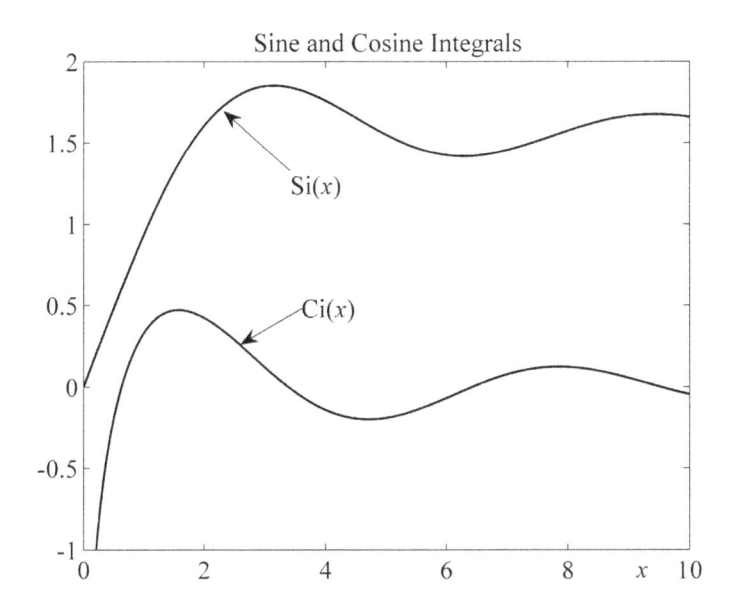

Sine and Cosine Integrals

2.3.5. The Elliptic Integral Functions:

$$F_1(x,k) = \int_0^x \frac{d\theta}{\sqrt{1-k^2\sin^2\theta}}, \quad 0<k<1$$

$$F_2(x,k) = \int_0^x \sqrt{1-k^2\sin^2\theta}\, d\theta, \quad 0<k<1$$

$$F_3(x,k,m) = \int_0^x \frac{d\theta}{\sqrt{(1-k^2\sin^2\theta)(1+m^2\sin^2\theta)}}, \quad 0<k<1,\ m\neq k$$

Special Functions and Orthogonal Polynomials

2.3.6. The Fresnel Sine and Cosine Integral Functions:

Fresnel Sine Integral Function is given by $S(x) = \int_0^x \sin\left(\frac{\pi}{2}\theta^2\right) d\theta$

Fresnel Cosine Integral Function is $C(x) = \int_0^x \cos\left(\frac{\pi}{2}\theta^2\right) d\theta$

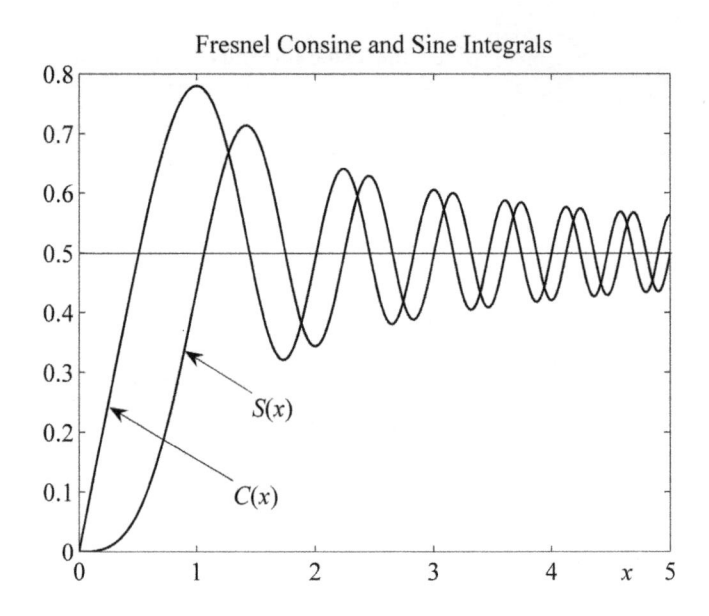

Fresnel Consine and Sine Integrals

2.3.7. The Debye Function:

$$D_n(x) = \int_0^x \frac{t^n}{e^t - 1} dt$$

Exercise 2.1

1. Evaluate the following integrals:

   i) $\int_0^1 x^7 (1-x)^8 \, dx$ ii) $\int_0^{\pi/2} \sin^4 \theta \cos^6 \theta \, d\theta$ iii) $\int_0^{\pi/2} \frac{d\theta}{\sqrt{\cos \theta}}$

   iv) $\int_0^1 \frac{dx}{\sqrt{1-x^3}}$ v) $\int_0^1 x^3 (1-\sqrt{x}) \, dx$ vi) $\int_0^2 \frac{x^2 dx}{\sqrt{2-x}}$

   vii) $\int_0^{2\pi} \sin^8 \theta \, d\theta$ viii) $\int_0^{\infty} \frac{dt}{1+t^4}$ ix) $\int_0^3 x \sqrt[3]{27-x^3} \, dx$

2. Show that $\int_0^1 \frac{dx}{\sqrt{1-x^n}} = \frac{\sqrt{\pi}}{n} \cdot \frac{\Gamma(1/n)}{\Gamma[(2+n)/(2n)]}$. Hint: Let $x^n = \sin^2 \theta$

3. Show that $\int_0^{\infty} \frac{\cos x}{x^n} \, dx = \frac{\pi}{2\Gamma(n) \cos n\pi/2}$, $0 < n < 1$.

   Hint: $\frac{1}{x^n} = \frac{1}{\Gamma(n)} \int_0^{\infty} t^{n-1} e^{-nt} \, dt$

4. Show that $\int_0^1 \frac{dx}{\sqrt{x \ln(1/x)}} = \sqrt{2\pi}$.   Hint: Let $\ln(1/x) = t$

5. Show that $\int_0^1 \frac{x^2 \, dx}{\sqrt{1-x^4}} \cdot \int_0^1 \frac{dx}{\sqrt{1+x^4}} = \frac{\pi}{4\sqrt{2}}$.

   Hint: let $x^2 = \sin \theta$ in the first integral, and $x^2 = \tan \phi$ in the second.

6. Evaluate the integrals:

   i) $\int_0^{\infty} e^{-\alpha x} x^{n-1} \cos \beta x \, dx$ ii) $\int_0^{\infty} e^{-\alpha x} x^{n-1} \sin \beta x \, dx$

   Hint: Consider the integral $\int_0^{\infty} e^{-x(\alpha - i\beta)} x^{n-1} dx$

7. Show that $\int_0^\infty \cos(x^{1/n}) dx = \Gamma(n+1) \cos n\pi/2$ **Hint**: Let $t = x^{1/n}$

8. Show that: $\int_{-\infty}^\infty e^{-\alpha|x|+ikx} dx = \dfrac{2\alpha}{k^2 + \alpha^2}$, $\alpha > 0$.

9. Show that $\int_0^1 x^m \left(\ln\dfrac{1}{x}\right)^n dx = \dfrac{1}{(m+1)^{n+1}} \Gamma(n+1)$, and hence deduce that

   i) $\int_0^1 \left(\ln\dfrac{1}{x}\right)^{1/3} dx = \sqrt{\pi}/2$ ii) $\int_0^1 \left(\ln\dfrac{1}{x}\right)^{-1/2} dx = \sqrt{\pi}$.

10. Show that: $\dfrac{2\Gamma'(2n)}{\Gamma(2n)} - \dfrac{\Gamma'(n)}{\Gamma(n)} - \dfrac{\Gamma'(n+1/2)}{\Gamma(n+1/2)} = 2\ln 2$.

11. Show that: $\beta(p,p) \cdot \beta(p+1/2, p+1/2) = \dfrac{\pi}{2^{4p-1}}$.

12. Show that: $\beta(p,q) \cdot \beta(p+q,m) = \beta(q,m) \cdot \beta(q+m,p)$.

13. Show that: $\int_0^1 \dfrac{x^m (1-x)^{n-1}}{(a+x)^{m+n}} dx = \dfrac{1}{a^n (1+a)^m} \beta(m,n)$.

    Hint: $\dfrac{x(1+a)}{a+x} = y \Rightarrow x = \dfrac{ay}{1+a-y}$

14. Show that: $\int_0^{\pi/2} \dfrac{d\theta}{\sqrt{\sin\theta}} \cdot \int_0^{\pi/2} \sqrt{\sin\theta}\, d\theta = \pi$.

15. Show that: $\Gamma\left(\tfrac{1}{2}+x\right) \cdot \Gamma\left(\tfrac{1}{2}-x\right) = \dfrac{\pi}{\cos \pi x}$.

16. Show that: $\dfrac{\Gamma\left(1+\tfrac{m}{2}\right) \cdot \Gamma\left(1-\tfrac{m}{2}\right)}{\Gamma(1+m) \cdot \Gamma(1-m)} = \cos\dfrac{\pi m}{2}$.

Chapter Three

Legendre Polynomials

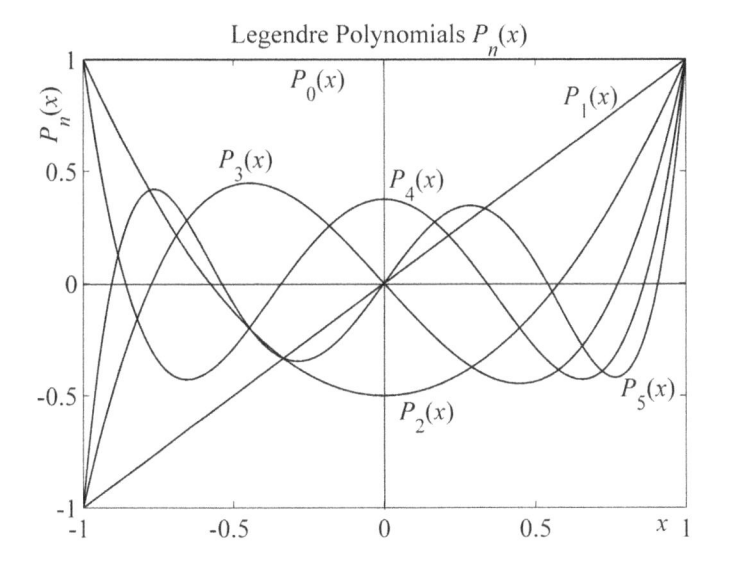

$$P_n(x) = \sum_{k=0}^{N} (-1)^k \cdot \frac{(2n-2k)!}{2^n \, k!(n-2k)!(n-k)!} \, x^{n-2k}$$

Chapter 3
Legendre Polynomials

3.1. Introduction

Legendre[11] *Polynomials*, also called *Legendre functions* are solutions of *Legendre differential equation*. They are *Orthogonal Polynomials* and are special case of the *Ultraspherical Functions* and *Jacobi*[12] *Polynomials*. They arise in numerous problems especially in those involving spheres or spherical coordinates or exhibiting spherical symmetry. In spherical polar coordinates, the angular dependence is always best handled by spherical harmonics that are defined in terms of Legendre functions. We start our study with the Legendre differential equation.

3.2. Legendre Differential Equation and its Solutions

Legendre Differential equation is given by:

$$(1-x^2)y'' - 2xy' + n(n+1)y = 0 \qquad (1)$$

where n is a real number. The solutions of this equation are called *Legendre Functions* of order n. When n is a non-negative integer, the Legendre Functions are often referred to as Legendre Polynomials as we will establish. Since the Legendre differential equation is a second order ordinary differential equation, it has two linearly independent solutions.

It is sometimes useful to write Legendre Differential equation in the form

$$\frac{d}{dx}\left[(1-x^2)\frac{dy}{dx}\right] + n(n+1)y = 0. \qquad (2)$$

It is clear that the only finite singularities of the differential equation are at $x = \pm 1$. In fact $x = \pm 1$ are regular points of the differential equation. Also, it is easy to show that the point at infinity is a regular point.

In order to investigate the solutions of Legendre differential equation, let us first consider the simple case where $n = 0$. In this case, we have

---

11. Adrien-Marie Legendre (1752-1833 Paris-France).
12. Carl Gustav Jacob Jacobi (1804 Potsdam Prussia – 1851 Berlin Germany).

$$\frac{d}{dx}\left[(1-x^2)\frac{dy}{dx}\right] = 0. \tag{3}$$

This is a simple differential equation whose solution is obtained easily. Integrating once, we get

$$(1-x^2)\frac{dy}{dx} = A. \tag{4}$$

Now, by separating the variables, we obtain

$$\int dy = \int \frac{A}{(1-x^2)} dx, \tag{5}$$

giving the solution

$$y = A \cdot \frac{1}{2}\ln\left(\frac{1+x}{1-x}\right) + B. \tag{6}$$

Since this Legendre differential equation is a second order differential equation, its solution contains two arbitrary constants, written here as A and B. And we shall denote the two linearly independent solutions by

$$P_0(x) = 1 \text{ and } Q_0(x) = \frac{1}{2}\ln\left(\frac{1+x}{1-x}\right), \tag{7}$$

where the subscript 0 represents the value of n. It is to be noticed that the second solution $Q_0(x)$ diverges at $x = 1$.

Since the Legendre equation is homogeneous, its general solution is the superposition of $P_0(x)$ and $Q_0(x)$,

$$y = AP_0(x) + BQ_0(x) \tag{8}$$

For the general case where $n \neq 0$, we will use series methods to solve the Legendre differential equation.

To obtain a series solution about $x = 0$ (an ordinary point), we use the method of undetermined coefficients by assuming a solution of the form:

$$y = \sum_{k=0}^{\infty} c_k x^k \tag{9}$$

Differentiating twice, we obtain

Legendre Polynomials

$$y' = \sum_{k=1}^{\infty} kc_k x^{k-1} = \sum_{k=0}^{\infty} (k+1)c_{k+1} x^k \tag{10}$$

$$y'' = \sum_{k=1}^{\infty} k(k+1)c_{k+1} x^{k-1} = \sum_{k=0}^{\infty} (k+1)(k+2)c_{k+2} x^k \tag{11}$$

If this assumption is true, then y must satisfy the differential equation. Substituting in the differential equation, we obtain:

$$(1-x^2) \sum_{k=0}^{\infty} (k+1)(k+2)c_{k+2} x^k$$

$$- 2x \sum_{k=0}^{\infty} (k+1)c_{k+1} x^k + n(n+1) \sum_{k=0}^{\infty} c_k x^k = 0$$

The coefficients of each power of x on the left hand side of this equation must be zero, then equating the coefficients of x^k to zero, we obtain

$$(k+1)(k+2)c_{k+2} - k(k-1)c_k - 2kc_k + n(n+1)c_k = 0. \tag{12}$$

Rearranging, we get

$$\boxed{c_{k+2} = -\frac{(k+n+1)(n-k)}{(k+1)(k+2)} c_k} \tag{13}$$

This is the *Recurrence Relation for the Coefficients*. Since $x = 0$ is an ordinary point of the differential equation, the two linearly independent solutions can be obtained with the help of this relation. Moreover, if we let c_0 and c_1 be the two arbitrary constants of the solution, then each c_k, $k \geq 2$ can be obtained in terms of either c_0 or c_1.

Now, for even values of k, we have

$$c_2 = -\frac{(n+1)n}{1 \cdot 2} c_0$$

$$c_4 = -\frac{(n+3)(n-2)}{3 \cdot 4} c_2 = \frac{(n+3)(n+1) \cdot n(n-2)}{1 \cdot 2 \cdot 3 \cdot 4} c_0$$

$$c_6 = -\frac{(n+5)(n-4)}{5 \cdot 6} c_4 = -\frac{(n+5)(n+3)(n+1) \cdot n(n-2)(n-4)}{1 \cdot 2 \cdot 3 \cdot 4 \cdot 5 \cdot 6} c_0$$

and so on. In general for the even-suffixed coefficients, we have

$$c_{2k} = (-1)^k \frac{(n+2k-1)(n+2k-3)\cdots(n+1)\cdot n(n-2)\cdots(n-2k+2)}{(2k)!} c_0$$

Similarly, for odd-suffixed coefficients, we obtain

$$c_3 = -\frac{(n+2)(n-1)}{2\cdot 3} c_1$$

$$c_5 = -\frac{(n+4)(n-3)}{4\cdot 5} c_3 = \frac{(n+4)(n+2)\cdot(n-1)(n-3)}{2\cdot 3\cdot 4\cdot 5} c_1$$

$$c_7 = -\frac{(n+6)(n-5)}{6\cdot 7} c_5 = -\frac{(n+6)(n+4)(n+2)\cdot(n-1)(n-3)(n-5)}{2\cdot 3\cdot 4\cdot 5\cdot 6\cdot 7} c_1$$

Then,

$$c_{2k+1} = (-1)^k \frac{(n+2k)(n+2k-2)\cdots(n+2)\cdot(n-1)(n-3)\cdots(n-2k+1)}{(2k+1)!} c_1$$

The solutions can now be obtained by making choices for c_0 and c_1. If $c_0 = 1$ and $c_1 = 0$, the first solution will be

$$y_1(x) = \sum_{k=0}^{\infty} (-1)^k \frac{(n+2k-1)(n+2k-3)\cdots(n+1)\cdot n(n-2)\cdots(n-2k+2)}{(2k)!} x^{2k}$$

(14)

On the other hand, If $c_0 = 0$ and $c_1 = 1$, the second solution will be

$$y_2(x) = \sum_{k=0}^{\infty} (-1)^k \frac{(n+2k)(n+2k-2)\cdots(n+2)\cdot(n-1)(n-3)\cdots(n-2k+1)}{(2k+1)!} x^{2k+1}$$

(15)

And since the nearest singularities are at $x = \pm 1$, then these two series solutions converge for all $x \in (-1, 1)$. Moreover, since the first solution is in even powers of x, while the second is in odd powers of x, the two solutions are in fact linearly independent.

The general solution of Legendre differential equation is given as a linear combination of the two linearly independent series solutions. Writing the first few terms in the general solution

$$y(x) = A\left\{1 - \frac{n\cdot(n+1)}{2!}x^2 + \frac{n(n-2)\cdot(n+1)(n+3)}{4!}x^4 - \cdots\right\}$$
$$+ B\left\{x - \frac{(n-1)\cdot(n+2)}{3!}x^3 + \frac{(n-1)(n-3)\cdot(n+2)(n+4)}{5!}x^5 - \cdots\right\} \quad (16)$$

Legendre Polynomials

Now, looking closely at each of the two solutions, we can observe that if n is an even integer, the first series reduces to an even polynomial of degree n. For example, if $n = 2$, the first solution, apart from the multiplicative constant, becomes

$$y_1(x) = 1 - 3x^2$$

While, if n is an odd integer, the second series reduces to an odd polynomial of degree n. For example, if $n = 3$, the second solution, apart from the multiplicative constant, becomes

$$y_2(x) = x - \frac{5}{3}x^3$$

These polynomial solutions, apart from the arbitrary constants, are called *Legendre Polynomials* or *Legendre Functions of the First Kind*. They are denoted by $P_n(x)$. We can say that $P_n(x)$ is the first solution of Legendre Differential Equation. For the second solution, if n is an even integer, $y_2(x)$ remains an infinite series, while if n is an odd integer, $y_1(x)$ is again an infinite series. In this case, these infinite series represent the second solution for Legendre Differential equation. They are denoted by $Q_n(x)$, and are called *Legendre Functions of the Second Kind*. We will get to them later. The general solution can now be written as

$$\boxed{y(x) = a P_n(x) + b Q_n(x)} \qquad (17)$$

The second solution $Q_n(x)$ can be expressed in terms of the first solution $P_n(x)$ from the analysis of ordinary differential equations as

$$Q_n(x) = P_n(x) \int \frac{1}{(1-x^2)[P_n(x)]^2} dx \ . \qquad (18)$$

We will study this later.

Example 1: Find the general solution of the Legendre equation

$$(1-x^2)y'' - 2xy' + 2y = 0.$$

Solution: Comparing this equation with Legendre Differential Equation

$(1-x^2)y'' - 2xy' + n(n+1)y = 0$, we have $n(n+1) = 2$,

then $n = 1$, and the first solution is $y_1 = P_1(x) = x$.

For the second solution, we have

Special Functions and Orthogonal Polynomials

$$y_2 = P_1(x) \int \frac{dx}{(1-x^2)[P_1(x)]^2} = x \int \frac{dx}{(1-x^2)x^2}$$

$$= \frac{x}{2} \int \left(\frac{1}{1+x} + \frac{1}{1-x} + \frac{2}{x^2} \right) dx = \left(\frac{x}{2} \ln\left[\frac{1+x}{1-x}\right] - 1 \right).$$

And the general solution will be

$$y = ax + b\left(\frac{x}{2} \ln\left[\frac{1+x}{1-x}\right] - 1 \right). \qquad \square$$

Example 2: Obtain one series solution for Legendre Differential Equation
$$(1-x^2)y'' - 2xy' + n(n+1)y = 0 \text{ about } x = 1.$$

Solution: $x = 1$ is a regular point of the differential equation, then we will use the method of Frobenius to obtain a series solution. But first, let us make the substitution $t = x - 1$. This will simplify somehow the procedure. Now,

$$\frac{dy}{dx} = \frac{dy}{dt} \cdot \frac{dt}{dx} = \frac{dy}{dt}, \text{ and}$$

$$\frac{d^2y}{dx^2} = \frac{d}{dx}\left(\frac{dy}{dt}\right) = \frac{d}{dt}\left(\frac{dy}{dt}\right) \cdot \frac{dx}{dt} = \frac{d^2y}{dt^2}.$$

Substituting in the differential equation, we get

$$[1-(1+t)^2]\frac{d^2y}{dt^2} - 2(1+t)\frac{dy}{dt} + n(n+1)y = 0, \text{ or}$$

$$t(t+2)\frac{d^2y}{dt^2} + 2(1+t)\frac{dy}{dt} - n(n+1)y = 0.$$

$t = 0$ is now a regular point of this differential equation, and we can assume a series solution of the form

$$y = \sum_{k=0}^{\infty} c_k t^{k+\alpha}$$

Differentiating twice, we obtain

$$\frac{dy}{dt} = \sum_{k=0}^{\infty} (k+\alpha) c_k t^{k+\alpha-1}, \text{ and}$$

Legendre Polynomials

$$\frac{d^2 y}{dt^2} = \sum_{k=0}^{\infty} (k+\alpha-1)(k+\alpha) c_k t^{k+\alpha-2}.$$

Substituting all these expressions in to differential equation, we obtain

$$t(t+2) \sum_{k=0}^{\infty} (k+\alpha-1)(k+\alpha) c_k t^{k+\alpha-2}$$

$$+ 2(t+1) \sum_{k=0}^{\infty} (k+\alpha) c_k t^{k+\alpha-1} + n(n+1) \sum_{k=0}^{\infty} c_k t^{k+\alpha} = 0$$

Equating the coefficients of the least powr of t ($t^{\alpha-1}$) to zero, we get $2\alpha(\alpha-1)c_0 + 2\alpha c_0 = 0$, and since $c_0 \neq 0$, we have $\boxed{\alpha = 0, 0}$ (a double root). Then Frobenius method will produce one solution.

Equating the coefficients of $t^{k+\alpha}$ to zero, we obtain

$$(k+\alpha-1)(k+\alpha) c_k + 2(k+\alpha)(k+\alpha+1) c_{k+1}$$

$$+ 2(k+\alpha) c_k + 2(k+\alpha+1) c_{k+1} - n(n+1) c_k = 0.$$

Re-arranging, we get the recurrence relation for the coefficients as

$$\boxed{c_{k+1} = -\frac{(k+\alpha)(k+\alpha+1) - n(n+1)}{2(k+\alpha+1)^2} c_k}.$$

For $\boxed{\alpha = 0}$, this recurrence relation becomes

$$\boxed{c_{k+1} = -\frac{k(k+1) - n(n+1)}{2(k+1)^2} c_k}$$

For various values of k, we have

$$k = 0: \quad c_1 = \frac{n(n+1)}{2} c_0$$

$$k = 1: \quad c_2 = \frac{1 \cdot 2 - n(n+1)}{2 \cdot 2^2} c_1 = \frac{n(n-1) \cdot (n+1)(n+2)}{2^2 \cdot (1 \cdot 2)^2} c_0$$

$$k = 2: \quad c_3 = \cdots = \frac{n(n-1)(n-2) \cdot (n+1)(n+2)(n+3)}{2^3 \cdot (1 \cdot 2 \cdot 3)^2} c_0.$$

And in general
$$c_k = \frac{n(n-1)\cdots(n-k+1)\cdot(n+1)(n+2)\cdots(n+k)}{2^k\cdot(k!)^2}c_0.$$

Using the Pochhammer[13] symbol, we can write

$$c_k = \frac{(n)_k\cdot(n+1)_k}{2^k\cdot(k!)^2}c_0,$$ and the series solution is now

$$y = c_0\left[1+\sum_{k=1}^{\infty}\frac{(n)_k\cdot(n+1)_k}{2^k\cdot(k!)^2}t^k\right].$$

Back-substituting, we obtain

$$y = c_0\left[1+\sum_{k=1}^{\infty}\frac{(n)_k\cdot(n+1)_k}{(k!)^2}\cdot\left(\frac{x-1}{2}\right)^k\right]. \qquad \square$$

---

13. Leo August Pochhammer (1841-1920). The Pochhammer symbol is given by
$$(n)_k = \frac{\Gamma(n+k)}{\Gamma(n)} = n(n+1)(n+2)\cdots(n+k-1),\quad (n)_0 = 1$$

Legendre Polynomials

3.3. Legendre Polynomials in Descending Powers of *x*

It is sometimes advantageous to write the polynomial solutions in descending powers of *x*. From the previous section, rewriting the recurrence relation for the coefficients as

$$c_k = -\frac{(k+2)(k+1)}{(k+n+1)(n-k)} c_{k+2}, \quad k = 0, 1, 2, \cdots \tag{19}$$

If $k = n$, then we have $c_{n+2} = c_{n+4} = c_{n+6} = \cdots = 0$.

Therefore for $k = n-2, n-4, n-6, \cdots$, we find that

$$c_{n-2} = -\frac{n(n-1)}{2(2n-1)} c_n$$

$$c_{n-4} = -\frac{(n-2)(n-3)}{4(2n-3)} c_{n-2} = \frac{n(n-1)(n-2)(n-3)}{2 \cdot 4 \cdot (2n-1)(2n-3)} c_n$$

$$c_{n-6} = \cdots = -\frac{n(n-1)(n-2)(n-3)(n-4)(n-5)}{2 \cdot 4 \cdot 6 \cdot (2n-1)(2n-3)(2n-5)} c_n$$

and so on. Then, the polynomial takes the form

$$y(x) = c_n \left\{ x^n - \frac{n(n-1)}{2(2n-1)} x^{n-2} + \frac{n(n-1)(n-2)(n-3)}{2 \cdot 4 \cdot (2n-1)(2n-3)} x^{n-4} \right. \\ \left. - \frac{n(n-1)(n-2)(n-3)(n-4)(n-5)}{2 \cdot 4 \cdot 6 \cdot (2n-1)(2n-3)(2n-5)} x^{n-6} + \cdots \right\} \tag{20}$$

To normalize these polynomials, we chose the arbitrary constant c_n, the coefficient of x^n, so that $P_n(1) = 1$. The value of c_n to achieve this is

$$c_n = \frac{(2n-1)(2n-3) \cdots 3 \cdot 1}{n!} \tag{21}$$

Then, the Legendre Polynomials are

$$P_n(x) = \frac{(2n-1)(2n-3) \cdots 3 \cdot 1}{n!} \left\{ x^n - \frac{n(n-1)}{2(2n-1)} x^{n-2} + \frac{n(n-1)(n-2)(n-3)}{2 \cdot 4 \cdot (2n-1)(2n-3)} x^{n-4} - \cdots \right\}$$

To find a compact form for $P_n(x)$, the general term in this last series is

$$(-1)^k \frac{1 \cdot 3 \cdot 5 \cdots (2n-1)}{n!} \cdot \frac{n(n-1)(n-2) \cdots (n-2k+1)}{2 \cdot 4 \cdot 6 \cdots 2k \cdot (2n-1)(2n-3) \cdots (2n-2k+1)} x^{n-2k}$$

Now, we do some tricks to make this general term more compact. We have

Special Functions and Orthogonal Polynomials

$$1 \cdot 3 \cdot 5 \cdots (2n-1) = \frac{1 \cdot 2 \cdot 3 \cdot 4 \cdots (2n-1) \cdot 2n}{2 \cdot 4 \cdot 6 \cdots 2n} = \frac{(2n)!}{2^n \, n!}$$

$n(n-1)(n-2)\cdots(n-2k+1)$

$$= \frac{n(n-1)(n-2)\cdots(n-2k+1)(n-2k)(n-2k-1)\cdots 3 \cdot 2 \cdot 1}{(n-2k)(n-2k-1)\cdots 3 \cdot 2 \cdot 1} = \frac{n!}{(n-2k)!}$$

$(2n-1)(2n-3)\cdots(2n-2k+1)$

$$= \frac{2n(2n-1)(2n-2)\cdots(2n-2k+2)(2n-2k+1)}{2n(2n-2)(2n-4)\cdots(2n-2k+2)} \times \frac{(2n-2k)!}{(2n-2k)!}$$

$$= \frac{(2n)!}{2^n \, n(n-1)(n-2)\cdots(n-k+1)(2n-2k)!}$$

$$= \frac{(2n)!}{2^n (2n-2k)!} \cdot \frac{(n-k)(n-k-1)\cdots 3 \cdot 2 \cdot 1}{n(n-1)(n-2)\cdots(n-k+1)(n-k-1)\cdots 3 \cdot 2 \cdot 1}$$

$$= \frac{(2n)!(n-k)!}{2^n (2n-2k)! \, n!}$$

Substituting all these values, the general term becomes

$$(-1)^k \cdot \frac{(2n-2k)!}{2^n \, k!(n-2k)!(n-k)!} \, x^{n-2k}$$

And since the polynomial is of degree n, k must be chosen so that $n-2k \geq 0$, i.e., $k \leq n/2$. Then, if n is even, k goes from 0 to $n/2$, while if n is odd, k goes from 0 to $(n-1)/2$. Hence $P_n(x)$ can now be written as

$$\boxed{P_n(x) = \sum_{k=0}^{N} (-1)^k \cdot \frac{(2n-2k)!}{2^n \, k!(n-2k)!(n-k)!} \, x^{n-2k}} \qquad (22)$$

where N is called the *Floor Function* and is given by

$$N = \begin{cases} n/2 & \text{if } n \text{ is even} \\ (n-1)/2 & \text{if } n \text{ is odd} \end{cases}$$

Here are few of these polynomials for different values of n:

$P_0(x) = 1$; $\qquad P_1(x) = x$;

$P_2(x) = \frac{1}{2}(3x^2 - 1)$; $\qquad P_3(x) = \frac{1}{2}(5x^3 - 3x)$;

$P_4(x) = \frac{1}{8}(35x^4 - 30x^2 + 3)$;

$P_5(x) = \frac{1}{8}(63x^5 - 70x^3 + 15x)$;

Legendre Polynomials

$$P_6(x) = \frac{1}{16}(231x^6 - 315x^4 + 105x^2 - 5).$$

Note: Although Legendre Polynomials are defined for all finite values of x, they are solutions of Legendre differential equation only for $x \in (-1, 1)$.

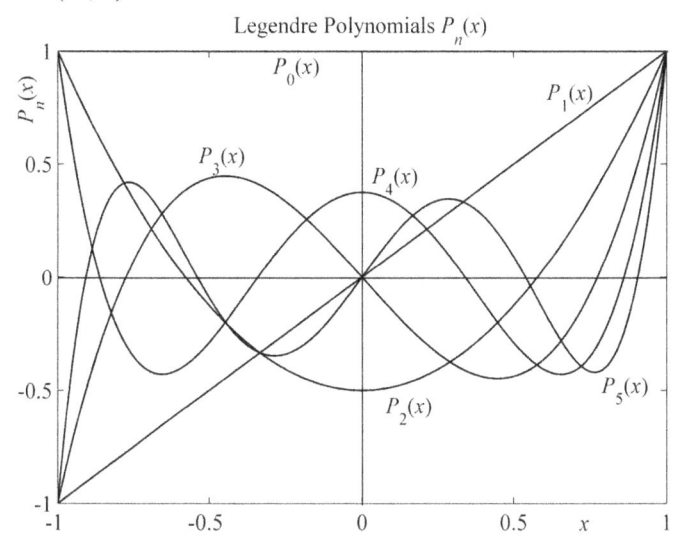

Legendre Polynomials can be generated using the *Gram-Schmidt*[14] *Orthogonalization Process* as follows: Given a set of linearly independent functions $\{1, x, x^2, x^3, \cdots\}$, we can construct Legendre Polynomials in the interval $(-1, 1)$. The steps are

$$P_0(x) = 1$$

$$P_1(x) = \left(x - \frac{\int_{-1}^{1} x P_0^2(x)\,dx}{\int_{-1}^{1} P_0^2(x)\,dx} \right) P_0(x) = x$$

$$P_k(x) = \left(x - \frac{\int_{-1}^{1} x P_{k-1}^2(x)\,dx}{\int_{-1}^{1} P_{k-1}^2(x)\,dx} \right) P_{k-1}(x) - \left(\frac{\int_{-1}^{1} P_{k-1}^2(x)\,dx}{\int_{-1}^{1} P_{k-2}^2(x)\,dx} \right) P_{k-2}(x),$$

$$k = 2, 3, 4, \cdots$$

---

14. Jørgen Pedersen Gram (1850–1916 Denmark), Erhard Schmidt (1876-1959 Germany)

Special Functions and Orthogonal Polynomials

Example 3: Express the function $f(x) = 4x^3 + 6x^2 + 7x + 2$ in terms of Legendre Polynomials.

Solution: We have $P_0(x) = 1$; $P_1(x) = x$; $P_2(x) = \frac{1}{2}(3x^2 - 1)$ and

$P_3(x) = \frac{1}{2}(5x^3 - 3x)$. Then

$f(x) = 4x^3 + 6x^2 + 7x + 2 = a_3 P_3 + a_2 P_2 + a_1 P_1 + a_0 P_0$

$= \frac{a_3}{2}(5x^3 - 3x) + \frac{a_2}{2}(3x^2 - 1) + a_1 x + a_0$.

Equating the coefficients of various powers of x in both sides, we obtain

coef. x^0: $\quad a_2 = 2$

coef. x^1: $\quad -\frac{3}{2}a_3 + a_1 = 7$

coef. x^2: $\quad \frac{3}{2}a_2 = 6 \Rightarrow a_2 = 4$

coef. x^3: $\quad \frac{5}{2}a_3 = 5 \Rightarrow a_3 = 2 \Rightarrow a_1 = 10$.

Substituting all these values, we get

$f(x) = 2P_3(x) + 4P_2(x) + 10P_1(x) + 2P_0(x)$. □

Example 4: Show that: $\sin^4 \theta = \frac{8}{15} P_0(\cos\theta) + \frac{16}{21} P_2(\cos\theta) + \frac{8}{35} P_4(\cos\theta)$.

Solution: We have

$\sin^4 \theta = (1 - \cos^2 \theta)^2 = 1 - 2\cos^2 \theta + \cos^4 \theta$

$= a P_0(\cos\theta) + b P_2(\cos\theta) + c P_4(\cos\theta)$

Also,

$P_0(\cos\theta) = 1$; $P_2(\cos\theta) = \frac{1}{2}(3\cos^2 \theta - 1)$

$$P_4(\cos\theta) = \frac{1}{8}(35\cos^4\theta - 30\cos^2\theta + 3).$$

Therefore,

$$\sin^4\theta = 1 - 2\cos^2\theta + \cos^4\theta$$
$$= a + \frac{b}{2}(3\cos^2\theta - 1) + \frac{c}{8}(35\cos^4\theta - 30\cos^2\theta + 3)$$

Equating various coefficients, we get

$$a = \frac{8}{15}, b = \frac{16}{21}, c = \frac{8}{35}.$$

Therefore,

$$\sin^4\theta = \frac{8}{15}P_0(\cos\theta) + \frac{16}{21}P_2(\cos\theta) + \frac{8}{35}P_4(\cos\theta). \qquad \square$$

In some applications, it is more convenient to use $P_n(\cos\theta)$. The graph is given below.

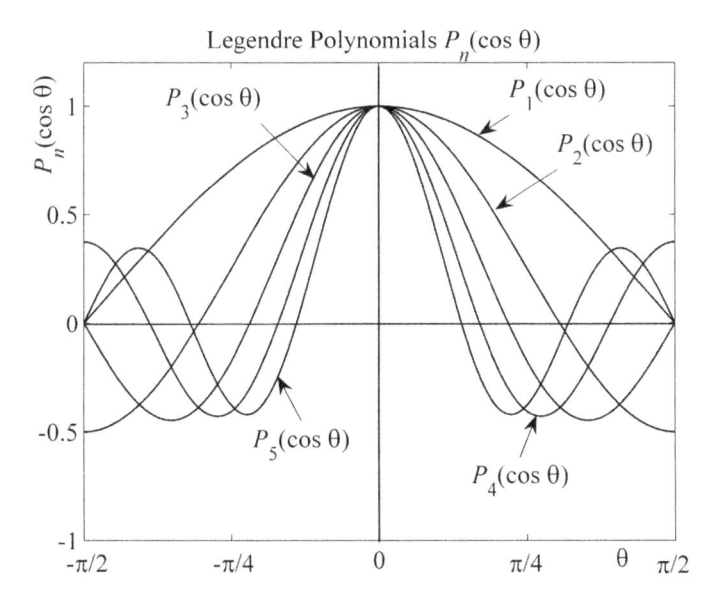

Legendre Polynomials $P_n(\cos\theta)$

Special Functions and Orthogonal Polynomials

Example 5: In Legendre differential equation, use the substitution $x = \cos\theta$ to show that:

i. $\dfrac{d^2y}{d\theta^2} + \cot\theta \dfrac{dy}{d\theta} + n(n+1)y = 0$.

ii. $\dfrac{1}{\sin\theta} \dfrac{d}{d\theta}\left(\sin\theta \dfrac{dy}{d\theta}\right) + n(n+1)y = 0$.

Solution: i. We have: $\dfrac{dy}{dx} = \dfrac{dy}{d\theta} \cdot \dfrac{d\theta}{dx} = -\csc\theta \cdot \dfrac{dy}{d\theta}$, and

$$\dfrac{d^2y}{dx^2} = \dfrac{d}{dx}\left(\dfrac{dy}{dx}\right) = \dfrac{d}{dx}\left(-\csc\theta \cdot \dfrac{dy}{d\theta}\right) = \dfrac{d}{d\theta}\left(-\csc\theta \cdot \dfrac{dy}{d\theta}\right) \cdot \dfrac{d\theta}{dx}$$

$$= \csc^2\theta \cdot \dfrac{d^2y}{d\theta^2} - \csc^2\theta \cot\theta \cdot \dfrac{dy}{d\theta}.$$

Now, from Legendre differential equation

$$(1-x^2)\dfrac{d^2y}{dx^2} - 2x\dfrac{dy}{dx} + n(n+1)y = 0,$$

we substitute all these values to obtain

$$\sin^2\theta \left(\csc^2\theta \cdot \dfrac{d^2y}{d\theta^2} - \csc^2\theta \cot\theta \cdot \dfrac{dy}{d\theta}\right)$$

$$-\cos\theta\left(-\csc\theta \cdot \dfrac{dy}{d\theta}\right) + n(n+1)y = 0$$

Rearranging, we get $\dfrac{d^2y}{d\theta^2} + \cot\theta \dfrac{dy}{d\theta} + n(n+1)y = 0$.

ii. We have

$$\dfrac{1}{\sin\theta} \dfrac{d}{d\theta}\left(\sin\theta \dfrac{dy}{d\theta}\right) + n(n+1)y$$

$$= \dfrac{1}{\sin\theta}\left(\sin\theta \dfrac{d^2y}{d\theta^2} + \cos\theta \dfrac{dy}{d\theta}\right) + n(n+1)y$$

$$= \dfrac{d^2y}{d\theta^2} + \cot\theta \dfrac{dy}{d\theta} + n(n+1)y = 0. \quad \square$$

Legendre Polynomials

Example 6: Show that all the roots of $P_n(x) = 0$ are distinct.

Solution: We will show this by contradiction. Suppose that we have a double root at $x = \alpha$, then from the theory of equations, this root must satisfy the two equations: $P_n(\alpha) = 0$ and $P_n'(\alpha) = 0$.

Since $P_n(x)$ satisfies Legendre differential equation, we have

$$(1-x^2)P_n''(x) - 2xP_n'(x) + n(n+1)P_n(x) = 0.$$

Differentiating m times and using Leibniz Theorem[15], we get

$$(1-x^2)\frac{d^{m+2}}{dx^{m+2}}P_n(x) - 2x(m+1)\frac{d^{m+1}}{dx^{m+1}}P_n(x)$$

$$+ \left[n(n+1) - m(m+1)\right]\frac{d^m}{dx^m}P_n(x) = 0$$

Letting $m = 0$ and $x = \alpha$ in this last equation, we get

$$(1-\alpha^2)P_n''(\alpha) - 2\alpha P_n'(\alpha) + n(n+1)P_n(\alpha) = 0.$$

But, $P_n(\alpha) = 0$ and $P_n'(\alpha) = 0$,

then we have, $(1-\alpha^2)P_n''(\alpha) = 0$ or $P_n''(\alpha) = 0$.

Similarly, Let $m = 1$ and $x = \alpha$, we conclude that $P_n'''(\alpha) = 0$.

If we continue with this process, we obtain for $m = n - 2$:

$$\frac{d^n}{dx^n}P_n(x)\bigg|_{x=\alpha} = 0.$$

But $P_n(x) = \frac{(2n-1)(2n-3)\cdots 3\cdot 1}{n!} \cdot \left\{ x^n - \frac{n(n-1)}{2(2n-1)} x^{n-2} + \cdots \right\}$.

Then $\frac{d^n}{dx^n}P_n(x)\bigg|_{x=\alpha} = \frac{(2n-1)(2n-3)\cdots 3\cdot 1}{n!} \cdot n!$.

This is a contradiction. Then our assumption is not true, and all the roots of $P_n(x) = 0$ must be distinct. □

---

15. **Leibniz Theorem:** If $u(x)$ and $v(x)$ are continuously differentiable, then the mth derivative of their product is given by:

$$(u \cdot v)_m = u \cdot v_m + m \cdot u_1 \cdot v_{m-1} + \frac{m(m-1)}{2!} \cdot u_2 \cdot v_{m-2} + \cdots + u_m \cdot v,$$

where u_m and v_m are the mth derivatives of $u(x)$ and $v(x)$.

Example 7: Show that $\int_x^1 P_n(x)\,dx = \dfrac{(1-x^2)P_n'(x)}{n(n+1)}$.

Solution: Since $P_n(x)$ satisfies Legendre Differential Equation, then

$$P_n = -\frac{1}{n(n+1)}\left\{(1-x^2)P_n'' - 2x\,P_n'\right\}.$$

Integrating with respect to x from x to 1, we get

$$I = \int_x^1 P_n\,dx = -\frac{1}{n(n+1)}\int_x^1\left\{(1-x^2)P_n'' - 2x\,P_n'\right\}dx$$

$$= -\frac{1}{n(n+1)}\left\{\int_x^1 (1-x^2)P_n''\,dx - 2\int_x^1 x\,P_n'\,dx\right\}$$

For the first integral, using integration by parts, we get

$$I = -\frac{1}{n(n+1)}\left\{\int_x^1 (1-x^2)\,dP_n' - 2\int_x^1 x\,P_n'\,dx\right\}$$

$$= -\frac{1}{n(n+1)}\left\{\left[(1-x^2)P_n'\right]_x^1 - \int_x^1 P_n'\,d(1-x^2) - 2\int_x^1 x\,P_n'\,dx\right\}$$

$$= -\frac{1}{n(n+1)}\left\{-(1-x^2)P_n' + 2\int_x^1 x\,P_n'\,dx - 2\int_x^1 x\,P_n'\,dx\right\}$$

$$= \frac{(1-x^2)P_n'}{n(n+1)}.$$

Therefore, $\int_x^1 P_n(x)\,dx = \dfrac{(1-x^2)P_n'(x)}{n(n+1)}$. □

Example 8: Solve Legendre differential equation in descending powers of x:

$$(1-x^2)y'' - 2xy' + n(n+1)y = 0.$$

Solution: Assume a solution of the form $y = \sum_{k=0}^{\infty} c_k x^{\alpha-k}$, then

Legendre Polynomials

$$y' = \sum_{k=0}^{\infty} (\alpha - k) c_k x^{\alpha - k - 1}$$

And $y'' = \sum_{k=0}^{\infty} (\alpha - k)(\alpha - k - 1) c_k x^{\alpha - k - 2}$

Substituting in the differential equation, we obtain

$$(1-x^2) \sum_{k=0}^{\infty} (\alpha - k)(\alpha - k - 1) c_k x^{\alpha - k - 2}$$

$$- 2x \sum_{k=0}^{\infty} (\alpha - k) c_k x^{\alpha - k - 1} + n(n+1) \sum_{k=0}^{\infty} c_k x^{\alpha - k} = 0$$

Equating the coefficients of the *highest power of* x, (x^{α}) to zero, we get

$-\alpha(\alpha - 1) c_0 - 2\alpha c_0 + n(n+1) c_0 = 0$, $c_0 \neq 0$, then

$(\alpha - n)(\alpha + n + 1) = 0$, giving $\boxed{\alpha = n}$ and $\boxed{\alpha = -(n+1)}$.

Equating the coefficients of the *next highest power of* x, $(x^{\alpha - 1})$ to zero, we get $-(\alpha - 1)(\alpha - 2) c_1 - 2(\alpha - 1) c_1 + n(n+1) c_1 = 0$, or

$c_1(n - \alpha + 1)(n + \alpha) = 0$, but $(n - \alpha + 1)(n + \alpha) = 0$, then $\boxed{c_1 = 0}$.

Now, equating the coefficients of the general term $x^{\alpha - k}$ to zero, we get

$(\alpha - k + 2)(\alpha - k + 1) c_{k-2}$

$- (\alpha - k)(\alpha - k - 1) c_k - 2(\alpha - k) c_k + n(n+1) c_k = 0$

Rearranging, we get

$$\boxed{c_k = \frac{(\alpha - k + 2)(\alpha - k + 1) - n(n+1)}{(\alpha - k + 2)(\alpha - k + 1)} c_{k-2}}$$

Since $c_1 = 0$, the it follows that $c_3 = c_5 = \cdots = 0$

For $\boxed{\alpha = n}$, we have

109

$$c_k = \frac{(n-k+2)(n-k+1) - n(n+1)}{(n-k+2)(n-k+1)} c_{k-2}$$

$$= -\frac{(n-k+2)(n-k+1)}{k(2n-k+1)} c_{k-2}$$

And the first solution is

$$y_1(x) = c_0 \left\{ \begin{array}{l} x^n - \dfrac{n(n-1)}{2(2n-1)} x^{n-2} + \dfrac{n(n-1)(n-2)(n-3)}{2 \cdot 4 \cdot (2n-1)(2n-3)} x^{n-4} \\ - \dfrac{n(n-1)(n-2)(n-3)(n-4)(n-5)}{2 \cdot 4 \cdot 6 \cdot (2n-1)(2n-3)(2n-5)} x^{n-6} + \cdots \end{array} \right\}$$

If we take, $c_0 = \dfrac{1 \cdot 3 \cdot 5 \cdots (2n-1)}{n!}$, This solution is in fact

Legendre Polynomial $P_n(x)$, therefore

$$P_n(x) = \frac{1 \cdot 3 \cdot 5 \cdots (2n-1)}{n!} \left\{ \begin{array}{l} x^n - \dfrac{n(n-1)}{2(2n-1)} x^{n-2} \\ + \dfrac{n(n-1)(n-2)(n-3)}{2 \cdot 4 \cdot (2n-1)(2n-3)} x^{n-4} \\ - \dfrac{n(n-1)(n-2) \cdots (n-5)}{2 \cdot 4 \cdot 6 \cdot (2n-1)(2n-3)(2n-5)} x^{n-6} + \cdots \end{array} \right\}$$

For $\boxed{\alpha = -(n+1)}$, we have $c_k = \dfrac{(n+k-1)(n+k)}{k(2n+k+1)} c_{k-2}$

And the second solution is

$$y_2(x) = c_0 \left\{ \begin{array}{l} x^{-n-1} + \dfrac{(n+1)(n+2)}{2(2n+3)} x^{-n-3} \\ + \dfrac{(n+1)(n+2)(n+3)(n+4)}{2 \cdot 4 \cdot (2n+3)(2n+5)} x^{-n-5} + \cdots \end{array} \right\}$$

If we take, $c_0 = \dfrac{n!}{1 \cdot 3 \cdot 5 \cdots (2n+1)}$, This solution is in fact

Legendre Function of the second kind $Q_n(x)$, therefore

$$Q_n(x) = \frac{n!}{1 \cdot 3 \cdot 5 \cdots (2n+1)} \left\{ \begin{array}{l} x^{-n-1} + \dfrac{(n+1)(n+2)}{2(2n+3)} x^{-n-3} \\ + \dfrac{(n+1)(n+2)(n+3)(n+4)}{2 \cdot 4 \cdot (2n+3)(2n+5)} x^{-n-5} + \cdots \end{array} \right\}$$

The general solution is then $y(x) = A P_n(x) + B Q_n(x)$. □

Note: From the previous example, Legendre Polynomials as well as Legendre Functions of the second kind can be put in a summation form as

$$P_n(x) = \sum_{k=0}^{N} \frac{(-1)^k (2n-2k)! x^{n-2k}}{2^n k!(n-k)!(n-2k)!}, \text{ where}$$

where N is the *Floor Function* $\quad N = \begin{cases} n/2 & \text{if } n \text{ is even} \\ (n-1)/2 & \text{if } n \text{ is odd.} \end{cases}$

This is exactly the expression for $P_n(x)$ as obtain by Equation (22).

For $Q_n(x)$, we have

$$Q_n(x) = \frac{2^n n!}{(2n+1)!} \sum_{k=0}^{\infty} \frac{(n+2k)! \cdot x^{-(n+2k+1)}}{2^k k!(2n+3)(2n+5)\cdots(2n+2k+1)} \qquad (23)$$

3.4. Generating Function for Legendre Polynomials

Generating functions are available for most special functions and polynomials. We start with a definition.

Definition: Let $f(x,t)$ be a function of x and t that can be expressed in a Taylor's series in t as $f(t,x) = \sum_{n=0}^{\infty} t^n C_n(x)$, then the function $f(x,t)$ is called a ***generating function*** of the functions $C_n(x)$.

For example, the function

$$f(x, t) = \frac{1}{1-xt} \tag{24}$$

is a generating function of the polynomials x^n since

$$\frac{1}{1-xt} = \sum_{n=0}^{\infty} t^n x^n, \quad |xt| < 1. \tag{25}$$

Now, the function

$$g(x, t) = (1 - 2xt + t^2)^{-1/2}, \text{ with } |x| \leq 1, \ |t| < 1 \tag{26}$$

is the generating function for Legendre polynomials. In fact, we have

$$\boxed{(1 - 2xt + t^2)^{-1/2} = \sum_{n=0}^{\infty} t^n P_n(x)} \tag{27}$$

To prove this relation, we expand the generating function to the left using the ***Binomial Expansion***[16], to obtain

$$(1 - 2xt + t^2)^{-1/2} = [1 + t(2x - t)]^{-1/2}$$

$$= 1 + \frac{1}{2}t(2x - t) + \frac{1 \cdot 3}{2 \cdot 4}t^2(2x - t)^2 + \cdots + \frac{1 \cdot 3 \cdot 5 \cdots (2n-1)}{2 \cdot 4 \cdot 6 \cdots 2n}t^n(2x - t)^n + \cdots$$

The coefficient of t^n in this expression is

---

16. Binomial Expansion:
$$(1+x)^m = 1 + mx + \frac{m(m-1)}{2!}x^2 + \frac{m(m-1)(m-2)}{3!}x^3 + \frac{m(m-1)(m-2)(m-3)}{4!}x^4 + \cdots$$

Legendre Polynomials

$$\frac{1\cdot 3\cdot 5\cdots (2n-1)}{2\cdot 4\cdot 6\cdots 2n}(2x)^n + \frac{1\cdot 3\cdot 5\cdots (2n-3)}{2\cdot 4\cdot 6\cdots (2n-2)}(n-1)(2x)^{n-2} + \cdots$$

$$= \frac{1\cdot 3\cdot 5\cdots (2n-1)}{n!}x^n + \frac{1\cdot 3\cdot 5\cdots (2n-1)}{n!}\cdot \frac{n(n-1)}{2(2n-1)}x^{n-2} + \cdots$$

$$= \frac{1\cdot 3\cdot 5\cdots (2n-1)}{n!}\left(x^n + \frac{n(n-1)}{2(2n-1)}x^{n-2} + \frac{n(n-1)(n-2)(n-3)}{2\cdot 4\cdot (2n-1)(2n-3)}x^{n-4} + \cdots\right)$$

$$= \sum_{k=0}^{N}(-1)^k \cdot \frac{(2n-2k)!}{2^k k!(n-2k)!(n-k)!} x^{n-2k} = P_n(x).$$

where N is the **Floor Function**

$$N = \begin{cases} n/2 & \text{if } n \text{ is even} \\ (n-1)/2 & \text{if } n \text{ is odd} \end{cases} \tag{28}$$

Using this generating function, we can derive many properties of Legendre Polynomials. But first, let us give some illustrative examples to show the use of this generating function.

Example 9: Using the generating function,

$$(1-2xt+t^2)^{-1/2} = \sum_{n=0}^{\infty} t^n P_n(x), \text{ or otherwise, show that:}$$

(1) $P_n(1) = 1$ (2) $P_n(-1) = (-1)^n$

(3) $P_n'(1) = \frac{n(n+1)}{2}$ (4) $P_n'(-1) = (-1)^{n+1}\frac{n(n+1)}{2}$

(5) $P_{2n}(0) = (-1)^n \frac{(2n)!}{2^{2n}(n!)^2}$

(6) $P_{2n+1}(0) = 0$ (7) $P_n(-x) = (-1)^n P_n(x)$

Solution: We will start from the generating function:

$$(1+2xt+t^2)^{-1/2} = \sum_{n=0}^{\infty} t^n P_n(x).$$

(1) Letting $x = 1$, we obtain

Special Functions and Orthogonal Polynomials

$$(1-2t+t^2)^{-1/2} = (1-t)^{-1} = \sum_{n=0}^{\infty} t^n P_n(1).$$

Using the Binomial expansion of $(1-t)^{-1}$, we obtain

$$1+t+t^2+\cdots+t^n+\cdots = \sum_{n=0}^{\infty} t^n P_n(1).$$

Equating the coefficients of t^n in both sides, we get

$P_n(1) = 1$. □

(2) Letting $x = -1$ in the generating function, we obtain

$$(1+2t+t^2)^{-1/2} = (1+t)^{-1} = \sum_{n=0}^{\infty} t^n P_n(-1).$$

Using the Binomial expansion of $(1+t)^{-1}$, we obtain

$$1-t+t^2-\cdots+(-1)^n t^n+\cdots = \sum_{n=0}^{\infty} t^n P_n(-1).$$

Equating the coefficients of t^n in both sides, we get

$P_n(-1) = (-1)^n$. □

(3) $P_n(x)$ satisfies the Legendre Differential equation, then

$$(1-x^2)P_n''(x) - 2xP_n'(x) + n(n+1)P_n(x) = 0.$$

Letting $x = 1$, we obtain $2P_n'(1) = n(n+1)P_n(1)$,
and since $P_n(1) = 1$, then

$P_n'(1) = \dfrac{n(n+1)}{2}$. □

(4) Letting $x = -1$ in the differential equation, we obtain

$2P_n'(-1) = -n(n+1)P_n(-1)$,

and since $P_n(-1) = (-1)^n$, then

$P_n'(-1) = (-1)^{n+1} \dfrac{n(n+1)}{2}$. □

Legendre Polynomials

(5) Letting $x = 0$ in the generating function, we obtain

$$(1+t^2)^{-1/2} = \sum_{n=0}^{\infty} t^n P_n(0)$$

Using the Binomial expansion of $(1+t^2)^{-1/2}$, we obtain

$$1 - \frac{1}{2} \cdot t^2 + \frac{1 \cdot 3}{2^2} \cdot \frac{t^4}{2!} - \cdots + (-1)^n \cdot \frac{1 \cdot 3 \cdot 5 \cdots (2n-1)}{2^n} \cdot \frac{t^{2n}}{n!} + \cdots$$

$$= \sum_{n=0}^{\infty} t^n P_n(0)$$

or $\sum_{n=0}^{\infty} (-1)^n \cdot \frac{1 \cdot 3 \cdot 5 \cdots (2n-1)}{2^n} \cdot \frac{t^{2n}}{n!} = \sum_{n=0}^{\infty} t^n P_n(0)$

or $\sum_{n=0}^{\infty} (-1)^n \cdot \frac{(2n)!}{2^{2n}(n!)^2} \cdot t^{2n} = \sum_{n=0}^{\infty} t^n P_n(0)$

Equating the coefficients of t^{2n} in both sides, we get

$$P_{2n}(0) = (-1)^n \cdot \frac{(2n)!}{2^{2n}(n!)^2}. \qquad \square$$

(6) From $\sum_{n=0}^{\infty} (-1)^n \cdot \frac{(2n)!}{2^{2n}(n!)^2} \cdot t^{2n} = \sum_{n=0}^{\infty} t^n P_n(0)$.

It is clear that the coefficient of t^{2n+1} in the right hand side is zero, then $P_{2n+1}(0) = 0$. $\qquad \square$

(7) Replacing x by $-x$ in the generating function, we obtain

$$(1 + 2xt + t^2)^{-1/2} = \sum_{n=0}^{\infty} t^n P_n(-x).$$

Now, replacing t by $-t$, we get

$$(1 - 2xt + t^2)^{-1/2} = \sum_{n=0}^{\infty} t^n P_n(x) = \sum_{n=0}^{\infty} (-1)^n t^n P_n(-x)$$

Equating the coefficients of t^n, we obtain

$$P_n(-x) = (-1)^n P_n(x). \qquad \square$$

115

Example 10: Using the generating function,
$$(1-2xt+t^2)^{-1/2} = \sum_{n=0}^{\infty} t^n P_n(x),\text{ show that:}$$

$$1+\frac{1}{2}P_1(\cos\theta)+\frac{1}{3}P_2(\cos\theta)+\cdots = \ln\left\{\frac{1+\sin(\theta/2)}{\sin(\theta/2)}\right\}.$$

Solution: From the generating function, integrating both sides with respect to t from 0 to 1, we obtain:
$$\int_0^1 \frac{dt}{\sqrt{1-2tx+t^2}} = \sum_{n=0}^{\infty} \int_0^1 t^n P_n(x)\, dt.$$

Now, letting $x = \cos\theta$, we get
$$\int_0^1 \frac{dt}{\sqrt{1-2t\cos\theta+t^2}} = \sum_{n=0}^{\infty} P_n(\cos\theta) \int_0^1 t^n\, dt\ ^{17}, \text{ or}$$

$$\int_0^1 \frac{dt}{\sqrt{(t-\cos\theta)^2+\sin^2\theta}} = \sum_{n=0}^{\infty} P_n(\cos\theta)\left[\frac{t^{n+1}}{n+1}\right]_0^1\ ^{18}.$$

$$\sum_{n=0}^{\infty} \frac{P_n(\cos\theta)}{n+1} = \left[\ln\left\{(t-\cos\theta)+\sqrt{(t-\cos\theta)^2+\sin^2\theta}\right\}\right]_0^1$$

$$= \ln\left[(1-\cos\theta)+\sqrt{(1-\cos\theta)^2+\sin^2\theta}\right]$$

$$-\ln\left[(-\cos\theta)+\sqrt{\cos^2\theta+\sin^2\theta}\right]$$

$$= \ln\left[\frac{1-\cos\theta+\sqrt{2}\sqrt{1-\cos\theta}}{1-\cos\theta}\right]_{19}$$

$$= \ln\left[\frac{\sqrt{1-\cos\theta}+\sqrt{2}}{\sqrt{1-\cos\theta}}\right]$$

---

17. $1-2t\cos\theta+t^2 = \sin^2\theta+\cos^2\theta-2t\cos\theta+t^2 = (t-\cos\theta)^2+\sin^2\theta$

18. $\int_0^1 \frac{dx}{\sqrt{x^2+a^2}} = \left[\ln\left\{x+\sqrt{x^2+a^2}\right\}\right]_0^1$

19. $1-\cos\theta = 2\sin^2\theta/2$

116

$$= \ln\left[\frac{\sqrt{2\sin^2\theta/2}+\sqrt{2}}{\sqrt{2\sin^2\theta/2}}\right] = \ln\left[\frac{1+\sin\theta/2}{\sin\theta/2}\right]. \quad \square$$

Example 11: From the generating function

$$(1-2xt+t^2)^{-1/2} = \sum_{n=0}^{\infty} t^n P_n(x), \text{ show that:}$$

$$P_n(\cos\theta) = \frac{1\cdot 3\cdot 5\cdots(2n-1)}{2^{n-1}n!}\left\{\cos n\theta + \frac{1\cdot n}{1\cdot(2n-1)}\cos[(n-2)\theta]\right.$$

$$\left.+\frac{1\cdot 3\cdot n(n-1)}{1\cdot 2\cdot(2n-1)(2n-3)}\cos[(n-4)\theta]+\cdots\right\}.$$

Solution: From the generating function, let $x = \cos\theta = \dfrac{e^{i\theta}+e^{-i\theta}}{2}$, then

$$\left[1-t(e^{i\theta}+e^{-i\theta})+t^2\right]= \left[(1-te^{i\theta})(1-te^{-i\theta})\right]^{-1/2} = \sum_{n=0}^{\infty}t^n P_n(\cos\theta)$$

Now, from the binomial expansion, we have

$$(1-te^{i\theta})^{-1/2} = 1 + \frac{1}{2}te^{i\theta} + \frac{1\cdot 3}{2\cdot 4}t^2 e^{2i\theta}+\cdots+\frac{1\cdot 3\cdots(2n-1)}{2\cdot 4\cdots(2n)}t^n e^{ni\theta}+\cdots$$

$$(1-te^{-i\theta})^{-1/2} = 1 + \frac{1}{2}te^{-i\theta} + \frac{1\cdot 3}{2\cdot 4}t^2 e^{-2i\theta}+\cdots+\frac{1\cdot 3\cdots(2n-1)}{2\cdot 4\cdots(2n)}t^n e^{-ni\theta}+\cdots$$

The coefficient of t^n in the product of these two series is

$$\frac{1\cdot 3\cdots(2n-1)}{2\cdot 4\cdots(2n)}\left(e^{ni\theta}+e^{-ni\theta}\right)+\frac{1}{2}\cdot\frac{1\cdot 3\cdots(2n-3)}{2\cdot 4\cdots(2n-2)}\left(e^{(n-2)i\theta}+e^{-(n-2)i\theta}\right)$$

$$+\frac{1\cdot 3}{2\cdot 4}\cdot\frac{1\cdot 3\cdots(2n-5)}{2\cdot 4\cdots(2n-4)}\left(e^{(n-4)i\theta}+e^{-(n-4)i\theta}\right)+\cdots$$

Therefore,

$$P_n(\cos\theta) = \frac{1\cdot 3\cdot 5\cdots(2n-1)}{2\cdot 4\cdots(2n)}\left\{2\cos n\theta + \frac{1\cdot n}{1\cdot(2n-1)}\cdot 2\cos[(n-2)\theta]\right.$$

$$\left.+\frac{1\cdot 3\cdot n(n-1)}{1\cdot 2\cdot(2n-1)(2n-3)}\cdot 2\cos[(n-4)\theta]+\cdots\right\}$$

or

$$P_n(\cos\theta) = \frac{1\cdot 3\cdot 5\cdots(2n-1)}{2^{n-1}n!}\left\{\cos n\theta + \frac{1\cdot n}{1\cdot(2n-1)}\cos[(n-2)\theta]\right.$$

$$\left.+\frac{1\cdot 3\cdot n(n-1)}{1\cdot 2\cdot(2n-1)(2n-3)}\cos[(n-4)\theta]+\cdots\right\} \quad \square$$

Special Functions and Orthogonal Polynomials

Note: If n is odd, the last term in the expansion of $P_n(\cos\theta)$ is a multiple of $\cos\theta$; while, if n is even, the last term is a constant. Using the expansion of $P_n(\cos\theta)$, we have

$$P_0(\cos\theta) = 1; \qquad P_1(\cos\theta) = \cos\theta;$$

$$P_2(\cos\theta) = \frac{1}{4}(3\cos 2\theta + 1)$$

$$P_3(\cos\theta) = \frac{1}{8}(5\cos 3\theta + 3\cos\theta);$$

$$P_4(\cos\theta) = \frac{1}{64}(35\cos 4\theta + 20\cos 2\theta + 9)$$

$$P_5(\cos\theta) = \frac{1}{128}(63\cos 5\theta + 35\cos 3\theta + 30\cos\theta).$$

Example 12: Show that $g(x,t) = (1 - 2xt + t^2)^{-1/2}$ satisfies:

i. $(1 - 2xt + t^2)\dfrac{\partial g}{\partial t} = (x - t)g$ ii. $(1 - 2xt + t^2)\dfrac{\partial g}{\partial x} = tg$.

Solution: $\dfrac{\partial g}{\partial t} = \dfrac{x - t}{(1 - 2xt + t^2)^{3/2}}$ or $(1 - 2xt + t^2)\dfrac{\partial g}{\partial t} = (x - t)g$.

$\dfrac{\partial g}{\partial x} = \dfrac{t}{(1 - 2xt + t^2)^{3/2}}$ or $(1 - 2xt + t^2)\dfrac{\partial g}{\partial x} = tg$. □

Example 13: Show that $g(x,t) = (1 - 2xt + t^2)^{-1/2}$ is a solution of the differential equation $t\dfrac{\partial^2}{\partial t^2}(tg) + \dfrac{\partial}{\partial x}\left\{(1-x^2)\dfrac{\partial g}{\partial x}\right\} = 0$.

Solution: From the generating function, we have

$$g = (1 - 2xt + t^2)^{-1/2} = \sum_{n=0}^{\infty} t^n P_n(x), \text{ then } tg = \sum_{n=0}^{\infty} t^{n+1} P_n(x).$$

Also, $\dfrac{\partial g}{\partial x} = \sum_{n=0}^{\infty} t^n P_n'(x)$, and

$$t\dfrac{\partial^2}{\partial t^2}(tg) = t\sum_{n=0}^{\infty} n(n+1)t^{n-1} P_n(x) = \sum_{n=0}^{\infty} n(n+1)t^n P_n(x)$$

118

Legendre Polynomials

$$\frac{\partial}{\partial x}\left\{(1-x^2)\frac{\partial g}{\partial x}\right\} = \frac{\partial}{\partial x}\left\{(1-x^2)\sum_{n=0}^{\infty} t^n P_n{}'(x)\right\}$$

$$= (1-x^2)\sum_{n=0}^{\infty} t^n P_n{}''(x) - 2x\sum_{n=0}^{\infty} t^n P_n{}'(x)$$

Substituting all these expressions in the given differential equation, we obtain

$$t\frac{\partial^2}{\partial t^2}(tg) + \frac{\partial}{\partial x}\left\{(1-x^2)\frac{\partial g}{\partial x}\right\}$$

$$= (1-x^2)\sum_{n=0}^{\infty} t^n P_n{}''(x) - 2x\sum_{n=0}^{\infty} t^n P_n{}'(x) + \sum_{n=0}^{\infty} n(n+1)t^n P_n(x)$$

$$= \sum_{n=0}^{\infty} t^n \left[(1-x^2)P_n{}''(x) - 2xP_n{}'(x) + n(n+1)P_n(x)\right] = 0 \quad \square$$

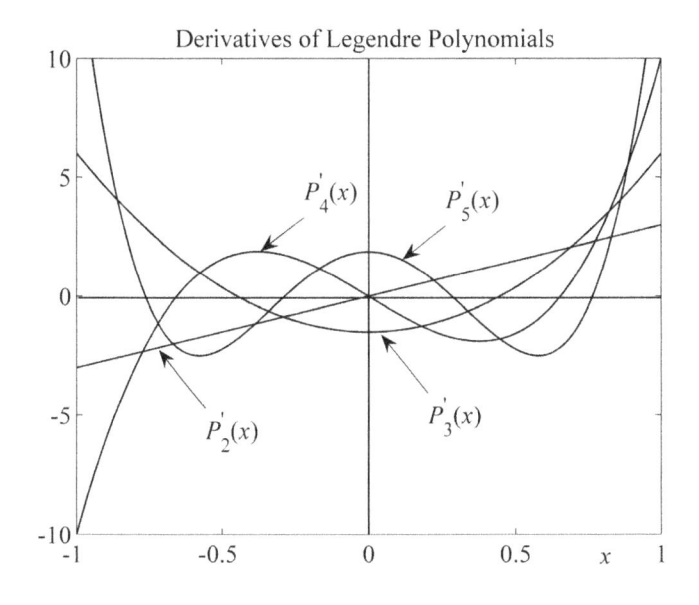

Derivatives of Legendre Polynomials

[Figure: Plot showing $P_2'(x)$, $P_3'(x)$, $P_4'(x)$, $P_5'(x)$ over $x \in [-1, 1]$]

3.5. Recurrence Relations for Legendre Polynomials

In the previous section we have studied the generating function for Legendre Polynomials. In this section, we will derive several recurrence relations from this generating function. These recurrence relations are particularly useful for computer evaluation of Legendre Polynomials and their derivatives.

I. $\boxed{(n+1)P_{n+1}(x) - (2n+1)xP_n(x) + nP_{n-1}(x) = 0}$ (29)

Proof: From the generating function $(1-2xt+t^2)^{-1/2} = \sum_{n=0}^{\infty} t^n P_n(x)$,

differentiating with respect to t, we get

$$-\frac{1}{2}(1-2xt+t^2)^{-3/2}(-2x+2t) = n\sum_{n=0}^{\infty} t^{n-1} P_n(x)$$ (30)

Multiplying both sides by $(1-2xt+t^2)$, we obtain

$$(x-t)(1-2xt+t^2)^{-1/2} = (1-2xt+t^2) \cdot n\sum_{n=0}^{\infty} t^{n-1} P_n(x), \text{ or}$$

$$(x-t)\sum_{n=0}^{\infty} t^n P_n(x) = (1-2xt+t^2) \cdot n\sum_{n=0}^{\infty} t^{n-1} P_n(x).$$

Equating the coefficients of t^n in both sides, we get

$$xP_n(x) - P_{n-1}(x) = (n+1)P_{n+1}(x) - 2nxP_n(x) + (n-1)P_{n-1}(x)$$

or

$$(n+1)P_{n+1}(x) - (2n+1)xP_n(x) + nP_{n-1}(x) = 0.$$

II. $\boxed{nP_n(x) = xP_n'(x) - P_{n-1}'(x)}$ (31)

Proof: From the generating function $(1-2xt+t^2)^{-1/2} = \sum_{n=0}^{\infty} t^n P_n(x)$,

differentiating with respect to x, we get

$$-\frac{1}{2}(1-2xt+t^2)^{-3/2}(-2t) = \sum_{n=0}^{\infty} t^n P_n'(x)$$

Multiplying both sides by $(2t-2x)$, we obtain

Legendre Polynomials

$$-\frac{1}{2}(-2x+2t)(1-2xt+t^2)^{-3/2}(-2t) = (-2x+2t)\sum_{n=0}^{\infty} t^n P_n'(x).$$

Using equation (30) in the previous page, and rearranging, we get

$$t\sum_{n=0}^{\infty} nt^{n-1} P_n(x) = (x-t)\sum_{n=0}^{\infty} t^n P_n'(x).$$

Equating the coefficients of t^n in both sides, we get

$$nP_n(x) = xP_n'(x) - P_{n-1}'(x).$$

III. $\boxed{(2n+1)P_n(x) = P_{n+1}'(x) - P_{n-1}'(x)}$ \hfill (32)

Proof: From recurrence relation **I**, we have

$$(2n+1)xP_n(x) = (n+1)P_{n+1}(x) + nP_{n-1}(x).$$

Differentiating with respect to x, we get

$$(2n+1)xP_n'(x) + (2n+1)P_n(x) = (n+1)P_{n+1}'(x) + nP_{n-1}'(x).$$

And from recurrence relation **II**, we have

$$xP_n'(x) = nP_n(x) + P_{n-1}'(x).$$

Then from these two equations, we finally obtain

$$(2n+1)P_n(x) = P_{n+1}'(x) - P_{n-1}'(x).$$

IV. $\boxed{(n+1)P_n(x) = P_{n+1}'(x) - xP_n'(x)}$ \hfill (33)

Proof: From recurrence relations **II** and **III**, we have

$$nP_n(x) = xP_n'(x) - P_{n-1}'(x)$$

$$(2n+1)P_n(x) = P_{n+1}'(x) - P_{n-1}'(x)$$

Subtracting these two equations, we get

$$(n+1)P_n(x) = P_{n+1}'(x) - xP_n'(x).$$

V. $\boxed{(1-x^2)P_n'(x) = n\ [P_{n-1}(x) - xP_n(x)\ ]}$ \hfill (34)

Proof: From recurrence relations **II** and **IV**, we have

$$nP_n(x) = xP_n'(x) - P_{n-1}'(x)$$

$$(n+1)P_n(x) = P_{n+1}'(x) - xP_n'(x)$$

Replacing n by n-1 in the last equation, we get

$$nP_{n-1}(x) = P'_n(x) - xP'_{n-1}(x)$$

Multiplying the first equation by x we get

$$nxP_n(x) = x^2 P'_n(x) - xP'_{n-1}(x).$$

Finally subtracting the last two equations, we obtain

$$(1-x^2)P'_n(x) = n\ [P_{n-1}(x) - xP_n(x)\ ].$$

VI. $\boxed{(1-x^2)P'_n(x) = (n+1)\ [xP_n(x) - P_{n+1}(x)\ ]}$ (35)

Proof: From recurrence relations **I** and **V**, we have

$$(2n+1)xP_n(x) = (n+1)P_{n+1}(x) + nP_{n-1}(x). \quad (i)$$

$$(1-x^2)P'_n(x) = n[P_{n-1}(x) - xP_n(x). \quad (ii)$$

Rewriting (*i*) as

$$[(n+1)+n]xP_n(x) = (n+1)P_{n+1}(x) + nP_{n-1}(x) \text{ or }$$

$$(n+1)[xP_n(x) - P_{n+1}(x)] = n[P_{n-1}(x) - xP_n(x)] \quad (iii)$$

From (*ii*) and (*iii*), we get

$$(1-x^2)P'_n(x) = (n+1)\ [xP_n(x) - P_{n+1}(x)\ ].$$

Example 14: Show that the coefficient of x^n in $P_n(x)$ is $\dfrac{1 \cdot 3 \cdot 5 \cdots (2n-1)}{n!}$.

Solution: Let c_n be the coefficient of x^n in $P_n(x)$, then from Recurrence Relation **I**, we have $(n+1)P_{n+1}(x) - (2n+1)xP_n(x) + nP_{n-1}(x) = 0$

Equating the coefficient of x^{n+1} to zero, we get

$$(n+1)c_{n+1} = (2n+1)c_n \text{ or } c_{n+1} = \frac{2n+1}{n+1}c_n.$$

Working backward, we obtain

$$c_n = \frac{2n-1}{n}c_{n-1} = \frac{(2n-1)(2n-3)}{n(n-1)}c_{n-2} = \cdots$$

$$= \frac{(2n-1)(2n-3)\cdots 5 \cdot 3 \cdot 1}{n!}c_0$$

But c_0 is the coefficient of x^0 in $P_0(x)$ which is 1.

Then $c_n = \dfrac{1 \cdot 3 \cdot 5 \cdots (2n-1)}{n!}.$ □

Legendre Polynomials

***Example* 15**: If $P_n(x)$ is Legendre polynomial of degree n, and if a is such that $P_n(a) = 0$, i.e. a is a root of $P_n(x) = 0$, show that $P_{n-1}(a)$ and $P_{n+1}(a)$ are of opposite signs.

Solution: From Recurrence Relation **I**, we have

$$(n+1)P_{n+1}(x) - (2n+1)xP_n(x) + nP_{n-1}(x) = 0.$$

Letting $x = a$, then $P_n(a) = 0$ and the last equation becomes

$$(n+1)P_{n+1}(a) - (2n+1)a \cdot 0 + nP_{n-1}(a) = 0, \text{ or}$$

$$\frac{P_{n+1}(a)}{P_{n-1}(a)} = -\frac{n}{n+1}.$$

And since n is a positive integer, then $P_{n-1}(a)$ and $P_{n+1}(a)$ are of opposite signs. □

***Example* 16**: Show that: (Beltrami's[20] Formula):

$$(2n+1)(x^2 - 1)P'_n(x) = n(n+1)[P_{n+1}(x) - P_{n-1}(x)]$$

Solution: From Recurrence Relation **V**, we have

$$xP'_n(x) = -\frac{(1-x^2)}{n}P'_n(x) + P'_{n-1}(x).$$

Substituting for $xP'_n(x)$ in Recurrence Relation **VI**, we get

$$(1-x^2)P'_n(x) = (n+1)\left[-\frac{(1-x^2)}{n}P'_n(x) + P'_{n-1}(x) - P_{n+1}(x)\right].$$

Re-arranging, we obtain

$$(2n+1)(x^2 - 1)P'_n(x) = n(n+1)[P_{n+1}(x) - P_{n-1}(x)]. \quad □$$

***Example* 17**: Show that: (Christoffel's[21] Expansion):

$$P'_n(x) = (2n-1)P_{n-1}(x) + (2n-5)P_{n-3}(x) + \cdots + \begin{cases} 3P_1(x) & n \text{ even} \\ P_0(x) & n \text{ odd} \end{cases}$$

Solution: In Recurrence Relation **III**, $(2n+1)P_n(x) = P'_{n+1}(x) - P'_{n-1}(x)$, replacing n by $n-1$ and re-arranging, we get

---

20. Eugenio Beltrami (1835-1900)
21. Elwin Christoffel (1829-1900)

Special Functions and Orthogonal Polynomials

$P'_n(x) = (2n-1)P_{n-1}(x) + P'_{n-2}(x)$.

<u>For n even</u>, replacing n by $n-2, n-4, \cdots, 4, 2$, we get

$P'_{n-2}(x) = (2n-5)P_{n-3}(x) + P'_{n-4}(x)$

$P'_{n-4}(x) = (2n-9)P_{n-5}(x) + P'_{n-6}(x)$

$P'_{n-6}(x) = (2n-13)P_{n-7}(x) + P'_{n-8}(x)$

...

$P'_4(x) = 7P_3(x) + P'_2(x)$

$P'_2(x) = 3P_1(x) + P'_0(x) = 3P_1(x)$.

Adding, we get

$P'_n(x) = (2n-1)P_{n-1}(x) + (2n-5)P_{n-3}(x) + \cdots + 7P_3(x) + 3P_1(x)$

<u>For n odd</u>, replacing n by $n-2, n-4, \cdots, 5, 3$, we get

$P'_{n-2}(x) = (2n-5)P_{n-3}(x) + P'_{n-4}(x)$

$P'_{n-4}(x) = (2n-9)P_{n-5}(x) + P'_{n-6}(x)$

$P'_{n-6}(x) = (2n-13)P_{n-7}(x) + P'_{n-8}(x)$

...

$P'_5(x) = 9P_4(x) + P'_3(x)$

$P'_3(x) = 5P_1(x) + P'_1(x) = 5P_2(x) + P_0(x)$.

Adding, we get

$P'_n(x) = (2n-1)P_{n-1}(x) + (2n-5)P_{n-3}(x) + \cdots + 5P_2(x) + P_0(x)$.

Combining the results, we finally obtain

$P'_n(x) = (2n-1)P_{n-1}(x) + (2n-5)P_{n-3}(x) + \cdots + \begin{cases} 3P_1(x) & n \text{ even} \\ P_0(x) & n \text{ odd} \end{cases}$

Example 18: Show that: (Christoffel's Summation):

$$\sum_{k=0}^{n} (2k+1)P_k(x)P_k(y) = \frac{n+1}{x-y}[P_{n+1}(x)P_n(y) - P_{n+1}(y)P_n(x)]$$

Solution: In Recurrence Relation **I**,

$(n+1)P_{n+1}(x) - (2n+1)xP_n(x) + nP_{n-1}(x) = 0$, we have

$(2k+1)xP_k(x) = (k+1)P_{k+1}(x) - kP_{k-1}(x)$ (i)

Legendre Polynomials

$$(2k+1)yP_k(y) = (k+1)P_{k+1}(y) - kP_{k-1}(y) \qquad (ii)$$

Multiplying (i) by $P_k(y)$ and (ii) by $P_k(x)$ and subtracting, we get

$$(2k+1)(x-y)P_k(x)P_k(y) =$$
$$(k+1)[P_{k+1}(x)P_k(y) - P_{k+1}(y)P_k(x)]$$
$$- k[P_{k-1}(y)P_k(x) - P_{k-1}(x)P_k(y)]$$

Letting $k = 0, 1, 2, \cdots, n$ and adding, we get

$$(x-y)\sum_{k=0}^{n}(2k+1)P_k(x)P_k(y) = (n+1)[P_{n+1}(x)P_n(y) - P_{n+1}(y)P_n(x)]$$

Note that all the other terms in the right hand side will cancel out in the addition process, then

$$\sum_{k=0}^{n}(2k+1)P_k(x)P_k(y) = \frac{n+1}{x-y}[P_{n+1}(x)P_n(y) - P_{n+1}(y)P_n(x)] \quad \square$$

Example 19: For $x > 1$, show that $P_n(x) < P_{n+1}(x)$.

Solution: First, $1 < x$ implies that $P_0(x) < P_1(x)$.

We will use mathematical induction: Suppose that this result is true for n, i.e., $P_{n-1}(x) < P_n(x)$, then $\dfrac{P_{n-1}(x)}{P_n(x)} < 1$.

From Recurrence Relation **I**, we have
$$(n+1)P_{n+1}(x) - (2n+1)xP_n(x) + nP_{n-1}(x) = 0, \text{ or}$$

$$\frac{P_{n+1}(x)}{P_n(x)} = \frac{2n+1}{n+1}\cdot x - \frac{n}{n+1}\cdot\frac{P_{n-1}(x)}{P_n(x)}$$

$$> \frac{2n+1}{n+1}\cdot x - \frac{n}{n+1} \qquad \left(\text{since } \frac{P_{n-1}(x)}{P_n(x)} < 1\right)$$

$$> \frac{2n+1}{n+1} - \frac{n}{n+1} = 1 \qquad (\text{since } x > 1)$$

Therefore, $P_n(x) < P_{n+1}(x)$. $\qquad \square$

Example 20: Show that: $P'_{n+1}(x) + P'_n(x) = \sum_{k=0}^{n}(2k+1)P_k(x)$

Solution: From Recurrence Relation **III**,
$$(2k+1)P_k(x) = P'_{k+1}(x) - P'_{k-1}(x),$$
let $k = 1, 2, 3, \cdots, n$, we get

$3P_1(x) = P_2'(x) - P_0'(x)$,

$5P_2(x) = P_3'(x) - P_1'(x)$,

$7P_3(x) = P_4'(x) - P_2'(x)$,

...

$(2n-3)P_{n-2}(x) = P_{n-1}'(x) - P_{n-3}'(x)$,

$(2n-1)P_{n-1}(x) = P_n'(x) - P_{n-2}'(x)$,

$(2n+1)P_n(x) = P_{n+1}'(x) - P_{n-1}'(x)$.

Adding all these equations, we get

$3P_1(x) + 5P_2(x) + 7P_3(x) + \cdots + (2n+1)P_n(x)$

$= P_{n+1}'(x) + P_n'(x) - P_0'(x) - P_1'(x) = P_{n+1}'(x) + P_n'(x) - P_0(x)$

Therefore,

$$P_{n+1}'(x) + P_n'(x) = \sum_{k=0}^{n} (2k+1)P_k(x). \qquad \square$$

Legendre Polynomials

3.6. Orthogonality Properties of Legendre Polynomials

Two functions $f(x)$ and $g(x)$ are said to be *orthogonal* to each other in the interval $[a, b]$ with respect to the weighting function $w(x)$ if and only if

$$\int_a^b w(x) f(x) g(x) dx = 0 \tag{36}$$

Legendre Polynomials among many other polynomials have this orthogonality property. This is given by the following theorem.

Theorem: If m and n are non-negative integers, then

$$\int_{-1}^{1} P_m(x) P_n(x) dx = \begin{cases} 0 & \text{if } m \neq n \\ \dfrac{2}{2n+1} & \text{if } m = n \end{cases} \tag{37}$$

The weighting function in this case is 1.

Proof: Case 1: $m \neq n$

Since $P_m(x)$ and $P_n(x)$ are solutions of Legendre Differential Equation, then omitting the argument x for convenience,

$$(1-x^2) P_m'' - 2x P_m' + m(m+1) P_m = 0,$$

$$(1-x^2) P_n'' - 2x P_n' + n(n+1) P_n = 0.$$

Multiplying the first equation by P_n and the second by P_m and subtracting, we obtain

$$(1-x^2)[P_n P_m'' - P_m P_n''] - 2x[P_n P_m' - P_m P_n']$$

$$+ [m(m+1) - n(n+1)] P_n P_m = 0$$

This can be written as

$$\frac{d}{dx}\left\{(1-x^2)[P_n P_m' - P_m P_n']\right\} = [(n-m)(n+m+1)] P_n P_m.$$

Integrating both sides of this last equation with respect to x form -1 to 1, we get

$$\left\{(1-x^2)[P_n P_m' - P_m P_n']\right\}_{-1}^{1} = [(n-m)(n+m+1)] \int_{-1}^{1} P_n P_m dx.$$

127

The expression on the left hand side vanishes at both limits, and since $m \neq n$ then $\int_{-1}^{1} P_n(x) P_m(x) dx = 0$, $m \neq n$.

Case 2: $m = n$.

Starting from the generating function, we have

$$(1 - 2xt + t^2)^{-1/2} = \sum_{n=0}^{\infty} t^n P_n(x), \text{ also}$$

$$(1 - 2xt + t^2)^{-1/2} = \sum_{m=0}^{\infty} t^m P_m(x).$$

Then, upon multiplication, we get

$$(1 - 2xt + t^2)^{-1} = \sum_{n=0}^{\infty} \sum_{m=0}^{\infty} t^{n+m} P_n(x) P_m(x).$$

Integration both sides with respect to x from -1 to 1, we obtain

$$\int_{-1}^{1} \frac{dx}{(1 - 2xt + t^2)} = \sum_{n=0}^{\infty} \sum_{m=0}^{\infty} \left\{ \int_{-1}^{1} P_n(x) P_m(x) dx \right\} t^{n+m},$$

And since $\int_{-1}^{1} P_n P_m dx = 0$, $m \neq n$, then the double summation to the right reduces to

$$\sum_{n=0}^{\infty} \left\{ \int_{-1}^{1} P_n^2(x) dx \right\} t^{2n}.$$

On the other hand, we have

$$\int_{-1}^{1} \frac{dx}{(1 - 2xt + t^2)} = -\frac{1}{2t} \left[\ln(1 - 2xt + t^2) \right]_{-1}^{1} = \frac{1}{t} [\ln(1+t) - \ln(1-t)]$$

$$= \frac{1}{t} \left\{ \left(t - \frac{t^2}{2} + \frac{t^3}{3} - \cdots \right) - \left(-t - \frac{t^2}{2} - \frac{t^3}{3} - \cdots \right) \right\}$$

$$= 2 \left\{ 1 + \frac{t^2}{3} + \frac{t^4}{5} + \cdots \right\} = \sum_{n=0}^{\infty} \frac{2}{2n+1} t^{2n}$$

Then, we have

Legendre Polynomials

$$\sum_{n=0}^{\infty}\left\{\int_{-1}^{1}P_n^2(x)dx\right\}t^{2n} = \sum_{n=0}^{\infty}\frac{2}{2n+1}t^{2n}.$$

Equating the coefficients of t^{2n} in both sides, we obtain

$$\int_{-1}^{1}P_n^2(x)dx = \frac{2}{2n+1}.$$

Note: If we let $x = \cos\theta$, the orthogonality property of Legendre Polynomials in trigonometric form will be

$$\int_0^{\pi} P_m(\cos\theta)P_n(\cos\theta)\sin\theta\, d\theta = \begin{cases} 0 & \text{if } m \neq n \\ \dfrac{2}{2n+1} & \text{if } m = n \end{cases} \qquad (38)$$

Example 21: Show that: $\int_{-1}^{1} x\, P_n(x)P_{n-1}(x)dx = \dfrac{2n}{4n^2-1}.$

Solution: From Recurrence Relation I, we have (omitting the argument for convenience)

$$x P_n = \frac{n+1}{2n+1}P_{n+1} + \frac{n}{2n+1}P_{n-1}.$$

Multiplying both sides by P_{n-1} and integrating with respect to x form -1 to 1, we obtain

$$\int_{-1}^{1} x P_n P_{n-1} dx = \frac{n+1}{2n+1}\int_{-1}^{1}P_{n+1}P_{n-1}dx + \frac{n}{2n+1}\int_{-1}^{1}P_{n-1}^2 dx.$$

From the orthogonality property of Legendre Polynomials, the first integral in the right hand side vanishes, while the second integral becomes

$$\frac{2}{2(n-1)+1} = \frac{2}{2n-1}.$$

Therofore, $\int_{-1}^{1} x\, P_n(x)P_{n-1}(x)dx = \dfrac{n}{2n+1}\cdot\dfrac{2}{2n-1} = \dfrac{2n}{4n^2-1}.$ □

129

Special Functions and Orthogonal Polynomials

Example 22: Show that

$$\int_{-1}^{1} (1-x^2) P'_m(x) P'_n(x) dx = \begin{cases} 0 & \text{if } m \neq n \\ \dfrac{2n(n+1)}{2n+1} & \text{if } m = n \end{cases}$$

Solution: Omitting the argument for convenience, we have

$$I = \int_{-1}^{1} (1-x^2) P'_m P'_n \, dx = \int_{-1}^{1} (1-x^2) P'_n \, dP_m .$$

Integrating by parts, we get

$$I = \int_{-1}^{1} (1-x^2) P'_m P'_n \, dx =$$

$$= \left[(1-x^2) P'_n P_m \right]_{-1}^{1} - \int_{-1}^{1} P_m \, d\left\{ (1-x^2) P'_n \right\}$$

$$= -\int_{-1}^{1} P_m \left\{ (1-x^2) P''_n - 2x P'_n \right\} dx$$

The term in the square brackets vanishes at both limits, and since P_n satisfies Legendre Differential Equation, we have

$$(1-x^2) P''_n - 2x \, P'_n = -n(n+1) P_n ,$$

and the integral becomes

$$I = \int_{-1}^{1} (1-x^2) P'_m P'_n \, dx = n(n+1) \int_{-1}^{1} P_n P_m \, dx$$

Then, from the Orthogonality Property, we obtain

$$\int_{-1}^{1} (1-x^2) P'_m(x) P'_n(x) dx = \begin{cases} 0 & \text{if } m \neq n \\ \dfrac{2n(n+1)}{2n+1} & \text{if } m = n \end{cases} \qquad \square$$

Example 23: Show that:

$$\int_{0}^{1} x^2 P_{n+1}(x) P_{n-1}(x) dx = \frac{n(n+1)}{(2n+3)(4n^2-1)} .$$

Solution: Since $x^2 P_{n+1} P_{n-1}$ is an even function, then

130

Legendre Polynomials

$$I = \int_0^1 x^2 P_{n+1} P_{n-1} dx = \frac{1}{2}\int_{-1}^1 x^2 P_{n+1} P_{n-1} dx$$

From Recurrence Relation **I**, we have

$$x P_n = \frac{n+1}{2n+1} P_{n+1} + \frac{n}{2n+1} P_{n-1}.$$

Replacing n once by $n+1$ and once by $n-1$, we obtain

$$x P_{n+1} = \frac{n+2}{2n+3} P_{n+2} + \frac{n+1}{2n+3} P_n \text{ and}$$

$$x P_{n-1} = \frac{n}{2n-1} P_n + \frac{n-1}{2n-1} P_{n-2}.$$

Multiplying these last two equations, we get

$$x^2 P_{n+1} P_{n-1} = \left(\frac{n+2}{2n+3} P_{n+2} + \frac{n+1}{2n+3} P_n\right)\left(\frac{n}{2n-1} P_n + \frac{n-1}{2n-1} P_{n-2}\right)$$

Integrating with respect to x from -1 to 1, and using the Orthogonality Property, we obtain

$$I = \frac{1}{2}\int_{-1}^1 x^2 P_{n+1} P_{n-1} dx = \frac{1}{2} \cdot \frac{n+1}{2n+3} \cdot \frac{n}{2n-1} \cdot \frac{2}{2n+1}$$

$$= \frac{n(n+1)}{(2n+3)(4n^2-1)}. \qquad \square$$

Example 24: Evaluate the following integrals:

i) $\int_{-1}^1 \frac{x^2 dx}{\sqrt{5-4x}}$; ii) $\int_{-1}^1 \frac{(1-x^3) dx}{(1-x)^{3/2}}$; iii) $\int_{-1}^1 \frac{P_1(x) P_5(x) dx}{\sqrt{2-2x}}$.

Solution: From the generating function $(1-2xt+t^2)^{-1/2} = \sum_{n=0}^{\infty} t^n P_n(x)$,

i) Letting $t = 1/2$, we obtain $(5/4-x)^{-1/2} = \sum_{n=0}^{\infty} \frac{1}{2^n} P_n(x)$, or

$$(5-4x)^{-1/2} = \sum_{n=0}^{\infty} \frac{1}{2^{n+1}} P_n(x),$$

Also, we know that $P_2(x) = \frac{1}{2}(3x^2 - 1)$, then

131

$$x^2 = \frac{1}{2}P_0(x) + \frac{2}{3}P_2(x).$$

The integral now becomes

$$\int_{-1}^{1}\frac{x^2 dx}{\sqrt{5-4x}} = \int_{-1}^{1}\left(\frac{1}{3}P_0(x) + \frac{2}{3}P_2(x)\right)\sum_{n=0}^{\infty}\frac{1}{2^{n+1}}P_n(x)dx.$$

From the orthogonality property, all terms in the integral to the right vanish except those for $n = 0$ and $n = 2$, then

$$\int_{-1}^{1}\frac{x^2 dx}{\sqrt{5-4x}} = \frac{1}{3} + \frac{1}{30} = \frac{11}{30}. \qquad \square$$

ii) $\int_{-1}^{1}\frac{(1-x^3)dx}{(1-x)^{3/2}} = \int_{-1}^{1}\frac{1+x+x^2}{\sqrt{1-x}}dx$

Letting $t = 1$, we obtain

$$\frac{1}{\sqrt{2-2x}} = \sum_{n=0}^{\infty}P_n(x), \text{ or } \frac{1}{\sqrt{1-x}} = \sqrt{2}\sum_{n=0}^{\infty}P_n(x),$$

also $1+x+x^2 = P_0 + P_1 + \frac{1}{3}P_0 + \frac{2}{3}P_2 = \frac{4}{3}P_0 + P + \frac{2}{3}P_2.$

The integral now becomes,

$$\int_{-1}^{1}\frac{(1-x^3)dx}{(1-x)^{3/2}} = \int_{-1}^{1}\left(\frac{4}{3}P_0 + P_1 + \frac{2}{3}P_2\right)\sum_{n=0}^{\infty}\sqrt{2}P_n \, dx.$$

From the orthogonality property, all terms in the integral to the right vanish except those for $n = 0$, $n = 1$ and $n = 2$, then

$$\int_{-1}^{1}\frac{(1-x^3)dx}{(1-x)^{3/2}} = \sqrt{2}\left(\frac{8}{3} + \frac{2}{3} + \frac{4}{15}\right) = \frac{18\sqrt{2}}{15}. \qquad \square$$

iii) From ii), we have

$$\frac{1}{\sqrt{2-2x}} = \sum_{n=0}^{\infty}P_n(x), \text{ also } P_1(x)P_5(x) = xP_5(x),$$

and from recurrence relation **I**, we have

$$xP_5(x) = \frac{5}{11}P_4(x) + \frac{6}{11}P_6(x),$$

Legendre Polynomials

then the integral becomes

$$\int_{-1}^{1}\frac{P_1(x)P_5(x)dx}{\sqrt{2-2x}} = \int_{-1}^{1}\left(\frac{5}{11}P_4(x)+\frac{6}{11}P_6(x)\right)\sum_{n=0}^{\infty}P_n(x)dx.$$

$$= \frac{10}{99}+\frac{12}{143} \approx 0.185. \qquad \Box$$

Example 25: If n is a positive integer, show that:

$$\int_{-1}^{1}(1-2xt+t^2)^{-1/2}P_n(x)dx = \frac{2t^n}{2n+1}.$$

Solution: From the generation function:

$$(1-2xt+t^2)^{-1/2} = \sum_{m=0}^{\infty} t^m P_m(x),$$

Substituting in the integral, we obtain

$$I = \int_{-1}^{1}(1-2xt+t^2)^{-1/2}P_n(x)dx = \int_{-1}^{1}P_n(x)\sum_{m=0}^{\infty}t^m P_m(x)dx.$$

From the Orthogonality Property of Legendre Polynomials, all integrals in the right hand side will vanish except when $m=n$, then

$$I = \int_{-1}^{1}(1-2xt+t^2)^{-1/2}P_n(x)dx = t^n \cdot \frac{2}{2n+1} = \frac{2t^n}{2n+1}. \qquad \Box$$

Example 26: Show that:

$$xP_n'(x) = nP_n(x)+(2n-3)P_{n-2}(x)+(2n-7)P_{n-4}(x)+ \\ +(2n-11)P_{n-6}(x)+\cdots$$

And hence deduce that:

$$\int_{-1}^{1}x P_n(x)P_n'(x)dx = \frac{2n}{2n+1}.$$

Solution: From Recurrence Relation **II**: $nP_n(x) = xP_n'(x)-P_{n-1}'(x)$, we have

$$xP_n'(x) = nP_n(x)+P_{n-1}'(x) \qquad (i)$$

133

Also, from Recurrence Relation **III**:

$$P'_{n+1}(x) = (2n+1)P_n(x) + P'_{n-1}(x),$$

Replacing n by $n-2, n-4, n-6, \cdots$, we get

$$\left.\begin{array}{l} P'_{n-1}(x) = (2n-3)P_{n-2}(x) + P'_{n-3}(x) \\ P'_{n-3}(x) = (2n-7)P_{n-4}(x) + P'_{n-5}(x) \\ P'_{n-5}(x) = (2n-11)P_{n-6}(x) + P'_{n-7}(x) \\ \cdots \end{array}\right\} \quad (ii)$$

Adding (*i*) and (*ii*), we obtain

$$xP'_n(x) = nP_n(x) + (2n-3)P_{n-2}(x) + (2n-7)P_{n-4}(x) + \\ + (2n-11)P_{n-6}(x) + \cdots \quad \square$$

If we multiply both sides by $P_n(x)$, integrate with respect to x from -1 to 1 and use the Orthogonality Property, we get

$$\int_{-1}^{1} x\, P_n(x) P'_n(x)\, dx = n \int_{-1}^{1} x\, P_n^2(x)\, dx = \frac{2n}{2n+1}. \quad \square$$

Example 27: Show that: $\int_{-1}^{1} [P'_n(x)]^2\, dx = n(n+1)$.

Solution: From Christoffel's Expansion:

$$P'_n(x) = (2n-1)P_{n-1}(x) + (2n-5)P_{n-3}(x) + \cdots + \begin{cases} 3P_1(x) & n \text{ even} \\ P_0(x) & n \text{ odd} \end{cases}$$

Squaring both sides, integrating with respect to x from -1 to 1 and using the Orthogonality property, we obtain

$$\int_{-1}^{1} [P'_n(x)]^2\, dx = (2n-1)^2 \int_{-1}^{1} P_{n-1}^2(x)\, dx + (2n-5)^2 \int_{-1}^{1} P_{n-3}^2(x)\, dx$$

$$+ \cdots + \begin{cases} 3^2 \int_{-1}^{1} P_1^2(x)\, dx & n \text{ even} \\ \int_{-1}^{1} P_0^2(x)\, dx & n \text{ odd} \end{cases}$$

Legendre Polynomials

$$= \frac{2(2n-1)^2}{2(n-1)+1} + \frac{2(2n-5)^2}{2(n-3)+1} + \frac{2(2n-9)^2}{2(n-5)+1} + \cdots + \begin{cases} \dfrac{2 \cdot 3^2}{3} & n \text{ even} \\ 2 & n \text{ odd} \end{cases}$$

$$= 2\left[(2n-1)+(2n-5)+(2n-9)+\cdots + \begin{cases} 3 & n \text{ even} \\ 1 & n \text{ odd} \end{cases}\right]$$

This is an arithmetic progession series, then:

<u>For n even</u>: the number of terms is $n/2$, and

$$\int_{-1}^{1} [P_n'(x)]^2 \, dx = 2 \cdot \frac{1}{2} \cdot \frac{n}{2} \cdot [(2n-1)+3] = n(n+1).$$

<u>For n odd</u>: the number of terms is $(n+1)/2$, and

$$\int_{-1}^{1} [P_n'(x)]^2 \, dx = 2 \cdot \frac{1}{2} \cdot \frac{n+1}{2} \cdot [(2n-1)+1] = n(n+1).$$

Therefore,

$$\int_{-1}^{1} [P_n'(x)]^2 \, dx = n(n+1) \cdot \qquad \square$$

Example 28: Show that: $\displaystyle\int_{-1}^{1} \frac{P_n(x) P_{n-1}(x)}{x} \, dx = \begin{cases} 0 & \text{if } n \text{ is even} \\ \dfrac{2}{n} & \text{if } n \text{ is odd} \end{cases}$.

Solution: Let: $u_n = \displaystyle\int_{-1}^{1} \frac{P_n(x) P_{n-1}(x)}{x} \, dx$, then from Recurrence Relation **I**,

$$\frac{P_n(x)}{x} = \frac{2n-1}{n} P_{n-1}(x) - \frac{n-1}{n} \frac{P_{n-2}(x)}{x}.$$

Substituting in the integral, we get,

$$u_n = \int_{-1}^{1} P_{n-1}(x) \left[\frac{2n-1}{n} P_{n-1}(x) - \frac{n-1}{n} \frac{P_{n-2}(x)}{x} \right] dx$$

$$= \frac{2n-1}{n} \cdot \frac{2}{2n-1} - \frac{n-1}{n} \int_{-1}^{1} \frac{P_{n-1}(x) P_{n-2}(x)}{x} \, dx$$

135

Special Functions and Orthogonal Polynomials

$$u_n = \frac{2}{n} - \frac{n-1}{n} u_{n-1}.$$

Substituting for u_{n-1}, we obtain $\boxed{u_n = \frac{n-2}{n} u_{n-2}}$.

Now, <u>For n even</u>: $u_n = \dfrac{(n-2)(n-4)\cdots 4 \cdot 2}{n(n-2)(n-4)\cdots 6 \cdot 4} u_2 = \dfrac{2}{n} u_2$, and

$$u_2 = \int_{-1}^{1} \frac{P_2(x) P_1(x)}{x} dx = \frac{1}{2} \int_{-1}^{1} \frac{x(x^2-1)}{x} dx = 0, \text{ therefore}$$

$$\int_{-1}^{1} \frac{P_n(x) P_{n-1}(x)}{x} dx = 0 \quad \text{if } n \text{ is even}.$$

<u>For n odd</u>: $u_n = \dfrac{(n-2)(n-4)\cdots 3 \cdot 1}{n(n-2)(n-4)\cdots 5 \cdot 3} u_1 = \dfrac{1}{n} u_1$, and

$$u_1 = \int_{-1}^{1} \frac{P_1(x) P_0(x)}{x} dx = \int_{-1}^{1} dx = 2, \text{ therefore}$$

$$\int_{-1}^{1} \frac{P_n(x) P_{n-1}(x)}{x} dx = \frac{2}{n} \quad \text{if } n \text{ is odd}. \qquad \square$$

3.7. Integral Form of Legendre Polynomials

If n is a positive integer, Legendre polynomials are given by

$$P_n(x) = \frac{1}{\pi} \int_0^\pi \left[x \pm \sqrt{x^2 - 1} \cos\phi \right]^n d\phi \qquad (39)$$

and

$$P_n(x) = \frac{1}{\pi} \int_0^\pi \frac{d\phi}{\left[x \pm \sqrt{x^2 - 1} \cos\phi \right]^{n+1}} \qquad (40)$$

These two forms are known as *Laplace's first and second integral* for Legendre polynomials. To prove form (39), we process as follows. It may be shown by elementary methods of integral calculus that

$$\int_0^\pi \frac{d\phi}{a \pm b \cos\phi} = \frac{\pi}{\sqrt{a^2 - b^2}}, \quad a^2 > b^2 \qquad (41)$$

Now, letting $a = 1 - tx$ and $b = t\sqrt{x^2 - 1}$, then

$$a^2 - b^2 = (1 - tx)^2 - t^2(x^2 - 1) = 1 - 2xt + t^2$$

Equation (41) becomes

$$\pi(1 - 2xt + t^2)^{-1/2} = \int_0^\pi \left[1 - tx \pm t\sqrt{x^2 - 1} \cos\phi \right]^{-1} d\phi$$

and from the generating function, we get

$$\pi \sum_{n=0}^\infty t^n P_n(x) = \int_0^\pi \left[1 - tx \pm t\sqrt{x^2 - 1} \cos\phi \right]^{-1} d\phi$$

Letting $z = x \pm \sqrt{x^2 - 1} \cos\phi$, then

$$\pi \sum_{n=0}^\infty t^n P_n(x) = \int_0^\pi [1 - tz]^{-1} d\phi = \int_0^\pi \left[1 + tz + t^2 z^2 + \cdots \right] d\phi$$

$$= \int_0^\pi \sum_{n=0}^\infty [tz]^n d\phi = \sum_{n=0}^\infty \int_0^\pi \left[x \pm \sqrt{x^2 - 1} \cos\phi \right]^n d\phi \cdot t^n$$

Equating the coefficients of t^n in both sides, we get

Special Functions and Orthogonal Polynomials

$$P_n(x) = \frac{1}{\pi}\int_0^\pi \left[x \pm \sqrt{x^2-1}\cos\phi\right]^n d\phi.\qquad \square$$

To prove the Laplace's second integral for Legendre polynomials $P_n(x)$, we let $a = tx - 1$ and $b = t\sqrt{x^2-1}$ in equation (41), we get

$$b^2 - a^2 = (tx-1)^2 - t^2(x^2-1) = 1 - 2tx + t^2$$

Equation (41) becomes

$$\pi(1-2xt+t^2)^{-1/2} = \int_0^\pi \left[-1+tx \pm t\sqrt{x^2-1}\cos\phi\right]^{-1} d\phi$$

or

$$\frac{\pi}{t}\left(1 - \frac{2x}{t} + \frac{1}{t^2}\right)^{-1/2} = \int_0^\pi \left[-1 + t\left\{x \pm \sqrt{x^2-1}\cos\phi\right\}\right]^{-1} d\phi \qquad (42)$$

From the generating function $(1-2xt+t^2)^{-1/2} = \sum_{n=0}^\infty t^n P_n(x)$

Replacing t by $1/t$, we get

$$\left(1 - \frac{2x}{t} + \frac{1}{t^2}\right)^{-1/2} = \sum_{n=0}^\infty \frac{1}{t^n} P_n(x)$$

Letting $z = x \pm \sqrt{x^2-1}\cos\phi$ equation (48) becomes

$$\frac{\pi}{t}\sum_{n=0}^\infty t^{-n} P_n(x) = \int_0^\pi [-1+tz]^{-1} d\phi = \int_0^\pi (tz)^{-1}\left[1 - \frac{1}{tz}\right]^{-1} d\phi$$

$$= \int_0^\pi \frac{1}{tz}\sum_{n=0}^\infty [tz]^{-n} d\phi = \sum_{n=0}^\infty \frac{1}{t^{n+1}}\int_0^\pi \frac{d\phi}{\left[x \pm \sqrt{x^2-1}\cos\phi\right]^{n+1}}$$

Equating the coefficients of $\dfrac{1}{t^{n+1}}$ in both sides, we get

$$P_n(x) = \frac{1}{\pi}\int_0^\pi \frac{d\phi}{\left[x \pm \sqrt{x^2-1}\cos\phi\right]^{n+1}} \qquad \square$$

138

Legendre Polynomials

3.8. Differential Form for Legendre Polynomials (Rodrigues'[22] Formula)

One of the fundamental identities involving Legendre polynomials is Rodrigues' Formula. This formula is given in differential form

$$P_n(x) = \frac{1}{2^n n!} \cdot \frac{d^n}{dx^n}\left[(x^2-1)^n\right] \tag{43}$$

To prove this formula, we let $v = (x^2-1)^n$, then

$$\frac{dv}{dx} = 2nx(x^2-1)^{n-1}$$

Multiplying by (x^2-1), we get $(x^2-1)\dfrac{dv}{dx} = 2nx(x^2-1)^n$

or
$$(1-x^2)\frac{dv}{dx} + 2nxv = 0 \tag{44}$$

Differentiating with respect to x, we get

$$(1-x^2)v'' - 2xv' + 2nxv' + 2nv = 0$$

or
$$(1-x^2)v'' + 2x(n-1)v' + 2nv = 0$$

Differentiating n times using Leibniz's theorem, we get

$$(1-x^2)v_{n+2} - 2nxv_{n+1} - 2\cdot\frac{n(n-1)}{2!}v_n$$

$$+ 2x(n-1)v_{n+1} + 2n(n-1)v_n + 2nv_n = 0$$

or $\quad (1-x^2)v_{n+2} - 2xv_{n+1} + n(n+1)v_n = 0$

Letting $u = v_n$, we get $(1-x^2)u'' - 2xu' + n(n+1)u = 0$

Then $u = v_n$ is a solution of Legendre differential equation.

But $u = v_n = \dfrac{d^n}{dx^n}\left[(x^2-1)^n\right]$. Hence u must be some multiple of $P_n(x)$, i.e.,

To determine the constant c, we know that for $x = 1$, $P_n(1) = 1$, therefore from Equation (45), we have

---

22. Benjamin Olinde Rodrigues (1794-1851 France)

Special Functions and Orthogonal Polynomials

$$c = \frac{d^n}{dx^n}\left[(x^2-1)^n\right]\bigg|_{x=1} \qquad (46)$$

But $\frac{d^n}{dx^n}\left[(x^2-1)^n\right] = \frac{d^n}{dx^n}\left[(x-1)^n\cdot(x+1)^n\right] = n!(x+1)^n + R$, where R contains $(x-1)$ as a factor, which makes $R=0$, as $x=1$. Then

$$c = \frac{d^n}{dx^n}\left[(x^2-1)^n\right]\bigg|_{x=1} = 2^n n!,$$

and
$$P_n(x) = \frac{1}{2^n n!}\cdot\frac{d^n}{dx^n}\left[(x^2-1)^n\right]. \qquad \square$$

Rodrigues' Formula can also be derived using the summation expression for Legendre Polynomials

$$P_n(x) = \sum_{k=0}^{N}(-1)^k \cdot \frac{(2n-2k)!}{2^n k!(n-2k)!(n-k)!}\, x^{n-2k} \qquad (47)$$

where N is the *Floor Function* $N = \begin{cases} n/2 & \text{if } n \text{ is even} \\ (n-1)/2 & \text{if } n \text{ is odd} \end{cases}$

$$P_n(x) = \frac{1}{2^n}\sum_{k=0}^{N}(-1)^k \cdot \frac{(2n-2k)(2n-2k-1)\cdots(n-2k+1)x^{n-2k}}{k!(n-k)!}.$$

Looking at the denominator in this expression, we can see that the factors in it allows us to replace the upper limit of the summation N by n, since one of the factors will vanish for every value of k greater that N. Therefore, we have

$$P_n(x) = \frac{1}{2^n}\sum_{k=0}^{n}(-1)^k \cdot \frac{(2n-2k)(2n-2k-1)\cdots(n-2k+1)x^{n-2k}}{k!(n-k)!}$$

$$= \frac{1}{2^n}\sum_{k=0}^{n}\frac{(-1)^k}{k!(n-k)!}\cdot\frac{d^n}{dx^n}\left(x^{2n-2k}\right)$$

$$= \frac{1}{2^n}\cdot\frac{d^n}{dx^n}\sum_{k=0}^{n}\frac{(-1)^k x^{2n-2k}}{k!(n-k)!}$$

$$= \frac{1}{2^n n!}\cdot\frac{d^n}{dx^n}\underbrace{\sum_{k=0}^{n}\frac{n!}{k!(n-k)!}\left(-\frac{1}{x^2}\right)^k x^{2n}}_{\left(1-\frac{1}{x^2}\right)^n \text{ (binomial expansion)}}$$

Legendre Polynomials

$$= \frac{1}{2^n n!} \cdot \frac{d^n}{dx^n}\left[\left(1-\frac{1}{x^2}\right)^n x^{2n}\right].$$

Hence $$P_n(x) = \frac{1}{2^n n!} \cdot \frac{d^n}{dx^n}\left[(x^2-1)^n\right].$$

Example 29: Starting from Rodrigues' Formula, show that:

$$P_n(x) = 1 + \sum_{k=1}^{\infty} \frac{(-1)^k \, n(n-1)\cdots(n-k+1)\cdot(n+1)(n+2)\cdots(n+k)}{2^k (k!)^2}(1-x)^k$$

Solution: From Rodrigues' Formula, we have

$$P_n(x) = \frac{1}{2^n n!} \cdot \frac{d^n}{dx^n}\left[(x^2-1)^n\right] = \frac{(-1)^n}{n!} \frac{d^n}{dx^n}\left[\frac{1}{2^n}(1-x^2)^n\right]$$

$$= \frac{(-1)^n}{n!} \frac{d^n}{dx^n}\left[(1-x)^n \cdot \frac{(1+x)^n}{2^n}\right] = \frac{(-1)^n}{n!}\frac{d^n}{dx^n}\left[(1-x)^n \cdot \left\{1-\left(\frac{1+x}{2}\right)\right\}^n\right]$$

$$= \frac{(-1)^n}{n!}\frac{d^n}{dx^n}\left[(1-x)^n \cdot \left\{1 - n\left(\frac{1-x}{2}\right) + \frac{n(n-1)}{2!}\left(\frac{1-x}{2}\right)^2 - \cdots\right\}\right]$$

$$= \frac{(-1)^n}{n!}\frac{d^n}{dx^n}\left[(1-x)^n - \frac{n}{2}(1-x)^{n+1} + \frac{n(n-1)}{2^2 2!}(1-x)^{n+2} - \cdots\right]$$

$$= \frac{(-1)^n}{n!}\left[(-1)^n n! - \frac{(-1)^n n}{2}\frac{(n+1)!}{1!}(1-x) + \frac{(-1)^n n(n-1)}{2^2 2!}\frac{(n+2)!}{2!}(1-x)^2 - \cdots\right]$$

$$= \left[1 - \frac{n}{2}\frac{(n+1)}{1!}(1-x) + \frac{n(n-1)}{2^2 2!}\frac{(n+1)(n+2)}{2!}(1-x)^2 - \cdots\right]$$

$$= 1 + \sum_{k=1}^{\infty}(-1)^n \frac{n(n-1)\cdots(n-k+1)\cdot(n+1)(n+2)\cdots(n+k)}{2^k (k!)}(1-x)^k \quad \square$$

Example 30: Starting from Rodrigues' Formula, prove the orthogonalty property of Legendre polynomials.

Solution: From Rodrigues' Formula, $P_n(x) = \frac{1}{2^n n!} \cdot \frac{d^n}{dx^n}\left[(x^2-1)^n\right]$, we have

141

Special Functions and Orthogonal Polynomials

$$I = \int_{-1}^{1} P_m(x) P_n(x) dx$$

$$= \frac{1}{2^{n+m} n! m!} \int_{-1}^{1} \frac{d^m}{dx^m}(x^2-1)^m \frac{d^n}{dx^n}(x^2-1)^n dx$$

$$= \frac{1}{2^{n+m} n! m!} \int_{-1}^{1} \frac{d^m}{dx^m}(x^2-1)^m d\left\{\frac{d^{n-1}}{dx^{n-1}}(x^2-1)^n\right\}$$

Integrating by parts, we get,

$$I = \frac{1}{2^{n+m} n! m!} \left[\frac{d^m}{dx^m}(x^2-1)^m \cdot \frac{d^{n-1}}{dx^{n-1}}(x^2-1)^n\right]_{-1}^{1}$$

$$- \frac{1}{2^{n+m} n! m!} \int_{-1}^{1} \frac{d^{n-1}}{dx^{n-1}}(x^2-1)^n d\left\{\frac{d^m}{dx^m}(x^2-1)^m\right\}$$

The first term in this expression will vanish at both limits, then

$$I = \frac{-1}{2^{n+m} n! m!} \int_{-1}^{1} \frac{d^{n-1}}{dx^{n-1}}(x^2-1)^n \frac{d^{m+1}}{dx^{m+1}}(x^2-1)^m dx \cdot$$

Now, integrating by parts $(m-1)$, we obtain

$$I = \frac{(-1)^m}{2^{n+m} n! m!} \int_{-1}^{1} \frac{d^{n-m}}{dx^{n-m}}(x^2-1)^n \frac{d^{2m}}{dx^{2m}}(x^2-1)^m dx \cdot$$

But $\dfrac{d^{2m}}{dx^{2m}}(x^2-1)^m = (2m)!$, then

$$I = \frac{(-1)^m (2m)!}{2^{n+m} n! m!} \int_{-1}^{1} \frac{d^{n-m}}{dx^{n-m}}(x^2-1)^n dx \cdot$$

If $m \neq n$ and $n > m$, then

$$I = \frac{(-1)^m (2m)!}{2^{n+m} n! m!} \left[\frac{d^{n-m-1}}{dx^{n-m-1}}(x^2-1)^n\right]_{-1}^{1} = 0 \cdot$$

This is the first part of the proof, for $m = n$, we have

Legendre Polynomials

$$I = \int_{-1}^{1} P_n^2(x)\,dx = \frac{(-1)^n (2n)!}{2^{2n}(n!)^2} \int_{-1}^{1} (x^2-1)^n\,dx \cdot$$

The integrand of the integral to the right is an even function, so, we may write

$$I = \frac{(-1)^n (2n)!}{2^{2n-1}(n!)^2} \int_{0}^{1} (x^2-1)^n\,dx \cdot$$

To evaluate this integral, let $x = \cos\theta$, then $dx = -\sin\theta\,d\theta$ and the limits of the integral will be from $\pi/2$ to 0, therefore

$$I = \frac{(-1)^n (2n)!}{2^{2n-1}(n!)^2} \int_{0}^{\pi/2} \sin^{2n}\theta \cdot \sin\theta\,d\theta$$

$$= \frac{(-1)^n (2n)!}{2^{2n-1}(n!)^2} \int_{0}^{\pi/2} \sin^{2n+1}\theta\,d\theta = \frac{(-1)^n (2n)!}{2^{2n-1}(n!)^2} \cdot \frac{\Gamma(n+1)\sqrt{\pi}}{2^{n+1}\Gamma(n+3/2)}$$

But $\dfrac{(2n)!}{n!} = 1 \cdot 3 \cdot 5 \cdots (2n-1)$, then

$$I = \frac{1 \cdot 3 \cdot 5 \cdots (2n-1)}{n!} \cdot \frac{2 \cdot n!\sqrt{\pi}}{(2n+1)(2n-1)\cdots 3 \cdot 1 \cdot \sqrt{\pi}} = \frac{2}{2n+1} \quad \square$$

Example 31: For any n continuously differentiable function $f(x)$, show that:

$$\int_{-1}^{1} f(x) P_n(x)\,dx = \frac{(-1)^n}{2^n n!} \int_{-1}^{1} (x^2-1)^n f^{(n)}(x)\,dx \cdot$$

Solution: Substituting from Rodrigues' Formula into the integral on the right hand side, $P_n(x) = \dfrac{1}{2^n n!} \cdot \dfrac{d^n}{dx^n}\left[(x^2-1)^n\right]$, we obtain

$$I = \int_{-1}^{1} f(x) P_n(x)\,dx = \frac{1}{2^n n!} \cdot \int_{-1}^{1} f(x) \frac{d^n}{dx^n}\left[(x^2-1)^n\right] dx$$

$$= \frac{1}{2^n n!} \cdot \int_{-1}^{1} f(x)\,d\left\{\frac{d^{n-1}}{dx^{n-1}}\left[(x^2-1)^n\right]\right\}.$$

Integrating by parts, we get

143

Special Functions and Orthogonal Polynomials

$$I = \frac{1}{2^n n!} \cdot \left[f(x) \frac{d^{n-1}}{dx^{n-1}} \left[(x^2-1)^n \right] \right]_{-1}^{1}$$

$$- \frac{1}{2^n n!} \int_{-1}^{1} f'(x) \frac{d^{n-1}}{dx^{n-1}} \left[(x^2-1)^n \right] dx.$$

The term in the square brackets vanishes at both limits, therefore

$$I = \frac{(-1)}{2^n n!} \int_{-1}^{1} f'(x) \frac{d^{n-1}}{dx^{n-1}} \left[(x^2-1)^n \right] dx.$$

Integrating by parts $(n-1)$ more times, we obtain

$$I = \int_{-1}^{1} f(x) P_n(x) dx = \frac{(-1)^n}{2^n n!} \int_{-1}^{1} (x^2-1)^n f^{(n)}(x) dx \cdot \qquad \square$$

Example 32: From Rodrigues' Formula, show that:

$$\int_{-1}^{1} (1-2xt+t^2)^{-n-1/2} (1-x^2)^n \, dx = \frac{2^{2n+1}(n!)^2}{(2n+1)!}.$$

Solution: From **Example 25**, we have

$$\int_{-1}^{1} (1-2xt+t^2)^{-1/2} P_n(x) dx = \frac{2t^n}{2n+1}.$$

Plugging in the Rodrigues' Fomula for $P_n(x)$, we obtain

$$\frac{1}{2^n n!} \int_{-1}^{1} (1-2xt+t^2)^{-1/2} \cdot \frac{d^n}{dx^n} \left[(x^2-1)^n \right] dx = \frac{2t^n}{2n+1}, \text{ or}$$

$$\int_{-1}^{1} (1-2xt+t^2)^{-1/2} \cdot \frac{d^n}{dx^n} \left[(x^2-1)^n \right] dx = \frac{2^{n+1} n! t^n}{2n+1} \text{ or}$$

$$\int_{-1}^{1} (1-2xt+t^2)^{-1/2} d \left\{ \frac{d^{n-1}}{dx^{n-1}} \left[(x^2-1)^n \right] \right\} = \frac{2^{n+1} n! t^n}{2n+1}.$$

Integrating by parts, we get

$$\left[(1-2xt+t^2)^{-1/2} \cdot \frac{d^{n-1}}{dx^{n-1}} \left[(x^2-1)^n \right] \right]_{-1}^{1}$$

$$- \int_{-1}^{1} \frac{d^{n-1}}{dx^{n-1}} \left[(x^2-1)^n \right] d\left\{ (1-2xt+t^2)^{-1/2} \right\} = \frac{2^{n+1} n! t^n}{2n+1}$$

The term in the square brackets vanishes at both end limits, then

$$- \int_{-1}^{1} \frac{d^{n-1}}{dx^{n-1}} \left[(x^2-1)^n \right] \left[t(1-2xt+t^2)^{-3/2} \right] dx = \frac{2^{n+1} n! t^n}{2n+1}, \text{ or}$$

$$- \int_{-1}^{1} \frac{d^{n-1}}{dx^{n-1}} \left[(x^2-1)^n \right] \left[(1-2xt+t^2)^{-3/2} \right] dx = \frac{2^{n+1} n! t^{n-1}}{1 \cdot (2n+1)}.$$

Integrating by parts one more time, we obtain

$$(-1)^2 \int_{-1}^{1} \frac{d^{n-1}}{dx^{n-1}} \left[(x^2-1)^n \right] \left[(1-2xt+t^2)^{-5/2} \right] dx = \frac{2^{n+1} n! t^{n-2}}{1 \cdot 3 \cdot (2n+1)}$$

Repeating this process $n-2$ times and simplifying, we get

$$\int_{-1}^{1} (x^2-1)^n (1-2xt+t^2)^{-n-1/2} dx = \frac{2^{n+1} n!}{1 \cdot 3 \cdot 5 \cdots (2n-1)(2n+1)}$$

$$= \frac{2^{n+1} n! \, 2 \cdot 4 \cdot 6 \cdots (2n)}{1 \cdot 2 \cdot 3 \cdots (2n+1)} = \frac{2^{2n+1} (n!)^2}{(2n+1)!} \quad \Box$$

Example 33: If $m < n$, show that:

(i) $\int_{-1}^{1} x^m P_n(x) dx = 0$, (ii) $\int_{-1}^{1} x^n P_n(x) dx = \frac{2^{n+1} (n!)^2}{(2n+1)!}$.

Solution: Substituting from Rodrigues' Formula, we get

$$I = \int_{-1}^{1} x^m P_n(x) dx = \frac{1}{2^n n!} \cdot \int_{-1}^{1} x^m \frac{d^n}{dx^n} \left[(x^2-1)^n \right] dx$$

$$= \frac{1}{2^n n!} \cdot \int_{-1}^{1} x^m \, d \frac{d^{n-1}}{dx^{n-1}} \left[(x^2-1)^n \right].$$

Integrating by parts, we get

145

$$I = \frac{1}{2^n n!} \cdot \left[x^m \cdot \frac{d^{n-1}}{dx^{n-1}} \left[(x^2-1)^n \right] \right]_{-1}^{1}$$

$$- \frac{m}{2^n n!} \int_{-1}^{1} x^{m-1} \frac{d^{n-1}}{dx^{n-1}} \left[(x^2-1)^n \right] dx .$$

The term in the square brackets vanishes at both limits, therefore

$$I = \frac{(-1)m}{2^n n!} \int_{-1}^{1} x^{m-1} \frac{d^{n-1}}{dx^{n-1}} \left[(x^2-1)^n \right] dx .$$

Integrating by parts one more times, we obtain

$$I = \frac{(-1)^2 m(m-1)}{2^n n!} \int_{-1}^{1} x^{m-2} \frac{d^{n-2}}{dx^{n-2}} \left[(x^2-1)^n \right] dx .$$

Integrating by parts $n-2$ more times, we obtain

$$I = \frac{(-1)^n m!}{2^n n!} \int_{-1}^{1} \frac{d^{n-m}}{dx^{n-m}} \left[(x^2-1)^n \right] dx$$

$$= \frac{(-1)^{n+m} m!}{2^n n!} \int_{-1}^{1} \frac{d^{n-m}}{dx^{n-m}} \left[(1-x)^n (1+x)^n \right] dx$$

(*i*) <u>For $m < n$</u>, and recalling Leibniz Theorem, we notice that the integral will vanish and

$$I = \int_{-1}^{1} x^m P_n(x) dx = 0 .$$

(*ii*) <u>For $m = n$</u>, we have

$$I = \int_{-1}^{1} x^n P_n(x) dx = \frac{1}{2^n} \int_{-1}^{1} (1-x^2)^n dx = \frac{1}{2^{n-1}} \int_{0}^{1} (1-x^2)^n dx ,$$

Let $x^2 = t$, $dx = \dfrac{dt}{2\sqrt{t}}$, and the limits for the dummy variable t will also be from 0 to 1, therefore

Legendre Polynomials

$$I = \frac{1}{2^{n-1}} \int_0^1 (1-x^2)^n \, dx = \frac{1}{2^{n-1}} \cdot \frac{1}{2} \int_0^1 (1-t)^n t^{-1/2} \, dx$$

$$= \frac{1}{2^n} \cdot \beta(n+1, 1/2) = \frac{1}{2^n} \cdot \frac{\Gamma(n+1)\Gamma(1/2)}{\Gamma(n+3/2)}$$

$$= \frac{1}{2^n} \cdot \frac{n!\sqrt{\pi}}{(n+1/2)(n-1/2)(n-3/2)\cdots 3/2 \cdot 1/2 \cdot \sqrt{\pi}}$$

$$= \frac{1}{2^n} \cdot \frac{2n!}{(2n+1)(2n-1)(2n-3)\cdots 3 \cdot 1} \cdot \frac{2 \cdot 4 \cdot 6 \cdots 2n}{2 \cdot 4 \cdot 6 \cdots 2n} = \frac{2^{n+1}(n!)^2}{(2n+1)!} \square$$

3.9. Schläfli's[23] Integral for Legendre Polynomials

From complex analysis, the nth derivative of the function $f(z)$ is given by Cauchy's Integral formula

$$\frac{d^n}{d\zeta^n}f(\zeta) = \frac{n!}{2\pi i}\int_C \frac{f(z)}{(z-\zeta)^{n+1}}dz \ . \tag{48}$$

Replacing ζ by x and z by t, we obtain

$$\frac{d^n}{dx^n}f(x) = \frac{n!}{2\pi i}\int_C \frac{f(t)}{(t-x)^{n+1}}dt \ . \tag{49}$$

Let $f(x) = (x^2-1)^n$, then

$$\frac{d^n}{dx^n}f(x) = \frac{d^n}{dx^n}\left[(x^2-1)^n\right] = \frac{n!}{2\pi i}\int_C \frac{(t^2-1)^n}{(t-x)^{n+1}}dt \ ,$$

From Rodrigues' Formula:

$$P_n(x) = \frac{1}{2^n n!}\cdot\frac{d^n}{dx^n}\left[(x^2-1)^n\right], \tag{50}$$

we get,

$$P_n(x) = \frac{1}{2^n n!}\cdot\frac{n!}{2\pi i}\int_C \frac{(t^2-1)^n}{(t-x)^{n+1}}dt = \frac{1}{2\pi i}\int_C \frac{(t^2-1)^n}{2^n(t-x)^{n+1}}dt$$

Legendre polynomial $P_n(x)$ are also defined by the contour integral[24]

$$P_n(x) = \frac{1}{2\pi i}\oint_C (1-2zx+z^2)^{-1/2}z^{-n-1}dz \ . \tag{51}$$

where C is the contour enclosing the origin and traversing in an anticlockwise direction.

---

23. Ludwig Schläfli (1814-1895 Switzerland)
24. See Arfken, G., *Mathematical Methods for Physicists, 3rd ed.*, Academic Press, Orlando, FL, 1985.

Legendre Polynomials

3.10. Associated Legendre Functions

Consider again the Legendre differential equation

$$(1-x^2)y'' - 2xy' + n(n+1)y = 0 \tag{52}$$

$P_n(x)$ is a solution. Now, differentiating m times with respect to x using Leibniz theorem, we get

$$(1-x^2)y_{m+2} - 2xmy_{m+1} - 2\frac{m(m-1)}{2!}y_m$$

$$- 2xy_{m+1} - 2my_m + n(n+1)y_m = 0.$$

Or $\quad (1-x^2)y_{m+2} - 2x(m+1)y_{m+1} + [n(n+1) - m(m+1)]y_m = 0$

Letting $v = y_m$, then

$$(1-x^2)v'' - 2x(m+1)v' + [n(n+1) - m(m+1)]v = 0 \tag{53}$$

Since $P_n(x)$ is a solution of Legendre differential equation, then (53) will have a solution v given by

$$v = \frac{d^m}{dx^m} P_m(x)$$

If we let $w = v(1-x^2)^{m/2}$, then $v = w(1-x^2)^{-m/2}$.

Differentiating twice we obtain

$$v' = mxw(1-x^2)^{-1-m/2} + w'(1-x^2)^{-m/2}, \text{ and}$$

$$v'' = m(m+2)x^2 w(1-x^2)^{-2-m/2}$$

$$+ 2mxw'(1-x^2)^{-1-m/2} + w''(1-x^2)^{-m/2}$$

Substituting all these values in the differential equation (53) and simplifying, we get

$$(1-x^2)w'' - 2xw' + \left[n(n+1) - \frac{m}{1-x^2}\right]w = 0 \tag{54}$$

Equation (54) differs from the original Legendre differential equation in that an additional term involving m appears in it. This equation is called the *Associated*

149

Legendre Differential Equation whose solution

$$P_n^m(x) = (1-x^2)^{m/2} \frac{d^m}{dx^m}[P_n(x)], \quad m \leq n \qquad (55)$$

is called the *Associated Legendre Function of the First Kind*. These functions are not necessarily polynomials. But, since $P_n(x)$ is a polynomial with degree n, one can differentiate it only n times before it vanishes. This shows that $P_n^m(x)$ is defined for $m \leq n$, otherwise it is zero.

It can be easily verified that

$$P_n^0(x) = P_n(x) \text{ and } P_n^m(x) = 0 \text{ if } m > n.$$

If we use Rodrigues' Formula for $P_n(x)$, and allow m to take negative values, we might write

$$P_n^m(x) = (1-x^2)^{m/2} \frac{d^m}{dx^m} \left\{ \frac{1}{2^n n!} \cdot \frac{d^n}{dx^n}\left[(x^2-1)^n\right] \right\}$$

$$= \frac{(1-x^2)^{m/2}}{2^n n!} \cdot \frac{d^{m+n}}{dx^{m+n}}\left[(x^2-1)^n\right]; \quad -n \leq m \leq n.$$

It can be shown that

$$P_n^{-m}(x) = (-1)^m \frac{(n-m)!}{(n+m)!} P_n^m(x). \qquad (56)$$

Also, if $Q_n(x)$ is the second solution of Legendre differential equation, then

$$Q_n^m(x) = (1-x^2)^{m/2} \frac{d^m}{dx^m}[Q_n(x)]$$

will be the second solution of the associated Legendre equation. It is called the *Associated Legendre Function of the Second Kind*. The general solution of the associated Legendre differential equation can now be written as

$$y = AP_n^m(x) + BQ_n^m(x) \qquad (57)$$

We state here, without proof, some of the properties of $P_n^m(x)$.

Orthogonality Property:

Legendre Polynomials

$$\int_{-1}^{1} P_n^m(x) P_k^m(x) dx = \begin{cases} 0 & \text{if } k \neq n \\ \dfrac{2(n+m)!}{(2n+1)(n-m)!} & \text{if } k = n \end{cases}$$

The angular form is given by:

$$\int_{0}^{\pi} \sin\theta \cdot P_n^m(\cos\theta) P_k^m(\cos\theta) dx = \begin{cases} 0 & \text{if } k \neq n \\ \dfrac{2(n+m)!}{(2n+1)(n-m)!} & \text{if } k = n \end{cases}$$

Recurrence Relations:

I. $P_n^{m+1}(x) - \dfrac{2mx}{\sqrt{1-x^2}} P_n^m(x) + [n(n+1) - m(m-1)] P_n^{m-1}(x) = 0$

II. $(2n+1) x P_n^m(x) = (n+m) P_{n-1}^m(x) + (n-m+1) P_{n+1}^m(x)$

III. $\sqrt{1-x^2} P_n^m(x) = \dfrac{1}{2n+1} \left[P_{n+1}^{m+1}(x) - P_{n-1}^{m+1}(x) \right]$

IV. $\sqrt{1-x^2} P_n^m(x) = \dfrac{1}{2n+1} \left[(n+m)((n+m-1) P_{n-1}^{m-1}(x) - (n-m+1)(n-m+2) P_{n+1}^{m-1}(x) \right]$

Example 34: Find the general solution of

$$\dfrac{d^2 y}{d\theta^2} + \cot\theta \dfrac{dy}{d\theta} + \left[n(n+1) - \dfrac{m^2}{\sin^2\theta} \right] y = 0.$$

Solution: Let $x = \cos\theta$, then $\dfrac{dx}{d\theta} = -\sin\theta = -\sqrt{1-x^2}$,

$$\dfrac{dy}{d\theta} = \dfrac{dy}{dx} \cdot \dfrac{dx}{d\theta} = -\sqrt{1-x^2} \dfrac{dy}{dx}, \text{ and}$$

$$\dfrac{d^2 y}{d\theta^2} = \dfrac{d}{dx}\left[-\sqrt{1-x^2} \dfrac{dy}{dx} \right] \cdot \dfrac{dx}{d\theta} = (1-x^2) \dfrac{d^2 y}{dx^2} - x \dfrac{dy}{dx}$$

Substituting in the differential equation, we get

Special Functions and Orthogonal Polynomials

$$(1-x^2)\frac{d^2y}{dx^2} - x\frac{dy}{dx} + \frac{x}{\sqrt{1-x^2}}\left[-\sqrt{1-x^2}\frac{dy}{dx}\right]$$
$$+ \left[n(n+1) - \frac{m^2}{1-x^2}\right]y = 0$$

or $(1-x^2)\dfrac{d^2y}{dx^2} - 2x\dfrac{dy}{dx} + \left[n(n+1) - \dfrac{m}{1-x^2}\right]y = 0$.

But this is the associated Legendre differential equation whose solution is $y = AP_n^m(x) + BQ_n^m(x)$. Then the general solution of the given equation is

$$y = AP_n^m(\cos\theta) + BQ_n^m(\cos\theta). \qquad \square$$

3.11. Series of Legendre Polynomials

If $f(x)$ and $f'(x)$ are piecewise continuous in the interval $(-1, 1)$, then there exists a Legendre series expansion of $f(x)$ of the form

$$f(x) = \sum_{n=0}^{\infty} c_n P_n(x) \tag{58}$$

To obtain the coefficients c_n, $n = 0, 1, 2, \cdots$, we multiply equation (58) by $P_m(x)$ and integrate with respect to x from -1 to 1 to obtain

$$\int_{-1}^{1} f(x) P_m(x)\, dx = \sum_{n=0}^{\infty} c_n \int_{-1}^{1} P_n(x) P_m(x)\, dx$$

and from the orthogonality property of Legendre polynomials, we get

$$\int_{-1}^{1} f(x) P_n(x)\, dx = \frac{2}{2n+1} \cdot c_n.$$

Therefore
$$c_n = \frac{2n+1}{2} \int_{-1}^{1} f(x) P_n(x)\, dx; \quad n = 0, 1, 2, \cdots. \tag{59}$$

This representation is only valid in the interval $(-1, 1)$ since this interval is the interval of convergence of the series in equation (58).

Note: 1. If $f(x)$ is a polynomial of degree n, then

$$f(x) = \sum_{k=0}^{n} c_k P_k(x), \text{ where } c_k \text{ is given by}$$

$$c_k = \frac{2k+1}{2} \int_{-1}^{1} f(x) P_k(x)\, dx; \quad k = 0, 1, 2, \cdots, n$$

2. If $f(x)$ is a polynomial of degree less than k, then

$$\int_{-1}^{1} f(x) P_k(x)\, dx = 0$$

3. The series $\sum_{n=0}^{\infty} c_n P_n(x)$, where $c_n = \frac{2n+1}{2} \int_{-1}^{1} f(x) P_n(x)\, dx$,

Special Functions and Orthogonal Polynomials

converges to $f(x)$ if x is not a point of discontinuity of $f(x)$ and to $\frac{1}{2}\left[f(x^+)+f(x^-)\right]$ if x is a point of discontinuity.

Example 35: Find the Legendre series expansion for $f(x) = \begin{cases} -1 & -1 \leq x < 0 \\ 1 & 0 < x \leq 1 \end{cases}$.

Solution: At first glance, the function $f(x)$ is an odd function. Then we would expect that the series representation will contain only odd Legendre polynomials, then

$$c_n = \frac{2n+1}{2}\int_{-1}^{1} f(x)P_n(x)\,dx = 0, \text{ if } n = 0, 2, 4, \cdots, \text{ and}$$

$$c_n = (2n+1)\int_{0}^{1} f(x)P_n(x)\,dx, \text{ if } n = 1, 3, 5, \cdots.$$

Using the recurrence relation $P_n(x) = \frac{1}{2n+1}\left[P'_{n+1}(x) - P'_{n-1}(x)\right]$,

we get $c_n = \int_{0}^{1} \left[P'_{n+1}(x) - P'_{n-1}(x)\right]dx; \quad n = 1, 3, 5, \cdots.$

Let $n = 2k+1$, then

$$c_{2k+1} = \int_{0}^{1} \left[P'_{2k+2}(x) - P'_{2k}(x)\right]dx; \quad k = 1, 2, 3, \cdots$$

Integrating, we get

$$c_{2k+1} = P_{2k+2}(1) - P_{2k}(1) - P_{2k+2}(0) + P_{2k}(0)$$
$$= -P_{2k+2}(0) + P_{2k}(0)$$

Now, $P_{2k}(0) = \frac{(-1)^k (2k)!}{2^{2k}(k!)^2}$, then

154

Legendre Polynomials

$$c_{2k+1} = \frac{(-1)^k (2k)!}{2^{2k}(k!)^2} - \frac{(-1)^{k+1}(2k+2)!}{2^{2k+2}[(k+1)!]^2}$$

$$= \frac{(-1)^k (2k)!}{2^{2k}(k!)^2}\left[1 + \frac{(2k+2)(2k+1)}{2^2(k+1)^2}\right] = \frac{(-1)^k (2k)!(4k+3)}{2^{2k+1}(k!)^2(k+1)}$$

Then the Legendre series expansion of $f(x)$ is

$$f(x) = \sum_{k=0}^{\infty} \frac{(-1)^k (2k)!(4k+3)}{2^{2k+1}(k!)^2(k+1)} P_{2k+1}(x), \quad -1 < x < 1$$

Writing the first few terms, we have

$$f(x) = \tfrac{3}{2}P_1(x) - \tfrac{7}{4}P_3(x) + \tfrac{11}{4}P_5(x) + \cdots. \qquad \square$$

Note: The first few powers of x in terms of Legendre polynomials are

$x = P_1(x)$,

$x^2 = \tfrac{1}{3}\left[P_0(x) + 2P_2(x)\right]$,

$x^3 = \tfrac{1}{5}\left[3P_1(x) + 2P_3(x)\right]$,

$x^4 = \tfrac{1}{35}\left[7P_0(x) + 20P_2(x) + 8P_4(x)\right]$,

$x^5 = \tfrac{1}{63}\left[27P_1(x) + 28P_3(x) + 8P_5(x)\right]$,

$x^6 = \tfrac{1}{231}\left[33P_0(x) + 110P_2(x) + 72P_4(x) + 16P_6(x)\right]$.

3.12. Legendre Functions of the Second Kind $Q_n(x)$

Recalling the general solution of Legendre Differential equation

$$y(x) = A\left\{1 - \frac{n\cdot(n+1)}{2!}x^2 + \frac{n(n-2)\cdot(n+1)(n+3)}{4!}x^4 - \cdots\right\}$$
$$+ B\left\{x - \frac{(n-1)\cdot(n+2)}{3!}x^3 + \frac{(n-1)(n-3)\cdot(n+2)(n+4)}{5!}x^5 - \cdots\right\}$$

The *Legendre functions of the second kind* are the series solutions of Legendre differential equations that do not terminate. If n is even, then the second series does not terminate, while if n is odd the first series does not terminate.

These series solutions, apart from the multiplicative constants, define the Legendre functions of the second kind.

If we choose $B = \dfrac{(-1)^{n/2} 2^n [(n/2)!]^2}{n!}$, the Legendre function of the second kind for n even is

$$Q_n(x) = \frac{(-1)^{n/2} 2^n [(n/2)!]^2}{n!}\left\{x - \frac{(n-1)(n+2)}{3!}x^3 + \frac{(n-1)(n-3)(n+2)(n+4)}{5!}x^5 - \cdots\right\}$$

On the other hand, if we choose $A = \dfrac{(-1)^{(n+1)/2} 2^{n-1}\{[(n-1)/2]!\}^2}{1\cdot 3\cdot 5 \cdots n}$, the Legendre function of the second kind for n odd is

$$Q_n(x) = \frac{(-1)^{(n+1)/2} 2^{n-1}\{[(n-1)/2]!\}^2}{1\cdot 3\cdot 5 \cdots n}\left\{1 - \frac{n(n+1)}{2!}x^2 + \frac{n(n-2)(n+1)(n+3)}{4!}x^4 - \cdots\right\}$$

The values of A and B are chosen so that the recurrence formulas for $P_n(x)$ apply also to $Q_n(x)$.

The first few Legendre Functions of the Second Kind are

$$Q_0(x) = \frac{1}{2}\ln\left(\frac{1+x}{1-x}\right); \qquad Q_1(x) = \frac{x}{2}\ln\left(\frac{1+x}{1-x}\right) - 1;$$

$$Q_2(x) = \frac{3x^2 - 1}{4}\ln\left(\frac{1+x}{1-x}\right) - \frac{3x}{2};$$

Legendre Polynomials

$$Q_3(x) = \frac{5x^3 - 3x}{4} \ln\left(\frac{1+x}{1-x}\right) - \frac{5x^2}{2} + \frac{2}{3}.$$

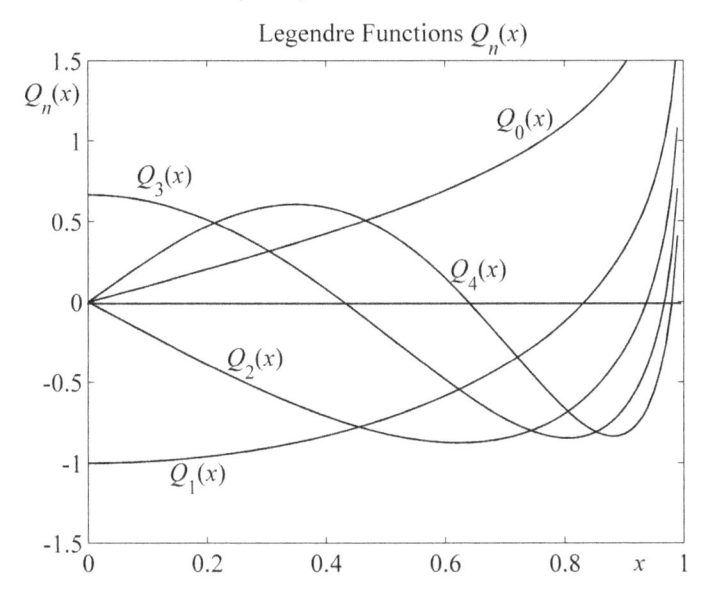

Legendre Functions $Q_n(x)$

The *Associated Legendre Functions of the Second Kind* are given by

$$Q_n^m(x) = (1-x^2)^{m/2} \frac{d^m}{dx^m} Q_n(x).$$

And satisfy the Associated Legendre Differential Equation

$$(1-x^2)y'' - 2x\, y' + \left[n(n+1) - \frac{m}{1-x^2}\right] y = 0.$$

3.12.1. Relation between $P_n(x)$ and $Q_n(x)$

Consider the Legendre differential equation

$$(1-x^2)y'' - 2xy' + n(n+1)y = 0$$

We have established that the Legendre polynomial $P_n(x)$ is a solution. Now, to obtain the second solution $Q_n(x)$, we proceed as follows. Since $P_n(x)$ and $Q_n(x)$ are the two linearly independent solutions of the Legendre differential equation, their Wronskian should not vanish, i.e.,

157

Special Functions and Orthogonal Polynomials

$$W = \begin{vmatrix} P_n(x) & Q_n(x) \\ P_n'(x) & Q_n'(x) \end{vmatrix} \neq 0$$

then

$$W = P_n(x)Q_n'(x) - Q_n(x)P_n'(x) = P_n^2(x)\frac{d}{dx}\left(\frac{Q_n(x)}{P_n(x)}\right)$$

or

$$\frac{d}{dx}\left(\frac{Q_n(x)}{P_n(x)}\right) = \frac{W}{P_n^2(x)}$$

Integrating with respect to x from ∞ to x, we get

$$Q_n(x) = P_n(x)\int_\infty^x \frac{W}{P_n^2(x)}dx$$

We have used that fact that $\lim_{x \to \infty} \frac{Q_n(x)}{P_n(x)} = 0$ (can you prove it?)

To obtain the Wronskian W, we know that $P_n(x)$ and $Q_n(x)$ satisfy the Legendre differential equation, then

$$(1-x^2)P_n''(x) - 2x\,P_n'(x) + n(n+1)P_n(x) = 0, \text{ and}$$

$$(1-x^2)Q_n''(x) - 2x\,Q_n'(x) + n(n+1)Q_n(x) = 0$$

Multiplying the first equation by $Q_n(x)$ and the second by $P_n(x)$ and subtracting, (dropping the argument for convenience) we get

$$(1-x^2)\{P_nQ_n'' - Q_nP_n''\} - 2x\{P_nQ_n' - Q_nP_n'\} = 0$$

or

$$(1-x^2)\frac{d}{dx}\{P_nQ_n' - Q_nP_n'\} - 2x\{P_nQ_n' - Q_nP_n'\} = 0$$

or

$$(1-x^2)\frac{dW}{dx} - 2x\,W = 0$$

This is a first order differential equation that is separable, then

$$W = \frac{c}{1-x^2}$$

It can be shown that $c = 1$, then

$$Q_n(x) = P_n(x)\int_x^\infty \frac{dx}{(x^2-1)P_n^2(x)}$$

We state here a theorem that gives $Q_n(x)$ in terms of $P_n(x)$.

158

Legendre Polynomials

Theorem[25]: The Legendre function of the second kind $Q_n(x)$ is given by

$$Q_n(x) = \frac{1}{2} P_n(x) \ln\left(\frac{1+x}{1-x}\right) - \sum_{k=0}^{N} \frac{(2n-4k-1)}{(2k+1)(n-k)} P_{n-2k-1}(x),$$

where $Q_0(x) = \frac{1}{2} P_0(x) \ln\left(\frac{1+x}{1-x}\right)$, and

$$N = \begin{cases} (n-1)/2 & \text{if } n \text{ is odd} \\ (n-2)/2 & \text{if } n \text{ is even} \end{cases}$$

3.12.2. Properties of Legendre Functions of the Second Kind

$Q_n(x)$ satisfies the same recurrence relation as $P_n(x)$, namely

I. $(n+1)Q_{n+1}(x) - (2n+1)x\, Q_n(x) + nQ_{n-1}(x) = 0$

II. $nQ_n(x) = xQ'_n(x) - Q'_{n-1}(x)$

III. $(2n+1)Q_n(x) = Q'_{n+1}(x) - Q'_{n-1}(x)$

IV. $(n+1)Q_n(x) = Q'_{n+1}(x) - x\, Q'_n(x)$

V. $(1-x^2)Q'_n(x) = n[Q_{n-1}(x) - x\, Q_n(x)]$

VI. $(1-x^2)Q'_n(x) = (n+1)[x\, Q_n(x) - Q_{n+1}(x)]$

It can also be shown that:

1. *Christoffel's Second Summation Formula*:

$$\frac{1}{y-x} = \sum_{n=0}^{\infty} (2n+1) P_n(x) Q_n(y), \quad x > 1 \text{ and } |y| \leq 1.$$

2. *Neumann's Integral Formula*:

$$Q_n(x) = \frac{1}{2} \int_{-1}^{1} \frac{P_n(x)}{x-y} dy, \quad |x| > 1.$$

---

25. The proof can be found in Bell, W.W., *Special Functions for Scientists and Engineers*, Van Nostrand, 1968.

Special Functions and Orthogonal Polynomials

Example 36: Evaluate $Q_0(x), Q_1(x)$ and $Q_2(x)$.

Solution: We know that $Q_n(x) = P_n(x) \int_x^\infty \dfrac{dx}{(x^2-1)P_n^2(x)}$, also $P_0(x) = 1$ and $P_1(x) = x$, then

$$Q_0(x) = \int_x^\infty \frac{dx}{(x^2-1)} = \frac{1}{2}\int_x^\infty \left[\frac{1}{1+x} - \frac{1}{1-x}\right] dx = \frac{1}{2}\ln\left(\frac{1+x}{1-x}\right).$$

$$Q_1(x) = x \int_x^\infty \frac{dx}{(1-x^2)x^2} = x\int_x^\infty \left[\frac{1}{1-x^2} + \frac{1}{x^2}\right] dx = \frac{x}{2}\ln\left(\frac{1+x}{1-x}\right) - 1$$

To obtain $Q_2(x)$, we use the recurrence relation **I**:

$$(n+1)Q_{n+1}(x) - (2n+1)x\, Q_n(x) + n Q_{n-1}(x) = 0$$

Let $n = 1$, we get $Q_2(x) = \frac{3}{2}x\, Q_1(x) - \frac{1}{2}Q_0(x)$

Substituting for $Q_1(x)$ and $Q_0(x)$, we get

$$Q_2(x) = \frac{3x^2-1}{4}\ln\left(\frac{1+x}{1-x}\right) - \frac{3x}{2}. \qquad \square$$

Example 37: Show that:

i) $n[Q_n(x)P_{n-1}(x) - Q_{n-1}(x)P_n(x)] =$
$$(n-1)[Q_{n-1}(x)P_{n-2}(x) - Q_{n-2}(x)P_{n-2}(x)]$$

ii) $P_n(x)Q_{n-1}(x) - P_{n-1}(x)Q_n(x) = \dfrac{1}{n}$

Solution: i) From the recurrence relations for $P_n(x)$ and $Q_n(x)$, we have

$$(n+1)P_{n+1}(x) - (2n+1)x\, P_n(x) + n\, P_{n-1}(x) = 0 \qquad (60)$$
$$(n+1)Q_{n+1}(x) - (2n+1)x\, Q_n(x) + n Q_{n-1}(x) = 0 \qquad (61)$$

Replacing n by $(n-1)$ in equations (60) and (61), we get

$$n P_n(x) - (2n-1)x\, P_{n-1}(x) + (n-1)P_{n-2}(x) = 0 \qquad (62)$$
$$n Q_n(x) - (2n-1)x\, Q_{n-1}(x) + (n-1)Q_{n-2}(x) = 0 \qquad (63)$$

Multiplying (62) by $Q_{n-1}(x)$ and (63) by $P_{n-1}(x)$, subtracting and rearranging, we get

Legendre Polynomials

$$n[Q_n(x)P_{n-1}(x)-Q_{n-1}(x)P_n(x)] =$$
$$(n-1)[Q_{n-1}(x)P_{n-2}(x)-Q_{n-2}(x)P_{n-2}(x)] \quad \square$$

ii) Let $u_n = n[Q_n(x)P_{n-1}(x)-Q_{n-1}(x)P_n(x)]$, then from i), we have $u_n = u_{n-1}$ and $u_n = u_{n-1} = u_{n-2} = \cdots = u_1$, then

$$n[Q_n(x)P_{n-1}(x)-Q_{n-1}(x)P_n(x)] = [Q_1(x)P_0(x)-Q_0(x)P_1(x)]$$

But $P_0(x)=1$, $P_1(x)=x$, $Q_0(x) = \dfrac{1}{2}\ln\left(\dfrac{1+x}{1-x}\right)$ and

$$Q_1(x) = \dfrac{x}{2}\ln\left(\dfrac{1+x}{1-x}\right) - 1.$$

Substituting for all these values, we get

$$n[Q_n(x)P_{n-1}(x)-Q_{n-1}(x)P_n(x)] = -1$$

Therefore,

$$P_n(x)Q_{n-1}(x) - P_{n-1}(x)Q_n(x) = \dfrac{1}{n}. \quad \square$$

3.13. Shifted Legendre Polynomials

The Shifted Legendre Polynomials $\tilde{P}_n(x)$ are orthogonal polynomials with respect to a weighing function of 1 in the interval (0, 1) and

$$\int_0^1 \tilde{P}_m(x)\tilde{P}_n(x)dx = \begin{cases} 0 & \text{if } m \neq n \\ \dfrac{1}{2n+1} & \text{if } m = n \end{cases}$$

Rodrigues' Formula is given by:

$$\tilde{P}_n(x) = \frac{1}{n!}\frac{d^n}{dx^n}\left[(x^2-x)^n\right].$$

Also it can be shown that:

$$\tilde{P}_n(x) = P_n(2x-1);$$

The first few Shifted Legendre Polynomials are:

$\tilde{P}_0(x) = 1;$ $\qquad \tilde{P}_1(x) = 2x - 1;$

$\tilde{P}_2(x) = 6x^2 - 6x + 1;$ $\qquad \tilde{P}_3(x) = 20x^3 - 30x^2 + 12x - 1.$

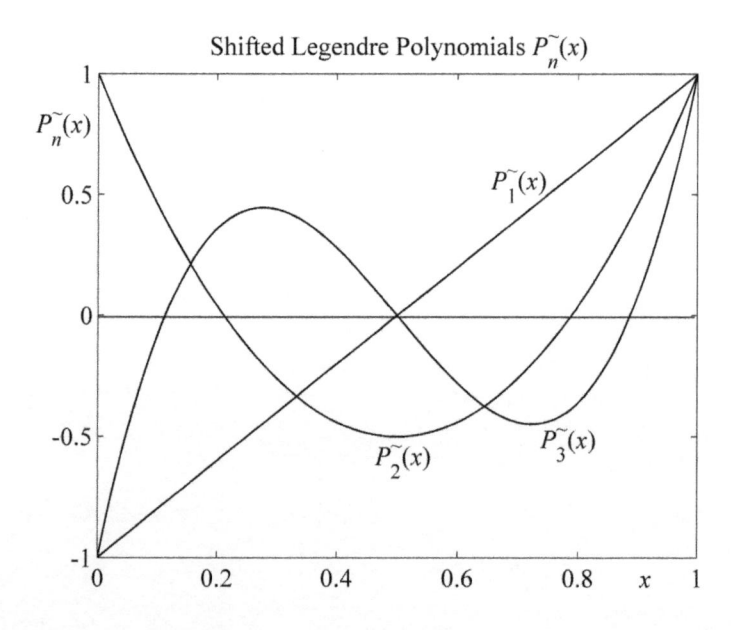

Legendre Polynomials

3.14. Summary of Legendre Polynomials and Functions

Legendre Differential equation:

$$(1-x^2)y'' - 2xy' + n(n+1)y = 0$$

Legendre Polynomials:

$$P_n(x) = \sum_{k=0}^{N} (-1)^k \cdot \frac{(2n-2k)!}{2^n k!(n-2k)!(n-k)!} x^{n-2k}$$

Generating Function:

$$(1 - 2xt + t^2)^{-1/2} = \sum_{n=0}^{\infty} t^n P_n(x)$$

Recurrence Relations for Legendre Polynomials:

I. $(n+1)P_{n+1}(x) - (2n+1)xP_n(x) + nP_{n-1}(x) = 0$

II. $nP_n(x) = xP_n'(x) - P_{n-1}'(x)$

III. $(2n+1)P_n(x) = P_{n+1}'(x) - P_{n-1}'(x)$

IV. $(n+1)P_n(x) = P_{n+1}'(x) - xP_n'(x)$

V. $(1-x^2)P_n'(x) = n[P_{n-1}(x) - xP_n(x)]$

VI. $(1-x^2)P_n'(x) = (n+1)[xP_n(x) - P_{n+1}(x)]$

Orthogonal Property:

$$\int_{-1}^{1} P_m(x) P_n(x) dx = \begin{cases} 0 & \text{if } m \neq n \\ \dfrac{2}{2n+1} & \text{if } m = n \end{cases}$$

$$\int_{0}^{\pi} P_m(\cos\theta) P_n(\cos\theta) \sin\theta\, d\theta = \begin{cases} 0 & \text{if } m \neq n \\ \dfrac{2}{2n+1} & \text{if } m = n \end{cases}$$

Integral Forms (Laplace Integrals):

$$P_n(x) = \frac{1}{\pi} \int_{0}^{\pi} \left[x \pm \sqrt{x^2 - 1} \cos\phi \right]^n d\phi$$

Special Functions and Orthogonal Polynomials

$$P_n(x) = \frac{1}{\pi}\int_0^\pi \frac{d\phi}{\left[x \pm \sqrt{x^2-1}\cos\phi\right]^{n+1}}$$

Differential Form (Rodrigues' Formula):

$$P_n(x) = \frac{1}{2^n n!} \cdot \frac{d^n}{dx^n}\left[(x^2-1)^n\right]$$

Associated Legendre Differential Equation:

$$(1-x^2)w'' - 2xw' + \left[n(n+1) - \frac{m}{1-x^2}\right]w = 0$$

Associated Legendre Functions:

$$P_n^m(x) = (1-x^2)^{m/2}\frac{d^m}{dx^m}[P_n(x)], \quad m \le n$$

$$Q_n^m(x) = (1-x^2)^{m/2}\frac{d^m}{dx^m}[Q_n(x)]$$

Recurrence Relations for the Associated Legendre Functions:

I. $P_n^{m+1}(x) - \dfrac{2mx}{\sqrt{1-x^2}}P_n^m(x) + [n(n+1)-m(m-1)]P_n^{m-1}(x) = 0$

II. $(2n+1)xP_n^m(x) = (n+m)P_{n-1}^m(x) + (n-m+1)P_{n+1}^m(x)$

III. $\sqrt{1-x^2}\,P_n^m(x) = \dfrac{1}{2n+1}\left[P_{n+1}^{m+1}(x) - P_{n-1}^{m+1}(x)\right]$

IV. $\sqrt{1-x^2}\,P_n^m(x) = \dfrac{1}{2n+1}\left[\begin{array}{l}(n+m)((n+m-1)P_{n-1}^{m-1}(x) \\ \quad -(n-m+1)(n-m+2)P_{n+1}^{m-1}(x)\end{array}\right]$

Orthogonality Property for the Associated Legendre Functions:

$$\int_{-1}^1 P_n^m(x)P_k^m(x)\,dx = \begin{cases}0 & \text{if } k \ne n \\ \dfrac{2(n+m)!}{(2n+1)(n-m)!} & \text{if } k = n\end{cases}$$

$$\int_0^\pi \sin\theta \cdot P_n^m(\cos\theta)P_k^m(\cos\theta)\,dx = \begin{cases}0 & \text{if } k \ne n \\ \dfrac{2(n+m)!}{(2n+1)(n-m)!} & \text{if } k = n\end{cases}$$

Legendre Polynomials

Series of Legendre Polynomials:

$$f(x) = \sum_{n=0}^{\infty} c_n P_n(x), \quad c_n = \frac{2n+1}{2} \int_{-1}^{1} f(x) P_n(x)\, dx; \quad n = 0, 1, 2, \cdots$$

Legendre Functions of the Second Kind:

$$Q_n(x) = P_n(x) \int_x^{\infty} \frac{dx}{(x^2-1) P_n^2(x)}$$

$$Q_n(x) = \frac{1}{2} P_n(x) \ln\left(\frac{1+x}{1-x}\right) - \sum_{k=0}^{N} \frac{(2n-4k-1)}{(2k+1)(n-k)} P_{n-2k-1}(x),$$

where $Q_0(x) = \frac{1}{2} P_0(x) \ln\left(\frac{1+x}{1-x}\right)$, and $N = \begin{cases} (n-1)/2 & \text{if } n \text{ is odd} \\ (n-2)/2 & \text{if } n \text{ is even} \end{cases}$

Recurrence Relations for Legendre Functions of the Second Kind:

I. $(n+1)Q_{n+1}(x) - (2n+1)x\, Q_n(x) + n Q_{n-1}(x) = 0$

II. $n Q_n(x) = x Q_n'(x) - Q_{n-1}'(x)$

III. $(2n+1) Q_n(x) = Q_{n+1}'(x) - Q_{n-1}'(x)$

IV. $(n+1) Q_n(x) = Q_{n+1}'(x) - x\, Q_n'(x)$

V. $(1-x^2) Q_n'(x) = n [Q_{n-1}(x) - x\, Q_n(x)]$

VI. $(1-x^2) Q_n'(x) = (n+1)[x\, Q_n(x) - Q_{n+1}(x)]$

Shifted Legendre Polynomials $\tilde{P}_n(x)$

Orthogonality Property for the Shifted Legendre Functions:

$$\int_0^1 \tilde{P}_m(x) \tilde{P}_n(x)\, dx = \begin{cases} 0 & \text{if } m \neq n \\ \dfrac{1}{2n+1} & \text{if } m = n \end{cases}$$

Rodrigues' Formula for the Shifted Legendre Functions:

$$\tilde{P}_n(x) = \frac{1}{n!} \frac{d^n}{dx^n}\left[(x^2 - x)^n\right].$$

Relation between $\tilde{P}_n(x)$ and $P_n(x)$:

$$\tilde{P}_n(x) = P_n(2x - 1)$$

165

Exercises

1. If $P_0(x) = 1$ and $P_1(x) = x$, find $P_2(x)$ and $P_3(x)$.

2. Show that $\dfrac{d}{dx} P_7(x) = 13 P_6(x) + 9 P_4(x) + 5 P_2(x) + P_0(x)$.

3. Prove that:

   a. $\displaystyle\int P_n(x)\, dx = \dfrac{1}{2n+1}[P_{n+1}(x) - P_{n-1}(x)] + c$

   b. $\displaystyle\int_x^1 P_n(x)\, dx = \dfrac{1}{n+1}[P_{n-1}(x) - P_{n+1}(x)]$

   c. $\displaystyle\sum_{k=0}^{n}(2k+1) P_k^2(x) = (n+1)^2 [P_n(x) P'_{n+1}(x) - P_{n+1}(x) P'_n(x)]$
   $= (n+1)^2 \left[P_n^2(x) - (x^2 - 1)\{P'_n(x)\}^2 \right]$

4. Starting from the recurrence relation:

   $(2n+1) P_n(x) = P'_{n+1}(x) - P'_{n-1}(x)$, show that:

   $P'_n(x) = (2n-1) P_{n-1}(x) + (2n-5) P_{n-3}(x) + (2n-9) P_{n-3}(x) + \cdots$
   $= \displaystyle\sum_{k=0}^{N}(2n - 4k - 1) P_{n-2k-1}(x)$

   where $N = \begin{cases} (n-2)/2 & \text{if } n \text{ is even} \\ (n-1)/2 & \text{if } n \text{ is odd} \end{cases}$

5. Starting from recurrence relation:

   $(n+1) P_{n+1}(x) - (2n+1) x P_n(x) + n P_{n-1}(x) = 0$, show that:

   $P'_n(x) + P'_{n-1}(x) = \displaystyle\sum_{k=0}^{n}(2k+1) P_k(x)$

6. Using the substitution $x = \cos\theta$, show that the equation:

   $\dfrac{1}{\sin\theta} \dfrac{d}{d\theta}\left(\sin\theta \dfrac{dy}{d\theta} \right) + n(n+1) y = 0$ reduces to Legendre differential equation.

Legendre Polynomials

7. Write down the general solution of the following differential equations in terms of Legendre functions:

   a. $(1-x^2)y'' - 2xy' + 2y = 0$

   b. $(1-x^2)y'' - 2xy' + 12y = 0$

   c. $\dfrac{d^2 y}{d\theta^2} + \cot\theta \dfrac{dy}{d\theta} + 2y = 0$

8. Show that: $P_n(x) = \dfrac{1}{n!} \dfrac{\partial^n}{\partial x^n}(1 - 2xt + t^2)^{-1/2}\Big|_{t=0}$.

9. Show that the roots of $P_n(x)$ lie between -1 and 1.

10. Show that:
$$P_n\left(-\tfrac{1}{2}\right) = P_0\left(-\tfrac{1}{2}\right)P_{2n}\left(\tfrac{1}{2}\right) + P_1\left(-\tfrac{1}{2}\right)P_{2n-1}\left(\tfrac{1}{2}\right) + P_2\left(-\tfrac{1}{2}\right)P_{2n-2}\left(\tfrac{1}{2}\right)$$
$$+ \cdots + P_{2n}\left(-\tfrac{1}{2}\right)P_0\left(\tfrac{1}{2}\right)$$

   Hint: Put $x = 1/2$ and $x = -1/2$ successively in the generating function, then replace t by t^2 and manipulate.

11. Show that:

   a. $P_{2n}(x) = \dfrac{(-1)^n}{2^{2n-1}} \displaystyle\sum_{k=0}^{n} \dfrac{(-1)^k (2n + 2k - 1)!}{(2k)!(n + k - 1)!(n - k)!} x^{2k}$

   b. $P_{2n+1}(x) = \dfrac{(-1)^n}{2^{2n}} \displaystyle\sum_{k=0}^{n} \dfrac{(-1)^k (2n + 2k + 1)!}{(2k + 1)!(n + k)!(n - k)!} x^{2k+1}$

12. Show that:

   a. $\displaystyle\int_0^{\pi} P_n(\cos\theta)\cos n\theta \, d\theta = \beta\left(n + \tfrac{1}{2}, \tfrac{1}{2}\right)$ if n is a positive integer.

   b. $\displaystyle\int_0^{\pi} P_n(\cos\theta)\cos n\theta \, d\theta = \dfrac{1 \cdot 3 \cdot 5 \cdots (2n-1)}{2 \cdot 4 \cdot 6 \cdots 2n}$.

c. $\displaystyle\sum_{n=0}^{\infty} P_n(\cos\theta) = \operatorname{cosec}\frac{\theta}{2}$.

13. Show that $|P_n(\cos\theta)| \le 1$ when θ is real.

14. Show that:

   a. $\displaystyle\int_{-1}^{1} x\, P_n(x)\,dx = \begin{cases} 0 & \text{if } n \ne 1 \\ 2/3 & \text{if } n = 1 \end{cases}$

   b. $\displaystyle\int_{-1}^{1} P_n(x) P'_{n+1}(x)\,dx = 2, \quad n = 0, 1, 2, 3, \cdots$

   c. $\displaystyle\int_{-1}^{1} x P_n(x) P'_{n+1}(x)\,dx = \frac{2n}{2n+1}, \quad n = 0, 1, 2, 3, \cdots$

   d. $\displaystyle\int_{-1}^{1} (1-x^2) P'_n(x) P'_m(x)\,dx = 0, \quad m \ne n$

   e. $\displaystyle\int_{-1}^{1} x^2 P_{n+1}(x) P_{n-1}(x)\,dx = \frac{2n(n+1)}{(2n-1)(2n+1)(2n+3)}$.

15. Show that $\displaystyle\frac{1-t^2}{(1-2xt+t^2)^{3/2}} = \sum_{n=0}^{\infty}(2n+1)t^n P_n(x)$.

16. Show that: $\displaystyle\int_0^1 P_{2n}(x) P_{2n+1}(x)\,dx = \int_0^1 P_{2n}(x) P_{2n-1}(x)\,dx$.

17. Use recurrence relation **V** and **VI** to show that:

$$(x^2-1)P'_n(x) = \frac{n(n+1)}{2n+1}[P_{n+1}(x) - P_{n-1}(x)],$$

Then deduce that $\displaystyle\int_{-1}^{1}(x^2-1)P_{n+1}(x)P'_n(x)\,dx = \frac{2n(n+1)}{(2n+1)(2n+3)}$.

18. Show that: $\displaystyle\sin^n\theta\, P_n(\sin\theta) = \sum_{k=0}^{n} \frac{(-1)^k\, n!}{m!(n-m)!} \cos^m\theta\, P_m(\cos\theta)$.

Legendre Polynomials

19. Show that:
$$(n+1)[P_n(x)P'_{n+1}(x) - P_{n+1}(x)P'_n(x)] = (n+1)^2 P_n^2(x) - (x^2-1)P'^2_n(x)$$

20. Show that:
$$P'_{2n+1}(x) = (2n+1)P_{2n}(x) + 2nxP_{2n-1}(x) + (2n-1)x^2 P_{2n-2}(x)$$
$$+ \cdots + 2x^{2n-1}P_1(x) + x^{2n}$$

21. Show that: $\displaystyle\sum_{k=0}^{\infty} \frac{x^{k+1}P_k(x)}{k+1} = \frac{1}{2}\ln\left(\frac{1+x}{1-x}\right).$

22. Show that $\dfrac{1+t}{t(1-2xt+t^2)^{1/2}} - \dfrac{1}{t} = \displaystyle\sum_{n=0}^{\infty} t^n [P_n(x) + P_{n+1}(x)].$

23. If $-1 < x < 1$ and n is a positive integer, show that:

    a. $|P_n(x)| < 1$ b. $|P_n(x)| < \sqrt{\dfrac{\pi}{2n(1-x^2)}}$

    Hint: Use Laplace first integral.

24. Using Rodrigue's formula, show that:
$$P'_{n+1}(x) - P'_{n-1}(x) = (2n+1)P_n(x) \text{ and hence deduce that:}$$
$$\int_x^1 P_n(x)\,dx = \frac{1}{2n+1}[P_{n-1}(x) - P_{n+1}(x)].$$

25. Show that $P_n(x)Q_{n-2}(x) - P_{n-2}(x)Q_n(x) = \dfrac{2n-1}{n(n-1)}.$

26. Expand $x^4 - 3x^2 + x$ in a series of Legendre polynomials.

27. Expand $f(x) = \begin{cases} 2x+1 & 0 < x \leq 1 \\ 0 & -1 \leq x < 0 \end{cases}$ in a series of Legendre polynomials.

28. Expand $f(x) = \begin{cases} 1 & 0 < x \leq 1 \\ 0 & -1 \leq x < 0 \end{cases}$ in a series of Legendre polynomials.

29. Expand $f(x) = x^2$ in a series of Legendre polynomials.

30. If $f(x) = \sum_{k=0}^{\infty} c_k P_k(x)$, show that $\int_{-1}^{1} [f(x)]^2 dx = \sum_{k=0}^{\infty} \frac{c_k^2}{2k+1}$.

31. Show that:

   a. $x^{2m} = \sum_{n=0}^{m} \frac{2^{2n}(4n+1)(2m)!(m+n)!}{(2m+2n+1)!(m-n)!} P_{2n}(x)$

   b. $x^{2m+1} = \sum_{n=0}^{m} \frac{2^{2n+1}(4n+3)(2m+1)!(m+n+1)!}{(2m+2n+3)!(m-n)!} P_{2n+1}(x)$

32. Find the Legendre series expansion for $f(x) = \cos^2 x$, $0 \le x \le \pi$.

33. Using Rodrigues' Formula, show that:
$(2n+1)P_n(x) = P'_{n+1}(x) - P'_{n-1}(x)$.

34. Prove Christoffel's Second Summation Formula:
$$\frac{1}{y-x} = \sum_{n=0}^{\infty} (2n+1)P_n(x)Q_n(y)$$

35. Prove Neumman's Integral Formula: $Q_n(x) = \frac{1}{2}\int_{-1}^{1} \frac{P_n(x)}{y-x} dx$, $y > 1$.

36. From the summation expression for $Q_n(x)$, show that:
$$\frac{d^{n+1}}{dx^{n+1}} Q_n(x) = \frac{(-1)^n 2^n n!}{(x^2-1)^{n+1}}.$$
Hint: Use Laplace first integral.

37. Show that: $Q_n(x) = \int_0^{\infty} \frac{dt}{\left[x + \Gamma(x^2-1)\cosh t\right]^{n+1}}$.

Chapter Four

Hermite Polynomials

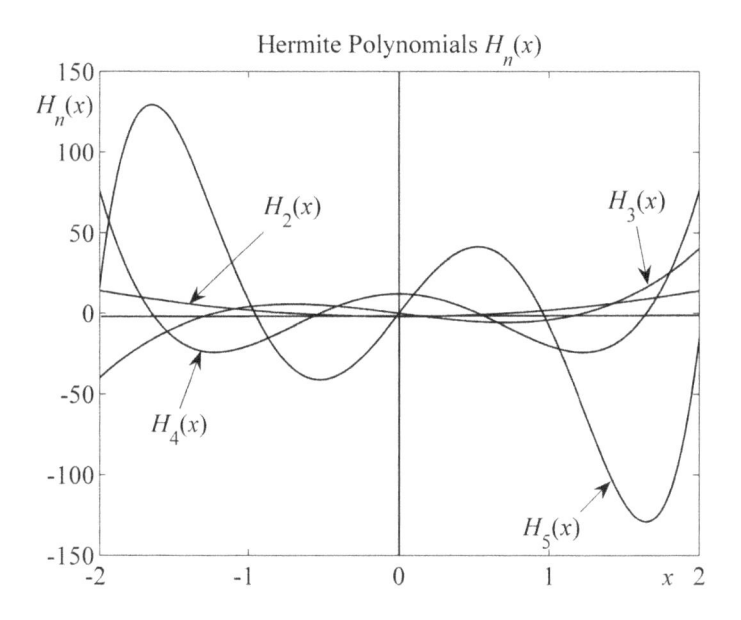

$$H_n(x) = \sum_{k=0}^{N} \frac{(-1)^k \, n!}{k!(n-2k)!} (2x)^{n-2k}$$

Chapter 4
Hermite Polynomials

4.1. Hermite[26] Differential Equation and Its Solutions

The differential equation of the form

$$y'' - 2xy' + 2ny = 0 \tag{1}$$

where n is a constant, is called *Hermite Differential Equation*. The point $x = 0$ is an ordinary point of the equation, and we can assume a series solution of the form

$$y = \sum_{k=0}^{\infty} c_k x^k \tag{2}$$

then
$$y' = \sum_{k=1}^{\infty} k\, c_k x^{k-1} \text{ and } y'' = \sum_{k=2}^{\infty} k(k-1) c_k x^{k-2} \tag{3}$$

Substituting in the differential equation, we get

$$\sum_{k=2}^{\infty} k(k-1) c_k x^{k-2} - 2x \sum_{k=1}^{\infty} k\, c_k x^{k-1} + 2n \sum_{k=0}^{\infty} c_k x^k = 0$$

Or $\sum_{k=0}^{\infty} (k+2)(k+1) c_{k+2} x^k - 2x \sum_{k=0}^{\infty} (k+1) c_{k+1} x^k + 2n \sum_{k=0}^{\infty} c_k x^k = 0$ (4)

Equating the coefficients of x^k to zero, we obtain

$$(k+2)(k+1) c_{k+2} - 2k\, c_k + 2n\, c_k = 0$$

or
$$\boxed{c_{k+2} = -\frac{2(n-k)}{(k+2)(k+1)} c_k, \quad k = 0, 1, 2, 3, \cdots}$$

This is the recurrence relation for the coefficients. For *even* values of k, we have

---

26. French mathematician Charles Hermite (1822-1901).

$k = 0: \quad c_2 = -\dfrac{2n}{2 \cdot 1} c_0$

$k = 2: \quad c_4 = -\dfrac{2(n-2)}{4 \cdot 3} c_2 = \dfrac{2^2 n(n-2)}{4!} c_0$

$k = 4: \quad c_6 = \cdots = -\dfrac{2^3 n(n-2)(n-4)}{6!} c_0$

and in general

$$c_{2k} = (-1)^k \dfrac{2^k n(n-2)(n-4)\cdots(n-2k+2)}{(2k)!} c_0 \qquad (5)$$

and a first solution for Hermite differential equation is

$$y_1(x) = c_0 \left[1 - \dfrac{2^2 n}{2!} x^2 + \dfrac{2^4 n(n-2)}{4!} x^4 - \cdots \right]$$

or $\qquad y_1(x) = c_0 \displaystyle\sum_{k=0}^{\infty} (-1)^k \dfrac{2^k n(n-2)(n-4)\cdots(n-2k+2)}{(2k)!} x^{2k} \qquad (6)$

For *odd* values of k, we have

$k = 1: \quad c_3 = -\dfrac{2(n-1)}{3 \cdot 2} c_1$

$k = 3: \quad c_5 = -\dfrac{2(n-3)}{5 \cdot 4} c_3 = \dfrac{2^2(n-1)(n-3)}{5!} c_1$

$k = 5: \quad c_7 = \cdots = -\dfrac{2^3(n-1)(n-3)(n-5)}{7!} c_1$

and in general

$$c_{2k+1} = (-1)^k \dfrac{2^k (n-1)(n-3)(n-5)\cdots(n-2k+1)}{(2k+1)!} c_1 \qquad (7)$$

and a second solution for Hermite differential equation is

$$y_2(x) = c_1 \left[x - \dfrac{2^3(n-1)}{3!} x^3 + \dfrac{2^5(n-1)(n-3)}{5!} x^5 - \cdots \right]$$

or $\qquad y_2(x) = c_1 \displaystyle\sum_{k=0}^{\infty} (-1)^k \dfrac{2^k (n-1)(n-3)(n-5)\cdots(n-2k+1)}{(2k+1)!} x^{2k+1} \qquad (8)$

Hermite Polynomials

Finally, the general solution of Hermite differential equation takes the form

$$y(x) = c_0\left[1 - \frac{2^2 n}{2!}x^2 + \frac{2^4 n(n-2)}{4!}x^4 - \cdots\right]$$

$$+ c_1\left[x - \frac{2^3(n-1)}{3!}x^3 + \frac{2^5(n-1)(n-3)}{5!}x^5 - \cdots\right] \quad (9)$$

Now, Hermite differential equation has no singularity for finite values of x, then these two series solutions converge for all finite values of x. Moreover, the two solutions are indeed linearly independent.

If n is even, the first series reduces to a polynomial of degree n. While, if n is odd the second series reduces to a polynomial of degree n. These polynomial solutions, apart from the arbitrary constants, are called Hermite polynomials. These polynomials can be represented in descending powers of x as we did for Legendre polynomials. From the recurrence relation for the coefficients, we have

$$c_k = -\frac{(k+2)(k+1)}{2(n-k)}c_{k+2} \quad (10)$$

Then for $k = n-2,\ n-4,\ \cdots$, we have

$$c_{n-2} = -\frac{(n-1)n}{2(n-n+2)}c_n = -\frac{(n-1)n}{2\cdot 2}c_n$$

$$c_{n-4} = -\frac{(n-3)(n-2)}{2(n-n+4)}c_{n-2} = \frac{n(n-1)(n-2)(n-3)}{2^2\cdot 2\cdot 4}c_n$$

$$c_{n-6} = \cdots = -\frac{n(n-1)(n-2)(n-3)(n-4)(n-5)}{2^3\cdot 2\cdot 4\cdot 6}c_n$$

and in general

$$c_{n-2k} = (-1)^k \frac{n(n-1)(n-2)\cdots(n-2k+1)}{2^{2k}\,k!}c_n$$

Multiplying both numerator and denominator by

$(n-2k)(n-2k-1)\cdots 3\cdot 2\cdot 1$, we get

$$c_{n-2k} = (-1)^k \frac{n!}{2^{2k}\,k!(n-2k)!}c_n \quad (11)$$

and the solution will be

175

Special Functions and Orthogonal Polynomials

$$y(x) = \sum_{k=0}^{N} \frac{(-1)^k n!}{2^{2k} k!(n-2k)!} x^{n-2k} \qquad (12)$$

where $N = n/2$ if n is even and $N = (n-1)/2$ if n is odd.

The standard form of the solution of Hermite differential equation is obtained by choosing the arbitrary constant as $c_n = 2^n$, then Hermite polynomials are given by

$$\boxed{H_n(x) = \sum_{k=0}^{N} \frac{(-1)^k n!}{k!(n-2k)!} (2x)^{n-2k}} \qquad (13)$$

For various values of n, we have

$H_0(x) = 1$

$H_1(x) = 2x$

$H_2(x) = 4x^2 - 2$

$H_3(x) = 8x^3 - 12x$

$H_4(x) = 16x^4 - 48x^2 + 12$

$H_5(x) = 32x^5 - 160x^3 + 120x$

$H_6(x) = 64x^6 - 480x^4 + 720x^2 - 120$

$H_7(x) = 128x^7 - 1344x^5 + 3360x^3 - 1680x$

$H_8(x) = 256x^8 - 3584x^6 + 13440x^4 - 13440x^2 + 1680$

$H_9(x) = 512x^9 - 9216x^7 + 48384x^5 - 80640x^3 + 30240x$

$H_{10}(x) = 1024x^{10} - 23040x^8 + 161280x^6 - 403200x^4 + 302400x^2 - 30240$

It can also be shown that:

$$H_{2n}(x) = (-1)^n 2^n (2n-1)!! \left[1 + \sum_{k=1}^{n} \frac{(-4n)(-4n+4)\cdots(-4n+4k-4)}{(2k)!} x^{2k} \right]$$

And

176

Hermite Polynomials

$$H_{2n+1}(x) = (-1)^n 2^{n+1}(2n+1)!! \left[1 + \sum_{k=1}^{n} \frac{(-4n)(-4n+4)\cdots(-4n+4k-4)}{(2k+1)!} x^{2k+1} \right]$$

Where $(2n-1)!!$ and $(2n+1)!!$ are the double factorial notation,

$$(k)!! = k(k-2)(k-4)\cdots \begin{cases} \cdot 3 \cdot 1 & k \text{ odd} \\ \cdot 4 \cdot 2 & k \text{ even} \end{cases}$$

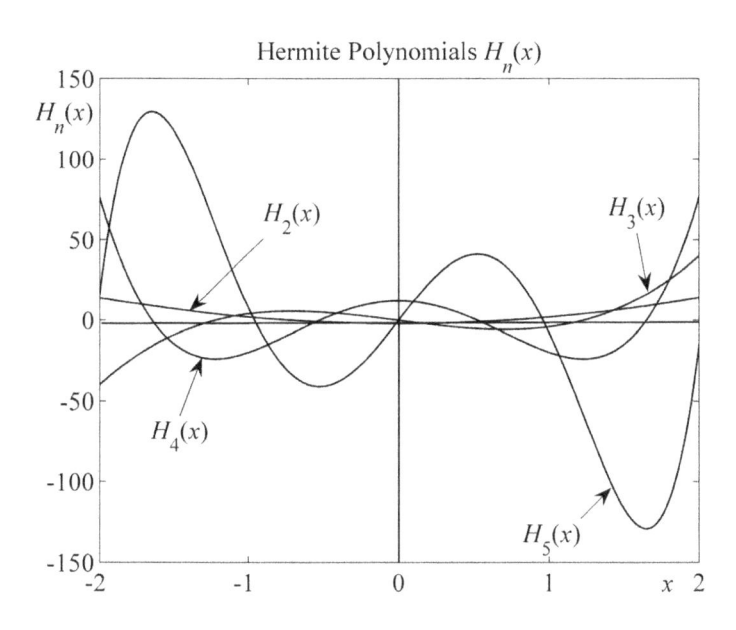

Hermite Polynomials $H_n(x)$

4.2. Generating Function for Hermite Polynomials

Hermite polynomials satisfy the generating equation

$$e^{2tx - t^2} = \sum_{n=0}^{\infty} \frac{t^n}{n!} H_n(x) \qquad (14)$$

To prove this, using Maclaurin's expansion, we get

$$e^{2tx - t^2} = e^{2tx} \cdot e^{-t^2} = \left[1 + 2tx + \frac{(2tx)^2}{2!} + \cdots\right] \cdot \left[1 - t^2 + \frac{t^4}{2!} - \cdots\right]$$

$$= \sum_{s=0}^{\infty} \frac{(2tx)^s}{s!} \cdot \sum_{r=0}^{\infty} \frac{(-1)^r t^{2r}}{r!} = \sum_{s=0}^{\infty} \sum_{r=0}^{\infty} \frac{(-1)^r (2x)^s}{r!\,s!} t^{s+2r}$$

Letting $s + 2r = n$ so that $s = n - 2r$, then for a fixed value of r, the coefficient of t^n is

$$\frac{(-1)^r (2x)^{n-2r}}{r!(n-2r)!} \qquad (15)$$

Now, $s \leq 0$ implies that $n - 2r \geq 0$ or $n \geq 2r$ or $r \leq n/2$, which gives all values of r for which (2) is the coefficient of t^n.

If n is even, then r varies from 0 to $n/2$. While if n is odd, r varies from 0 to $(n-1)/2$. Then, the total coefficient of t^n in the expansion of $e^{2tx - t^2}$ is

$$\sum_{r=0}^{N} \frac{(-1)^r (2x)^{n-2r}}{r!(n-2r)!}$$

where $N = n/2$ if n is even and $N = (n-1)/2$ if n is odd.

Equating the coefficients of t^n in both sides of equation (14), we get

$$\sum_{r=0}^{N} \frac{(-1)^r (2x)^{n-2r}}{r!(n-2r)!} = \frac{H_n(x)}{n!}$$

or

$$H_n(x) = \sum_{r=0}^{N} \frac{(-1)^r n!}{r!(n-2r)!} (2x)^{n-2r} \qquad (16)$$

Hermite Polynomials

which is true from the expression of $H_n(x)$ obtained in the previous section, then the generating equation is true.

The MacLaurin's expansion for the generating function $f(t) = e^{2tx - t^2}$ is given by

$$f(t) = e^{2tx - t^2} = f(0) + tf'(0) + \frac{t^2}{2!}f''(0) + \cdots + \frac{t^n}{n!}f^{(n)}(0) + \cdots$$

But we have

$$e^{2tx - t^2} = \sum_{n=0}^{\infty} \frac{t^n}{n!} H_n(x)$$

$$= \frac{H_0(x)}{(0)!} + \frac{tH_1(x)}{(1)!} + \frac{t^2 H_2(x)}{(2)!} + \cdots + \frac{t^n H_n(x)}{(n)!} + \cdots$$

Therefore, by comparing coefficients, we obtain

$$f^{(n)}(0) = H_n(x) = \left[\frac{\partial^n}{\partial t^n} e^{2tx - t^2} \right]_{t=0}.$$

And from Cauchy's integral formula for the nth derivative, we have

$$f^{(n)}(\zeta) = \frac{n!}{2\pi i} \oint_C \frac{f(t)\,dt}{(t - \zeta)^{n+1}}.$$

Let $\zeta = 0$, we get

$$f^{(n)}(0) = \frac{n!}{2\pi i} \oint_C \frac{f(t)\,dt}{t^{n+1}}$$

Therefore, the Hermite Polynomials can be defined by the contour integral

$$\boxed{H_n(x) = \frac{n!}{2\pi i} \oint_C \frac{e^{2tx - t^2}}{t^{n+1}}\,dt}$$

4.3. Recurrence Relations for Hermite Polynomials

Hermite polynomials satisfy the following recurrence relations:

I. $\boxed{H_{n+1}(x) = 2xH_n(x) - 2nH_{n-1}(x), \quad n \geq 1}$, and $\boxed{H_1(x) = 2xH_0(x)}$

II. $\boxed{H'_n(x) = 2nH_{n-1}(x), \quad n \geq 1}$, and $\boxed{H'_0(x) = 0}$

To prove this, we start from the generating function

$$e^{2tx - t^2} = \sum_{n=0}^{\infty} \frac{t^n}{n!} H_n(x)$$

Differentiating with respect to t, we get

$$2(x-t)e^{2tx-t^2} = \sum_{n=1}^{\infty} \frac{nt^{n-1}}{n!} \cdot H_n(x)$$

or

$$2(x-t) \sum_{n=0}^{\infty} \frac{t^n}{n!} \cdot H_n(x) = \sum_{n=0}^{\infty} \frac{t^n}{n!} \cdot H_{n+1}(x)$$

Equating the coefficients of t^0 in both sides, we get

$$H_1(x) = 2xH_0(x) \tag{17}$$

Again, equating the coefficients of t^n in both sides and re-arranging, we get

$$H_{n+1}(x) = 2xH_n(x) - 2nH_{n-1}(x), \quad n \geq 1 \tag{18}$$

For the second recurrence relation, differentiating the generating function with respect to x, we get

$$2t\, e^{2tx - t^2} = \sum_{n=0}^{\infty} \frac{t^n}{n!} \cdot H'_n(x)$$

or

$$2t \sum_{n=0}^{\infty} \frac{t^n}{n!} \cdot H_n(x) = \sum_{n=0}^{\infty} \frac{t^n}{n!} \cdot H'_n(x)$$

Equating the coefficients of t^0 in both sides, we get

$$H'_0(x) = 0 \tag{19}$$

Equating the coefficients of t^n in both sides and re-arranging, we get

$$H'_n(x) = 2nH_{n-1}(x), \quad n \geq 1 \tag{20}$$

4.4. Rodrigues' Formula for Hermite Polynomials

An alternative formula for Hermite polynomials is given in differential form as

$$H_n(x) = (-1)^n e^{x^2} \frac{d^n}{dx^n}\left[e^{-x^2}\right] \qquad (21)$$

To prove this, we start from the generating function

$$e^{2tx-t^2} = \sum_{n=0}^{\infty} \frac{t^n}{n!} H_n(x)$$

Using Taylor's expansion about $t = 0$, we get

$$\sum_{n=0}^{\infty}\left[\frac{\partial^n}{\partial t^n} e^{2tx-t^2}\right]_{t=0} \cdot \frac{t^n}{n!} = \sum_{n=0}^{\infty} \frac{t^n}{n!} H_n(x)$$

Equating the coefficients of t^n in both sides, we get

$$H_n(x) = \left[\frac{\partial^n}{\partial t^n} e^{2tx-t^2}\right]_{t=0} = \left[\frac{\partial^n}{\partial t^n} e^{x^2-(x-t)^2}\right]_{t=0}$$

$$= e^{x^2}\left[\frac{\partial^n}{\partial t^n} e^{-(x-t)^2}\right]_{t=0}$$

Using the fact that:

$$\frac{\partial^n}{\partial t^n} f(x-t) = (-1)^n \frac{\partial^n}{\partial x^n} f(x-t),$$

then

$$H_n(x) = e^{x^2}\left[(-1)^n \frac{\partial^n}{\partial x^n} e^{-(x-t)^2}\right]_{t=0} = e^{x^2}\left[(-1)^n \frac{\partial^n}{\partial x^n} e^{-x^2}\right] \quad \square$$

4.5. Orthogonality Property of Hermite Polynomials

Hermite polynomials satisfy the orthogonality property

$$\int_{-\infty}^{\infty} e^{-x^2} H_m(x) H_n(x) dx = \begin{cases} 0 & m \neq n \\ 2^n n! \sqrt{\pi} & m = n \end{cases} \quad (22)$$

The function e^{-x^2} is called the weighing function. To prove the orthogonality property, we start once again from the generation function

$$e^{2tx - t^2} = \sum_{n=0}^{\infty} \frac{t^n}{n!} H_n(x) \quad (23)$$

also, we may write

$$e^{2sx - s^2} = \sum_{m=0}^{\infty} \frac{s^m}{m!} H_m(x) \quad (24)$$

Multiplying equations (23) and (24), we get

$$e^{2tx - t^2 + 2sx - s^2} = \sum_{n=0}^{\infty} \sum_{m=0}^{\infty} \frac{t^n s^m}{n! m!} H_n(x) H_m(x)$$

Multiplying both sides by e^{-x^2} and integrating with respect to x from $-\infty$ to ∞, we get

$$\int_{-\infty}^{\infty} e^{[(x+s+t)^2 - 2st]} dx = \sum_{n=0}^{\infty} \sum_{m=0}^{\infty} \frac{t^n s^m}{n! m!} \int_{-\infty}^{\infty} e^{-x^2} H_n(x) H_m(x) dx$$

The integral to the left of this equation is equal to

$$\int_{-\infty}^{\infty} e^{-[(x+s+t)^2 - 2st]} dx = e^{2st} \int_{-\infty}^{\infty} e^{-(x+s+t)^2} dx$$

$$= e^{2st} \underbrace{\int_{-\infty}^{\infty} e^{-u^2} du}_{\sqrt{\pi}} = e^{2st} \sqrt{\pi} = \sqrt{\pi} \sum_{m=0}^{\infty} \frac{2^m s^m t^m}{m!}$$

or

$$\sqrt{\pi} \sum_{m=0}^{\infty} \frac{2^m s^m t^m}{m!} = \sum_{n=0}^{\infty} \sum_{m=0}^{\infty} \frac{t^n s^m}{n! m!} \int_{-\infty}^{\infty} e^{-x^2} H_n(x) H_m(x) dx \quad (25)$$

Hermite Polynomials

For $m = n$, we get upon equating the coefficients of $s^n t^n$,

$$\frac{2^n \sqrt{\pi}}{n!} = \frac{1}{(n!)^2} \int_{-\infty}^{\infty} e^{-x^2} H_n^2(x)\, dx \quad \text{or} \quad \int_{-\infty}^{\infty} e^{-x^2} H_n^2(x)\, dx = 2^n n! \sqrt{\pi}$$

For $m \neq n$, we note that the powers of t and s are always equal in each term on the right hand side of equation (25). Hence equating the coefficients of $t^n t^m$, $m \neq n$, in both sides of (25), we obtain

$$0 = \frac{1}{n! m!} \int_{-\infty}^{\infty} e^{-x^2} H_n(x) H_m(x)\, dx$$

then

$$\int_{-\infty}^{\infty} e^{-x^2} H_n(x) H_m(x)\, dx = 0, \quad m \neq n \qquad \square$$

4.6. Weber-Hermite Functions

These are solutions of the differential equation

$$\frac{d^2y}{dx^2} + (\alpha - x^2)y = 0 \tag{26}$$

Making the substitution $y = z e^{-x^2/2}$, we get

$$\frac{dy}{dx} = e^{-x^2/2}\left(\frac{dz}{dx} - xz\right) \text{ and } \frac{d^2y}{dx^2} = e^{-x^2/2}\left(\frac{d^2z}{dx^2} - 2x\frac{dz}{dx} - (1-x^2)z\right)$$

The differential equation becomes

$$\frac{d^2z}{dx^2} - 2x\frac{dz}{dx} - (\alpha - 1)z = 0$$

which is Hermite differential equation of order n, with $n = \frac{\alpha - 1}{2}$. Then the solution of equation (26) is

$$\Psi_n(x) = e^{-x^2/2} H_n(x)$$

and is called the *Weber-Hermite function* of order n.

This function satisfies the following recurrence relations:

I. $\Psi_{n+1}(x) = 2x\Psi_n(x) - 2n\Psi_{n-1}(x), \; n \geq 1$

II. $\Psi'_n(x) = 2n\Psi_{n-1}(x) - x\Psi_n(x)$

III. $\Psi'_n(x) = x\Psi_n(x) - \Psi_{n+1}(x)$

IV. $\Psi'_n(x) = n\Psi_n(x) - \frac{1}{2}\Psi_{n+1}(x)$

To prove recurrence relation **I**, we have

$$2x\Psi_n(x) - 2n\Psi_{n-1}(x) = 2xe^{-x^2/2}H_n(x) - 2ne^{-x^2/2}H_{n-1}(x)$$

$$= e^{-x^2/2}[2xH_n(x) - 2nH_{n-1}(x)] = e^{-x^2/2}H_{n+1}(x) = \Psi_{n+1}(x)$$

To prove recurrence relations **II**, **III** and **IV**, we proceed as follows.

$$\Psi'_n(x) = \frac{d}{dx}\left[e^{-x^2/2}H_n(x)\right] = e^{-x^2/2}H'_n(x) - xe^{-x^2/2}H_n(x)$$

$$= e^{-x^2/2}[H'_n(x) - xH_n(x)] = e^{-x^2/2}[2nH_{n-1}(x) - xH_n(x)]$$

$$= 2n\Psi_{n-1}(x) - x\Psi_n(x)$$

Substituting from recurrence relation **I** into **II**, we get

Hermite Polynomials

$$\Psi'_n(x) = 2x\,\Psi_n(x) - \Psi_{n+1}(x) - x\,\Psi_n(x) = x\,\Psi_n(x) - \Psi_{n+1}(x)$$

Finally, adding recurrence relations **II** and **III** then dividing by 2, we get

$$\Psi'_n(x) = n\,\Psi_n(x) - \frac{1}{2}\Psi_{n+1}(x) \qquad \square$$

For the orthogonality of the Weber-Hermite functions, we have

$$\int_{-\infty}^{\infty} \Psi_m(x)\Psi_n(x)\,dx = \int_{-\infty}^{\infty} e^{-x^2} H_m(x) H_n(x)\,dx = \begin{cases} 0 & m \neq n \\ 2^n n! \sqrt{\pi} & m = n \end{cases}$$

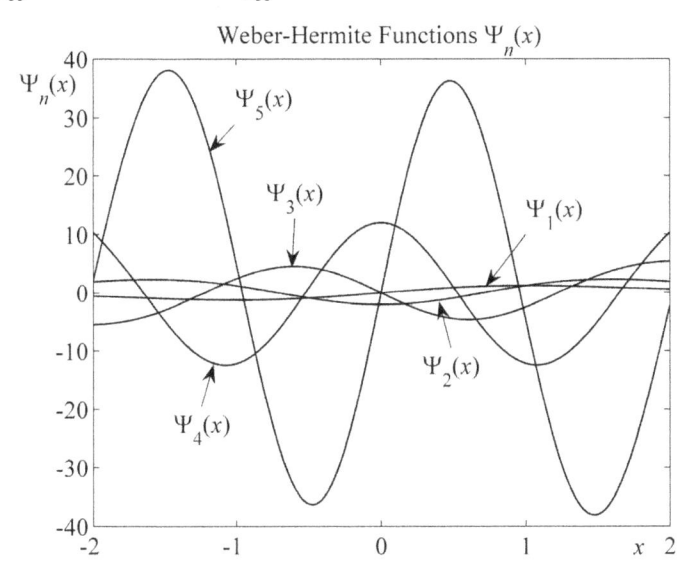

Weber-Hermite Functions $\Psi_n(x)$

Example 1: Show that: (Symmetric property) $H_n(-x) = (-1)^n H_n(x)$.

Solution: From the expression for $H_n(x)$, we have

$$H_n(x) = \sum_{r=0}^{N} \frac{(-1)^r n!}{r!(n-2r)!} (2x)^{n-2r}.$$

Replacing x by $-x$, we get

$$H_n(-x) = \sum_{r=0}^{N} \frac{(-1)^r n! (-1)^{n-2r}}{r!(n-2r)!} (2x)^{n-2r}$$

$$= (-1)^n \sum_{r=0}^{N} \frac{(-1)^r n!}{r!(n-2r)!} (2x)^{n-2r} = (-1)^n H_n(x). \qquad \square$$

Example 2: Show that $H_{2n}(0) = \dfrac{(-1)^n (2n)!}{n!}$ and $H_{2n+1}(0) = 0$.

Solution: From the generating function $e^{2tx-t^2} = \sum_{n=0}^{\infty} \dfrac{t^n}{n!} H_n(x)$.

Letting $x = 0$, we obtain $e^{-t^2} = \sum_{n=0}^{\infty} \dfrac{t^n}{n!} H_n(0)$.

Expanding the left hand side in powers of t, we get

$$\sum_{n=0}^{\infty} (-1)^n \dfrac{t^{2n}}{n!} = \sum_{n=0}^{\infty} \dfrac{t^n}{n!} H_n(0)$$

Equating the coefficients of the corresponding powers of t in both sides, we get $H_n(0) = 0$ if n is odd, or equivalently $H_{2n+1}(0) = 0$, n is a non-negative integer.

And, $\dfrac{(-1)^n}{n!} = \dfrac{H_{2n}(0)}{(2n)!}$ or $H_{2n}(0) = \dfrac{(-1)^n (2n)!}{n!}$. □

Example 3: Show that $H'_{2n}(0) = 0$ and $H'_{2n+1}(0) = \dfrac{2(-1)^n (2n+1)!}{n!}$.

Solution: From the summation expression for Hermite polynomials, we have,

$$H_{2n}(x) = \sum_{k=0}^{n} \dfrac{(-1)^k (2n)!}{k!(2n-2k)!} (2x)^{2n-2k}. \qquad (i)$$

Differentiating with respect to x, we get

$$H'_{2n}(x) = \sum_{k=0}^{n} \dfrac{(-1)^k (2n)!(2n-2k) \cdot 2}{k!(2n-2k)!} (2x)^{2n-2k-1}$$

Letting $x = 0$, all terms in the summation will vanish and $H'_{2n}(0) = 0$

Replacing $2n$ by $2n+1$, we get

$$H'_{2n+1}(x) = \sum_{k=0}^{(2n+1)/2} \dfrac{(-1)^k (2n+1)!(2n-2k+1) \cdot 2}{k!(2n-2k+1)!} (2x)^{2n-2k}.$$

Letting $x = 0$, all terms in the summation will vanish except the one where $k = n$, therefore

$$H'_{2n+1}(0) = \dfrac{2 \cdot (-1)^n (2n+1)!}{n!}.$$

□

Hermite Polynomials

Example 4: Show that $P_n(x) = \dfrac{2}{\sqrt{\pi} n!} \displaystyle\int_0^\infty t^n e^{-t^2} H_n(xt)\, dt$.

Solution: From the expression for $H_n(x)$, we have

$$H_n(xt) = \sum_{k=0}^{N} \frac{(-1)^k n!}{k!(n-2k)!}(2xt)^{n-2k}$$

Then,

$$\frac{2}{\sqrt{\pi} n!}\int_0^\infty t^n e^{-t^2} H_n(xt)\, dt$$

$$= \frac{2}{\sqrt{\pi} n!}\int_0^\infty t^n e^{-t^2} \sum_{k=0}^{N} \frac{(-1)^k n! 2^{n-2k} x^{n-2k}}{k!(n-2k)!} t^{n-2k}\, dt$$

$$= \sum_{k=0}^{N} \frac{(-1)^k 2^{n-2k+1} x^{n-2k}}{\sqrt{\pi}\, k!(n-2k)!} \int_0^\infty e^{-t^2} t^{2n-2k}\, dt$$

From the definition of Gamma function, we have

$$\int_0^\infty e^{-t^2} t^{2n-2k}\, dt = \frac{1}{2}\Gamma\!\left(n-k+\frac{1}{2}\right)$$

and from Legendre Duplication Formula, we have

$$\Gamma\!\left(n-k+\frac{1}{2}\right) = \frac{(2n-2k)!\sqrt{\pi}}{2^{2n-2k}(n-k)!}$$

Then

$$\frac{2}{\sqrt{\pi} n!}\int_0^\infty t^n e^{-t^2} H_n(xt)\, dt$$

$$= \sum_{k=0}^{N} \frac{(-1)^k 2^{n-2k+1} x^{n-2k}}{\sqrt{\pi}\, k!(n-2k)!} \cdot \frac{1}{2}\frac{(2n-2k)!\sqrt{\pi}}{2^{2n-2k}(n-k)!}$$

$$= \sum_{k=0}^{N} \frac{(-1)^k (2n-2k)! x^{n-2k}}{2^n k!(n-2k)!(n-k)!} = P_n(x) \qquad \square$$

Special Functions and Orthogonal Polynomials

Example 5: Use the generating function for Hermite polynomials to find $H_0(x)$, $H_1(x)$, $H_2(x)$ and $H_3(x)$.

Solution: From the generating function $e^{2tx-t^2} = \sum_{n=0}^{\infty} \frac{t^n}{n!} H_n(x)$, expanding the left hand side in powers of t, we get

$$e^{2tx-t^2} = 1+(2tx-t^2)+\frac{(2tx-t^2)^2}{2!}+\frac{(2tx-t^2)^3}{3!}+\cdots$$

$$= 1+(2x)t+(2x^2-1)t^2+\frac{1}{2}(4x^3-6x)t^3+\cdots$$

$$= \sum_{n=0}^{\infty} \frac{t^n}{n!} H_n(x)$$

Equating to coefficients of corresponding powers of t in both sides, we get

$H_0(x)=1$, $H_1(x)=2x$, $H_2(x)=4x^2-2$ and $H_3(x)=8x^3-12x$. □

Example 6: If $m < n$, show that $\frac{d^m}{dx^m} H_n(x) = \frac{2^m n!}{(n-m)!} H_{n-m}(x)$.

Solution: From the generating function $e^{2tx-t^2} = \sum_{n=0}^{\infty} \frac{t^n}{n!} H_n(x)$, differentiating m times with respect to x, we obtain

$$(2t)^m e^{2tx-t^2} = \sum_{n=0}^{\infty} \frac{t^n}{n!} \frac{d^m}{dx^m} H_n(x), \text{ or}$$

$$2^m t^m \sum_{n=0}^{\infty} \frac{t^n}{n!} H_n(x) = \sum_{n=0}^{\infty} \frac{t^n}{n!} \frac{d^m}{dx^m} H_n(x), \text{ or}$$

$$2^m \sum_{n=0}^{\infty} \frac{t^{n+m}}{n!} H_n(x) = \sum_{n=0}^{\infty} \frac{t^n}{n!} \frac{d^m}{dx^m} H_n(x)$$

Hermite Polynomials

Letting $n = k - m$ in the summation to the left, we obtain

$$2^m \sum_{k=m}^{\infty} \frac{t^k}{(k-m)!} H_{k-m}(x) = \sum_{n=0}^{\infty} \frac{t^n}{n!} \frac{d^m}{dx^m} H_n(x)$$

Equating the coefficients of t^n in both sides, we obtain

$$\frac{2^m}{(n-m)!} H_{n-m}(x) = \frac{1}{n!} \frac{d^m}{dx^m} H_n(x), \text{ or}$$

$$\frac{d^m}{dx^m} H_n(x) = \frac{2^m n!}{(n-m)!} H_{n-m}(x). \qquad \square$$

This example can be also solved starting from Recurrence Relation **II**. Let us see: From Recurrence Relation **II**, we have

$$\frac{d}{dx} H_n(x) = 2n H_{n-1}(x).$$

Differentiating with respect to x, we obtain

$$\frac{d^2}{dx^2} H_n(x) = 2n \frac{d}{dx} H_{n-1}(x) = 2n \left[2(n-1) H_{n-2}(x) \right] \text{ or}$$

$$\frac{d^2}{dx^2} H_n(x) = 2^2 n(n-1) H_{n-2}(x).$$

Differentiating one more time, we get

$$\frac{d^3}{dx^3} H_n(x) = 2^3 n(n-1)(n-2) H_{n-3}(x).$$

If we continue with this differentiation process $(m-3)$ times, we finally obtain

$$\frac{d^m}{dx^m} H_n(x) = 2^m n(n-1)(n-2) \cdots (n-m+1) H_{n-m}(x)$$

$$= \frac{2^m n!}{(n-m)!} H_{n-m}(x). \qquad \square$$

Example 7: Show that $H_n(x) = 2^n \left[e^{-\frac{1}{4}\frac{d^2}{dx^2}} \right] x^n$.

Solution: The exponential of a differential operator is defined by its power series expansion. Also, we know that $e^x = \sum_{n=0}^{\infty} \frac{x^n}{n!}$, then

$$\left[e^{-\frac{1}{4}\frac{d^2}{dx^2}} \right] e^{2tx} = \sum_{n=0}^{\infty} \frac{1}{n!} \left[-\frac{1}{4}\frac{d^2}{dx^2} \right]^n e^{2tx} = \sum_{n=0}^{\infty} \frac{(-1)^n}{n!} \left[\frac{1}{2}\frac{d}{dx} \right]^{2n} e^{2tx}$$

$$= \sum_{n=0}^{\infty} \frac{(-1)^n}{n! 2^{2n}} \frac{d^{2n}}{dx^{2n}} e^{2tx} = \sum_{n=0}^{\infty} \frac{(-1)^n}{n! 2^{2n}} (2t)^{2n} e^{2tx}$$

$$= e^{2tx} \sum_{n=0}^{\infty} \frac{(-1)^n}{n!} t^{2n} = e^{2tx} \sum_{n=0}^{\infty} \frac{(-t^2)^n}{n!} = e^{2tx} e^{-t^2} = \sum_{n=0}^{\infty} H_n(x) \frac{t^n}{n!}$$

Also, we have $e^{2tx} = \sum_{n=0}^{\infty} \frac{2^n x^n}{n!} t^n$, then

$$\sum_{n=0}^{\infty} \left[e^{-\frac{1}{4}\frac{d^2}{dx^2}} \right] \frac{2^n x^n}{n!} t^n = \sum_{n=0}^{\infty} \frac{H_n(x)}{n!} t^n.$$

Equating the coefficients of t^n in both sides, we obtain

$$H_n(x) = 2^n \left[e^{-\frac{1}{4}\frac{d^2}{dx^2}} \right] x^n. \qquad \square$$

Example 8: Show that $e^{-x^2} = \frac{2}{\sqrt{\pi}} \int_0^{\infty} e^{-t^2} \cos(2xt) dt$, then deduce that:

i. $H_{2n}(x) = \frac{(-1)^n 2^{2n+1} e^{x^2}}{\sqrt{\pi}} \int_0^{\infty} e^{-t^2} t^{2n} \cos(2xt) dt$

ii. $H_{2n+1}(x) = \frac{(-1)^n 2^{2n+2} e^{x^2}}{\sqrt{\pi}} \int_0^{\infty} e^{-t^2} t^{2n+1} \sin(2xt) dt$

Hermite Polynomials

Solution: From Taylor's expansion, we have

$$\cos(2xt) = 1 - \frac{(2xt)^2}{2!} + \frac{(2xt)^4}{4!} + \cdots = \sum_{n=0}^{\infty} (-1)^n \frac{(2xt)^{2n}}{(2n)!},$$

then

$$\frac{2}{\sqrt{\pi}} \int_0^{\infty} e^{-t^2} \cos(2xt) dt = \frac{2}{\sqrt{\pi}} \int_0^{\infty} e^{-t^2} \sum_{n=0}^{\infty} (-1)^n \frac{(2xt)^{2n}}{(2n)!} dt$$

$$= \sum_{n=0}^{\infty} (-1)^n \frac{2^{2n+1} x^{2n}}{\sqrt{\pi}(2n)!} \int_0^{\infty} e^{-t^2} t^{2n} dt$$

$$= \sum_{n=0}^{\infty} (-1)^n \frac{2^{2n+1} x^{2n}}{\sqrt{\pi}(2n)!} \Gamma\left(n + \frac{1}{2}\right)$$

$$= \sum_{n=0}^{\infty} (-1)^n \frac{2^{2n+1} x^{2n}}{\sqrt{\pi}(2n)!} \cdot \frac{(2n)!\sqrt{\pi}}{2^{2n} n!}$$

$$= \sum_{n=0}^{\infty} (-1)^n \frac{x^{2n}}{n!} = e^{-x^2}$$

We have used the Gamma function as well as Legendre Duplication Formula. Then

$$e^{-x^2} = \frac{2}{\sqrt{\pi}} \int_0^{\infty} e^{-t^2} \cos(2xt) dt . \tag{27}$$

i. Differentiating Equation (27) $2n$ times with respect to x, we get

$$\frac{d^{2n}}{dx^{2n}} e^{-x^2} = \frac{2}{\sqrt{\pi}} \int_0^{\infty} e^{-t^2} (-1)^n 2^{2n} t^{2n} \cos(2xt) dt$$

$$= \frac{(-1)^n 2^{2n+1}}{\sqrt{\pi}} \int_0^{\infty} e^{-t^2} t^{2n} \cos(2xt) dt$$

But, from Rodrigues' Formula, we have $H_{2n}(x) = e^{x^2} \dfrac{d^{2n}}{dx^{2n}} e^{-x^2}$.

Then $H_{2n}(x) = \dfrac{(-1)^n 2^{2n+1} e^{x^2}}{\sqrt{\pi}} \displaystyle\int_0^{\infty} e^{-t^2} t^{2n} \cos(2xt) dt$

191

ii. Differentiating Equation (27) (2n+1) times with respect to x, we get

$$\frac{d^{2n+1}}{dx^{2n+1}}e^{-x^2} = -\frac{2}{\sqrt{\pi}}\int_0^\infty e^{-t^2}(-1)^n 2^{2n+1} t^{2n+1} \sin(2xt)\,dt$$

$$= -\frac{(-1)^n 2^{2n+2}}{\sqrt{\pi}}\int_0^\infty e^{-t^2} t^{2n+1} \sin(2xt)\,dt$$

But, from Rodrigues' Formula, we have

$$H_{2n+1}(x) = -e^{x^2}\frac{d^{2n+1}}{dx^{2n+1}}e^{-x^2}.$$

Then

$$H_{2n+1}(x) = \frac{(-1)^n 2^{2n+2} e^{x^2}}{\sqrt{\pi}}\int_0^\infty e^{-t^2} t^{2n+1} \sin(2xt)\,dt. \quad \square$$

Example 9: Show that $\int_{-\infty}^{\infty} x^2 e^{-x^2} H_n^2(x)\,dx = \sqrt{\pi}\, 2^n\, n!(n+1/2)$.

Solution: From Recurrence Relation **I**, we have

$$xH_n(x) = nH_{n-1}(x) + \frac{1}{2}H_{n+1}(x)\,1.$$

Then the integral becomes

$$\int_{-\infty}^{\infty} x^2 e^{-x^2} H_n^2(x)\,dx = \int_{-\infty}^{\infty} e^{-x^2}\left[nH_{n-1}(x) + \frac{1}{2}H_{n+1}(x)\right]^2 dx$$

$$= \int_{-\infty}^{\infty} e^{-x^2}\left[n^2 H_{n-1}^2(x) + nH_{n-1}(x)H_{n+1}(x) + \frac{1}{2}H_{n+1}^2(x)\right]dx$$

And from the orthogonality property, we get

$$\int_{-\infty}^{\infty} x^2 e^{-x^2} H_n^2(x)\,dx = n^2 2^{n-1}(n-1)!\sqrt{\pi} + \frac{1}{4}2^{n+1}(n+1)!\sqrt{\pi}$$

$$= \sqrt{\pi}\, 2^n\, n!(n+1/2)! \quad \square$$

Hermite Polynomials

Example 10: Evaluate the integral $I = \int_{-\infty}^{\infty} xe^{-x^2} H_n(x) H_m(x) dx$.

Solution: From Recurrence Relation **I**, we have

$$xH_n(x) = nH_{n-1}(x) + \frac{1}{2} H_{n+1}(x).$$

Multiplying both sides by $e^{-x^2} H_m(x)$ and integrating with respect to x from $-\infty$ to ∞, we obtain

$$I = \int_{-\infty}^{\infty} xe^{-x^2} H_n(x) H_m(x) dx$$

$$= \int_{-\infty}^{\infty} e^{-x^2} H_m(x) \left[nH_{n-1}(x) + \frac{1}{2} H_{n+1}(x) \right] dx$$

$$= \int_{-\infty}^{\infty} e^{-x^2} H_m(x) H_{n-1}(x) dx + \frac{1}{2} \int_{-\infty}^{\infty} e^{-x^2} H_m(x) H_{n+1}(x) dx$$

From the orthogonality property, we have:

If $m = n-1$, then $I = \sqrt{\pi}\, 2^{n-1} (n-1)!$

If $m = n+1$, then $I = \frac{1}{2} \sqrt{\pi}\, 2^{n+1} (n+1)!$

and if $m \neq n-1$ and $m \neq n+1$, $I = 0$.

Therefore

$$I = \int_{-\infty}^{\infty} xe^{-x^2} H_n(x) H_m(x) dx = \begin{cases} \sqrt{\pi}\, 2^{n-1}(n-1)! & m = n-1 \\ \sqrt{\pi}\, 2^n (n+1)! & m = n+1 \\ 0 & \text{otherwise} \end{cases}$$

Example 11: Evaluate $I = \int_{-\infty}^{\infty} \Psi_m(x) \Psi'_n(x) dx$

Solution: Substituting from recurrence relation **IV** in the integral, we get

$$I = \int_{-\infty}^{\infty} \Psi_m(x) \Psi'_n(x) dx$$

$$= n \int_{-\infty}^{\infty} \Psi_m(x) \Psi_{n-1}(x) dx - \frac{1}{2} \int_{-\infty}^{\infty} \Psi_m(x) \Psi_{n+1}(x) dx$$

Special Functions and Orthogonal Polynomials

Then for $m = n-1$: $I = n \cdot 2^{n-1}(n-1)!\sqrt{\pi} = 2^{n-1}n!\sqrt{\pi}$

For $m = n+1$: $I = -\frac{1}{2}2^{n+1}(n+1)!\sqrt{\pi} = -2^n(n+1)!\sqrt{\pi}$

And for $m \neq n-1$ and $m \neq n+1$: $I = 0$.

Therefore

$$I = \int_{-\infty}^{\infty} \Psi_m(x)\Psi'_n(x)\,dx = \begin{cases} \sqrt{\pi}\, 2^{n-1}n! & m = n-1 \\ -\sqrt{\pi}\, 2^n(n+1)! & m = n+1 \\ 0 & \text{otherwise} \end{cases}$$

4.7. Summary of Hermite Polynomials

Hermite Differential Equation: $y'' - 2xy' + 2ny = 0$

Hermite Polynomials:

$$H_n(x) = \sum_{k=0}^{N} \frac{(-1)^k n!}{k!(n-2k)!} (2x)^{n-2k}$$

$$H_{2n}(x) = (-1)^n 2^n (2n-1)!! \left[1 + \sum_{k=1}^{n} \frac{4^k (-n)_k}{(2k)!} x^{2k} \right]$$

$$H_{2n+1}(x) = (-1)^n 2^{n+1} (2n+1)!! \left[1 + \sum_{k=1}^{n} \frac{4^k (-n)_k}{(2k+1)!} x^{2k+1} \right]$$

Generating Function: $\displaystyle e^{2tx - t^2} = \sum_{n=0}^{\infty} \frac{t^n}{n!} H_n(x)$

Recurrence Relations:

I. $H_{n+1}(x) = 2xH_n(x) - 2nH_{n-1}(x)$, $n \geq 1$, and $H_1(x) = 2xH_0(x)$

II. $H_n'(x) = 2nH_{n-1}(x)$, $n \geq 1$, and $H_0'(x) = 0$

Orthogonal Properties of Hermite Polynomials:

$$\int_{-\infty}^{\infty} e^{-x^2} H_m(x) H_n(x) dx = \begin{cases} 0 & m \neq n \\ 2^n n! \sqrt{\pi} & m = n \end{cases}$$

Differential Form of Hermite Polynomials (Rodrigue's Formula):

$$H_n(x) = (-1)^n e^{x^2} \frac{d^n}{dx^n} \left[e^{-x^2} \right]$$

Weber-Hermite Differential Equation: $\displaystyle \frac{d^2 y}{dx^2} + (\alpha - x^2) y = 0$

Weber-Hermite Function: $\Psi_n(x) = e^{-x^2/2} H_n(x)$

Recurrence Relations for Weber-Hermite Polynomials:

I. $\Psi_{n+1}(x) = 2x \Psi_n(x) - 2n \Psi_{n-1}(x)$, $n \geq 1$

II. $\Psi_n'(x) = 2n \Psi_{n-1}(x) - x \Psi_n(x)$

III. $\Psi'_n(x) = x\Psi_n(x) - \Psi_{n+1}(x)$

IV. $\Psi'_n(x) = n\Psi_n(x) - \frac{1}{2}\Psi_{n+1}(x)$

Orthogonal Property of Weber-Hermite Polynomials

$$\int_{-\infty}^{\infty} \Psi_m(x)\Psi_n(x)dx = \begin{cases} 0 & m \neq n \\ 2^n n!\sqrt{\pi} & m = n \end{cases}$$

Exercises:

1. Show that: $I = \int_{-\infty}^{\infty} x^2 e^{-x^2} H_n(x) dx = \begin{cases} \frac{1}{2}\sqrt{\pi} & n = 0 \\ 2\sqrt{\pi} & n = 2 \\ 0 & \text{otherwise} \end{cases}$

2. If $f(x)$ is a polynomial of degree m, show that it can be expressed in the form $f(x) = \sum_{n=0}^{m} c_n H_n(x)$, where

$$c_n = \frac{1}{2^n n!\sqrt{\pi}} \int_{-\infty}^{\infty} e^{-x^2} f(x) H_n(x) dx \text{ , then deduce that:}$$

$$\int_{-\infty}^{\infty} e^{-x^2} f(x) H_n(x) dx = 0 \text{ if } f(x) \text{ is a polynomial of degree less than } n.$$

3. Expand the function $f(x) = 8x^3 - 24x^2 + 16x$ into a series of the form $f(x) = \sum_{n=0}^{3} c_n H_n(x)$.

4. Show that $H_n(x) = \frac{2^n(-i)^n e^{x^2}}{\sqrt{\pi}} \int_{-\infty}^{\infty} e^{-t^2 + 2itx} t^n dt$.

5. Show that:

   i. $\sum_{k=0}^{n} \frac{H_k(x)H_k(y)}{2^k k!} = \frac{H_{n+1}(y)H_n(x) - H_{n+1}(x)H_n(y)}{2^{n+1} n!(y-x)}$

ii. $e^{2xt-t^2} H_n(x-t) = \sum_{k=0}^{\infty} \dfrac{H_{k+n}(x) t^k}{k!}$

iii. $H_n(x) = 2^{n+1} e^{x^2} \int_x^{\infty} e^{-t^2} t^{n+1} P_n\left(\dfrac{x}{t}\right) dt$

iv. $\Psi_n(-x) = (-1)^n e^{-x^2/2} H_n(x)$

v. $\dfrac{d}{dx}\left[e^{-x^2} H_n(x)\right] = -e^{-x^2} H_{n+1}(x)$

Special Functions and Orthogonal Polynomials

Chapter Five

Laguerre and Other Orthogonal Polynomials

Laguerre Polynomials $L_n(x)$

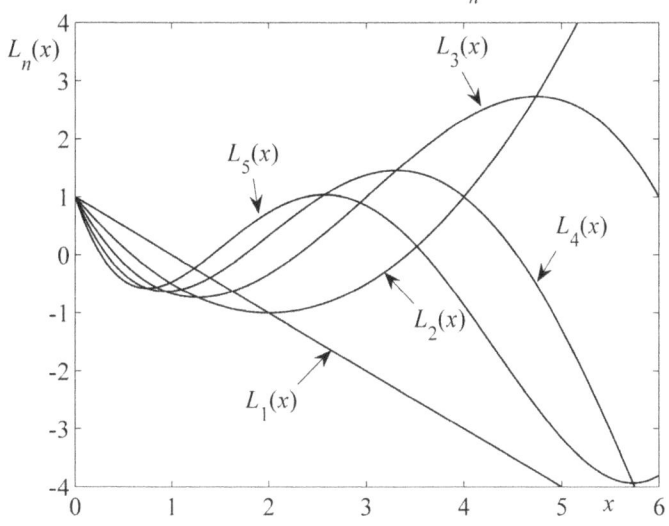

$$L_n(x) = \sum_{k=0}^{n} \frac{(-1)^k \, n!}{(n-k)!(k!)^2} x^k$$

Special Functions and Orthogonal Polynomials

Chapter 5
Laguerre and Other Orthogonal Polynomials

5.1. Laguerre[27] Differential Equation

Laguerre differential equation is

$$xy'' + (1-x)y' + ny = 0 \qquad (1)$$

where n is a positive integer. The point $x = 0$ is a regular point of the equation, so in order to obtain a series solution valid near the origin, we use the method of Frobenius. Assume a series solution of the form

$$y = \sum_{k=0}^{\infty} c_k x^{k+\alpha}, \qquad (2)$$

$$y' = \sum_{k=0}^{\infty} (k+\alpha) c_k x^{k+\alpha-1}, \quad y'' = \sum_{k=0}^{\infty} (k+\alpha)(k+\alpha-1) c_k x^{k+\alpha-2} \qquad (3)$$

Substituting these values in the differential equation, we get

$$x \sum_{k=0}^{\infty} (k+\alpha)(k+\alpha-1) c_k x^{k+\alpha-2}$$
$$+ (1-x) \sum_{k=0}^{\infty} (k+\alpha) c_k x^{k+\alpha-1} + n \sum_{k=0}^{\infty} c_k x^{k+\alpha} = 0 \qquad (4)$$

To obtain the indicial equation, we equate the coefficients of the least power of x ($x^{\alpha-1}$) to zero, to get

$$\alpha(\alpha-1) c_0 + \alpha c_0 = 0 \text{ or } \alpha^2 c_0 = 0 \qquad (5)$$

and since $c_0 \neq 0$, then $\alpha^2 = 0$, and we have a double root at $\alpha = 0$.

To obtain the recurrence relation for the coefficients, we equate the coefficient of $x^{k+\alpha}$ to zero

$$(k+\alpha+1)(k+\alpha) c_{k+1} + (k+\alpha+1) c_{k+1} - (k+\alpha) c_k + n c_k = 0$$

---

27. French mathematician Edmond Nicolas Laguerre (1834-1886 France)

or
$$c_{k+1} = \frac{k+\alpha-n}{(k+\alpha+1)^2} c_k \qquad (6)$$

The two independent solutions are $y\big|_{\alpha=0}$ and $\frac{\partial y}{\partial x}\big|_{\alpha=0}$. The second solution is known to contain a logarithmic term $\ln x$ (since we have a double root, see Chapter 1). If it is required that the solution be finite for all finite x, and since $\ln x$ in the second solution is infinite at the origin, only the first solution will be of interest.

For $\alpha = 0$, the recurrence relation for the coefficients is

$$c_{k+1} = -\frac{n-k}{(k+1)^2} c_k \qquad (7)$$

then for various values of k, we have

$k = 0:$ $c_1 = -\frac{n}{1^2} c_0,$

$k = 1:$ $c_2 = -\frac{n-1}{2^2} c_1 = \frac{n(n-1)}{(2!)^2} c_0$

$k = 2:$ $c_3 = -\frac{n-2}{3^2} c_2 = -\frac{n(n-1)(n-2)}{(3!)^2} c_0$

And in general,

$$c_k = (-1)^k \frac{n(n-1)(n-2)\cdots(n-k+1)}{(k!)^2} c_0$$

Then the series solution is

$$y = c_0 \left\{ 1 - \frac{n}{1^2} x + \frac{n(n-1)}{(2!)^2} x^2 - \frac{n(n-1)(n-2)}{(3!)^2} x^3 + \cdots \right. \\ \left. \cdots + (-1)^k \frac{n(n-1)(n-2)\cdots(n-k+1)}{(k!)^2} x^k + \cdots \right\} \qquad (8)$$

We can see from the recurrence relation for the coefficients (Equation (7)) that the highest power of x will be n, therefore

$$y = c_0 \sum_{k=0}^{n} (-1)^k \frac{n(n-1)(n-2)\cdots(n-k+1)}{(k!)^2} x^k = c_0 \sum_{k=0}^{n} \frac{(-1)^k n!}{(n-k)!(k!)^2} x^k$$

If we let $c_0 = 1$, we obtain the standard solution denoted by $L_n(x)$, called the **Laguerre Polynomial of Order n**, and is given by

Laguerre and Other Orthogonal Polynomials

$$L_n(x) = \sum_{k=0}^{n} \frac{(-1)^k n!}{(n-k)!(k!)^2} x^k \qquad (9)$$

Letting $n = 0, 1, 2, 3, 4$, we get some explicit expressions for Laguerre polynomials:

$L_0(x) = 1$;

$L_1(x) = 1 - x$

$L_2(x) = \frac{1}{2!}(2 - 4x + x^2)$

$L_3(x) = \frac{1}{3!}(6 - 18x + 9x^2 - x^3)$

$L_4(x) = \frac{1}{4!}(24 - 29x + 72x^2 - 16x^3 + x^4)$

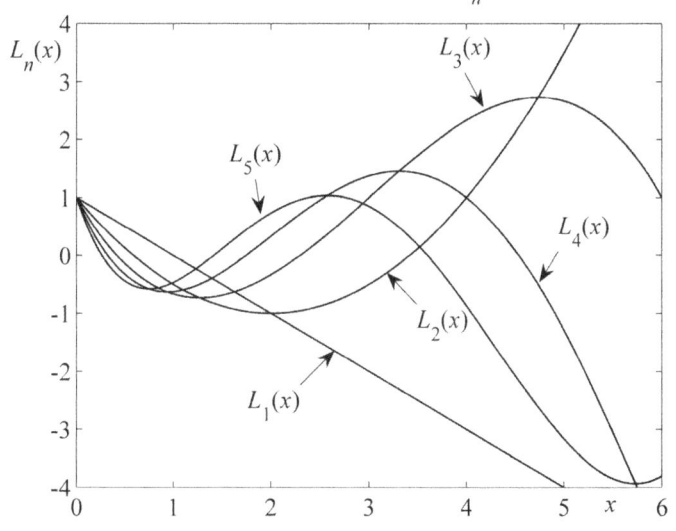

Laguerre Polynomials $L_n(x)$

Laguerre Polynomials are also defined by the contour integral

$$L_n(x) = \frac{n!}{2\pi i} \oint_C \frac{t^{-n-1}}{(1-t)} e^{-xt/(1-t)} dt \ .$$

203

5.2. Generating Function for Laguerre Polynomials

Laguerre polynomials as defined by Equation (9) satisfy the generating relation

$$\frac{1}{1-t}e^{-xt/(1-t)} = \sum_{n=0}^{\infty} L_n(x)t^n \qquad (10)$$

To prove this relation, we have from MacLaurin's expansion

$$\frac{1}{1-t}e^{-xt/(1-t)} = \frac{1}{1-t}\sum_{k=0}^{\infty}\frac{1}{k!}\left(\frac{-xt}{1-t}\right)^k = \sum_{k=0}^{\infty}\frac{(-1)^k x^k t^k}{k!(1-t)^{k+1}}$$

and from the Binomial theorem,

$$\frac{1}{(1-t)^{k+1}} = \sum_{m=0}^{\infty}\frac{(k+m)!}{k!m!}t^m$$

Therefore

$$\frac{1}{1-t}e^{-xt/(1-t)} = \sum_{k=0}^{\infty}\sum_{m=0}^{\infty}\frac{(-1)^k(k+m)!}{(k!)^2 m!}x^k t^{k+m}$$

It remains to show that the coefficient of t^n in this double summation is $L_n(x)$. For a fixed value of k, the coefficient of t^n is obtained by letting $k+m=n$, then

$$\text{coefficient of } t^n = \frac{(-1)^k n!}{(k!)^2(n-k)!}x^k$$

and the total coefficient of t^n in the double summation will be

$$\text{total coefficient of } t^n = \sum_{k=0}^{\infty}\frac{(-1)^k n!}{(k!)^2(n-k)!}x^k$$

which is nothing but $L_n(x)$.

Hence the proof. □

Laguerre and Other Orthogonal Polynomials

5.3. Differential Form of Laguerre Polynomials (Rodrigue's Formula)

Laguerre polynomials are given in differential form as

$$L_n(x) = \frac{e^x}{n!} \frac{d^n}{dx^n}\left[x^n e^{-x}\right] \tag{11}$$

Proof: Using Leibniz's theorem, we have

$$\frac{e^x}{n!} \frac{d^n}{dx^n}\left[x^n e^{-x}\right] = \frac{e^x}{n!} \sum_{k=0}^{n} \frac{n!}{k!(n-k)!} \frac{d^{n-k}}{dx^{n-k}} x^k \frac{d^k}{dx^k} e^{-x}$$

$$= \frac{e^x}{n!} \sum_{k=0}^{n} \frac{n!}{k!(n-k)!} \cdot \frac{n!}{k!} x^k (-1)^k e^{-x}$$

$$= \sum_{k=0}^{n} \frac{(-1)^k n!}{(k!)^2 (n-k)!} \cdot x^k = L_n(x) \qquad \square$$

5.4. Recurrence Relations for Laguerre Polynomials

We give here the recurrence relations for Laguerre polynomials together with their proof.

I. $\boxed{(n+1)L_{n+1}(x) = (2n+1-x)L_n(x) - nL_{n-1}(x)}$

Proof: From the generating function $\dfrac{1}{1-t} e^{-xt/(1-t)} = \sum\limits_{n=0}^{\infty} L_n(x) t^n$,

differentiating both sides with respect to t, we get

$$-\frac{1}{1-t} \cdot \frac{x}{(1-t)^2} e^{-xt/(1-t)} + \frac{1}{(1-t)^2} e^{-xt/(1-t)} = \sum_{n=0}^{\infty} n L_n(x) t^{n-1}$$

Using the generating function, we obtain

$$-\frac{x}{(1-t)^2} \sum_{n=0}^{\infty} L_n(x) t^n + \frac{1}{1-t} \sum_{n=0}^{\infty} L_n(x) t^n = \sum_{n=0}^{\infty} n L_n(x) t^{n-1}$$

Special Functions and Orthogonal Polynomials

Multiplying both sides by $(1-t)^2$, we get

$$-x\sum_{n=0}^{\infty} L_n(x)t^n + (1-t)\sum_{n=0}^{\infty} L_n(x)t^n = (1-2t+t^2)\sum_{n=0}^{\infty} nL_n(x)t^{n-1}$$

Equating the coefficients of t^n in both sides, and re-arranging, we get

$$(n+1)L_{n+1}(x) = (2n+1-x)L_n(x) - nL_{n-1}(x).\qquad\square$$

II. $\boxed{L_n'(x) - L_{n-1}'(x) = -L_{n-1}(x)}$

Proof: Differentiating the generating function with respect to x, we obtain

$$-\frac{1}{1-t}\cdot\frac{t}{1-t}e^{-xt/(1-t)} = \sum_{n=0}^{\infty} L_n'(x)t^n.$$

Using the generating function again, we get

$$-\frac{t}{1-t}\sum_{n=0}^{\infty} L_n(x)t^n = \sum_{n=0}^{\infty} L_n'(x)t^n \qquad (12)$$

Multiplying both sides by $(1-t)$, we get

$$-t\sum_{n=0}^{\infty} L_n(x)t^n = (1-t)\sum_{n=0}^{\infty} L_n'(x)t^n$$

Equating the coefficients of t^n in both sides, we get

$$L_n'(x) - L_{n-1}'(x) = -L_{n-1}(x).\qquad\square$$

III. $\boxed{L_n'(x) = \dfrac{n}{x}[L_n(x) - L_{n-1}(x)]}$

Proof: Differentiating Recurrence Relation **I** with respect to x, we get

$$(n+1)L_{n+1}'(x) = (2n+1-x)L_n'(x) - L_n(x) - nL_{n-1}'(x) \qquad (13)$$

From Recurrence Relation **II**, we have

$$L_{n-1}'(x) = L_n'(x) + L_{n-1}(x) \qquad (14)$$

Replacing n by $n+1$ in Equation (14), we obtain

$$L'_{n+1}(x) = L'_n(x) - L_n(x) \tag{15}$$

Substituting for $L'_{n-1}(x)$ and $L'_{n+1}(x)$ from Equations (14) and (15) into Equation (13) and simplifying, we get

$$L'_n(x) = \frac{n}{x}[L_n(x) - L_{n-1}(x)] \qquad \square$$

IV. $\boxed{L'_n(x) = -\sum_{k=0}^{n-1} L_k(x)}$

Proof: From the generating function

$$\sum_{n=0}^{\infty} L_n(x)t^n = \frac{1}{1-t} e^{-xt/(1-t)}.$$

Differentiating with respect to x, we get

$$\sum_{k=0}^{\infty} L'_k(x)t^k = -\frac{t}{(1-t)^2} e^{-xt/(1-t)} = -\frac{t}{1-t} \sum_{k=0}^{\infty} L_k(x)t^k$$

$$= -t(1-t)^{-1} \sum_{k=0}^{\infty} L_k(x)t^k = -\sum_{s=0}^{\infty} t^{s+1} \sum_{k=0}^{\infty} L_k(x)t^k$$

$$= -\sum_{s=0}^{\infty} \sum_{k=0}^{\infty} L_k(x) t^{k+s+1} \tag{16}$$

Let $s = n - k - 1$, then $s \geq 0$ or $k \leq n-1$, then

$$\sum_{k=0}^{\infty} L'_k(x)t^k = -\sum_{n=0}^{\infty} \sum_{k=0}^{n-1} L_k(x)t^n.$$

Equating the coefficients of t^n in both sides, we obtain

$$L'_n(x) = -\sum_{k=0}^{n-1} L_k(x). \qquad \square$$

5.5. Orthogonality Property of Laguerre Polynomials

The functions $e^{-x/2}L_n(x)$ and $e^{-x/2}L_m(x)$ are orthogonal in the interval from 0 to ∞. In fact

$$\int_0^\infty e^{-x} L_n(x) L_m(x) \, dx = \begin{cases} 0 & m \neq n \\ 1 & m = n \end{cases}$$

Proof: From the generating function, we have

$$\frac{1}{1-t} e^{-xt/(1-t)} = \sum_{n=0}^\infty L_n(x) t^n \, , \text{ and}$$

$$\frac{1}{1-s} e^{-xs/(1-s)} = \sum_{m=0}^\infty L_m(x) t^m \, .$$

Multiplying these two equations together with the weighing function e^{-x}, we obtain

$$\frac{1}{(1-t)(1-s)} e^{-x[1+t/(1-t)+s/(1-s)]} = \sum_{n=0}^\infty \sum_{m=0}^\infty e^{-x} L_n(x) L_m(x) t^n s^m$$

Integrating with respect to x from 0 to infinity, we obtain

$$\sum_{n=0}^\infty \sum_{m=0}^\infty t^n s^m \int_0^\infty e^{-x} L_n(x) L_m(x) \, dx$$

$$= \frac{1}{(1-t)(1-s)} \int_0^\infty e^{-x[1+t/(1-t)+s/(1-s)]} dx$$

$$= \frac{1}{(1-t)(1-s)+t(1-s)+s(1-t)} = \frac{1}{1-ts} = \sum_{n=0}^\infty t^n s^n$$

If $m \neq n$, the coefficient of $t^n s^m$ on the right hand side of this equation is zero and

$$\int_0^\infty e^{-x} L_n(x) L_m(x) \, dx = 0, \quad m \neq n \, .$$

But if $m = n$, we have $\int_0^\infty e^{-x} L_n^2(x) \, dx = 1$. $\square$

5.6. The Associated Laguerre Differential Equation

Starting from Laguerre differential equation of order $n+m$, we have

$$xy'' + (1-x)y' + (n+m)y = 0 \tag{17}$$

Differentiating this equation m times with respect to x using Leibniz's theorem and using the notation $y_m = \dfrac{d^m y}{dx^m}$, we obtain

$$xy_{m+2} + my_{m+1} + (1-x)y_{m+1} - my_m + (n+m)y_m = 0$$

or
$$xy_{m+2} + (m+1-x)y_{m+1} + ny_m = 0 \tag{18}$$

Letting $z = y_m$, we get

$$\boxed{xz'' + (m+1-x)z' + nz = 0} \tag{19}$$

This equation is called the *Associated Laguerre Differential Equation*. Since Laguerre polynomial $L_n(x)$ is a solution of Laguerre differential equation, then a solution of the associated Laguerre equation will be $\dfrac{d^m}{dx^m} L_{n+m}(x)$; and we define the **Associated Laguerre Polynomials** by this solution multiplied by the constant factor $(-1)^m$ as

$$\boxed{L_n^m(x) = (-1)^m \dfrac{d^m}{dx^m} L_{n+m}(x)} \tag{20}$$

The following results can be easily obtained from the properties of $L_n(x)$. We will state them and leave the proofs as an exercise.

1. Series Expansion of the Associated Laguerre Polynomials:

$$\boxed{L_n^m(x) = \sum_{k=0}^{n} \dfrac{(-1)^k (n+m)!}{(n-k)!\, k!\, (m+k)!} x^k} \tag{21}$$

2. Generating Function:

$$\boxed{\dfrac{1}{(1-t)^{m+1}} e^{-xt/(1-t)} = \sum_{n=0}^{\infty} L_n^m(x) t^n} \tag{22}$$

3. Recurrence Relations:

I. $L_{n-1}^m(x) + L_n^{m-1}(x) = L_n^m(x)$

II. $(n+1)L_{n+1}^m(x) = (2n+m+1-x)L_n^m(x) - (n+m)L_{n-1}^m(x)$

III. $x\dfrac{d}{dx}L_n^m(x) = nL_n^m(x) - (n+m)L_{n-1}^m(x)$

IV. $\dfrac{d}{dx}L_n^m(x) = -\sum_{k=0}^{n-1} L_k^m(x)$

V. $\dfrac{d}{dx}L_n^m(x) = -L_{n-1}^{m+1}(x)$

VI. $L_n^{m+1}(x) = \sum_{k=0}^{m} L_k^m(x)$

4. Orthogonality Property:

$$\int_0^\infty e^{-x} x^m L_n^m(x) L_k^m(x) dx = \begin{cases} 0 & k \neq n \\ \dfrac{(n+m)!}{n!} & k = n \end{cases} \qquad (23)$$

5. Rodrigue's Formula:

$$L_n^m(x) = \dfrac{e^x x^{-m}}{n!} \dfrac{d^n}{dx^n}\left[x^{n+m} e^{-x}\right] \qquad (24)$$

Example 1: Show that: i. $L_n(0) = 1$; ii. $L_n'(0) = -n$; iii. $L_n''(0) = \tfrac{1}{2}n(n-1)$.

Solution: From the generating function $\dfrac{1}{1-t} e^{-xt/(1-t)} = \sum_{n=0}^{\infty} L_n(x) t^n$:

i. Letting $x = 0$, we get

$$\dfrac{1}{1-t} = \sum_{n=0}^{\infty} L_n(0) t^n \text{ or } \sum_{n=0}^{\infty} t^n = \sum_{n=0}^{\infty} L_n(0) t^n.$$

Equating the coefficients of t^n in both sides, we get $L_n(0) = 1$.

ii. $L_n(x)$ satisfies the Laguerre differential equation, then

$$xL_n''(x) + (1-x)L_n'(x) + nL_n(x) = 0.$$

Letting $x = 0$, we get $L_n'(0) + nL_n(0) = 0$, but $L_n(0) = 1$, then $L_n'(0) = -n$.

iii. Differentiating the generating function twice with respect to x, we obtain

$$\frac{1}{1-t} \cdot \left(\frac{-t}{1-t}\right)^2 e^{-xt/(1-t)} = \sum_{n=0}^{\infty} L_n''(x) t^n$$

Letting $x = 0$, we get $\dfrac{t^2}{(1-t)^3} = \sum_{n=0}^{\infty} L_n''(0) t^n$.

From the binomial expansion of the left hand side, we have

$$t^2 \left[1 + 3t + \frac{3 \cdot 4}{2!} t^2 + \cdots + \frac{3 \cdot 4 \cdot 5 \cdots n}{(n-2)!} t^{n-2} \right] = \sum_{n=0}^{\infty} L_n''(0) t^n$$

Equating the coefficients of t^n in both sides, we get

$$L_n''(0) = \frac{3 \cdot 4 \cdot 5 \cdots n}{(n-2)!} = \frac{n!}{2(n-2)!} = \frac{n(n-1)}{2}.$$

5.7. Chebyshev[28] Polynomials

These polynomials are orthogonal polynomials obtained by solving the *Chebyshev Differential Equation*

$$(1-x^2)y'' - xy' + n^2 y = 0 \tag{25}$$

They are denoted by $T_n(x)$ and $U_n(x)$, where n is a non-negative integer. $T_n(x)$ is called the *Chebyshev Polynomial of the First Kind*, whereas $U_n(x)$ is called the *Chebyshev Polynomial of the Second Kind*. They are given by the identities

$$T_n(x) = \cos(n \cos^{-1} x) \tag{26}$$

and

$$U_n(x) = \sin(n \cos^{-1} x) \tag{27}$$

From Equations (26) and (27), it is easy to notice that:

$$T_n(1) = 1, \; T_n(-1) = (-1)^n \text{ and } U_n(1) = U_n(-1) = 0.$$

The summation expressions for $T_n(x)$ and $U_n(x)$ are given below, they are of course the solutions of Chebyshev differential equation (25):

$$T_n(z) = \frac{n}{2} \sum_{k=0}^{N} \frac{(-1)^k (n-k-1)!}{k!(n-2k)!} (2x)^{n-2k},$$

$$T_n(z) = \sum_{k=0}^{N} \frac{n!}{(2k)!(n-2k)!} x^{n-2k} (x^2-1)^k.$$

And

$$U_n(z) = \sum_{k=0}^{N} \frac{(-1)^k (n-k)!}{k!(n-2k)!} (2x)^{n-2k},$$

$$U_n(z) = \sum_{k=0}^{N} \frac{(n+1)!}{(2k+1)!(n-2k)!} x^{n-2k} (x^2-1)^k.$$

Where N is the Floor Function.

$T_n(z)$ is also defined by the contour integral

$$T_n(z) = \frac{1}{4\pi i} \oint_C \frac{(1-t^2)t^{-n-1}}{(1-2tz+t^2)} dt$$

---

28. Russian mathematician Pafnuti Chebyshev (1821-1894).

Laguerre and Other Orthogonal Polynomials

where C is the contour that encloses the origin and traverses in an anti-clockwise direction.

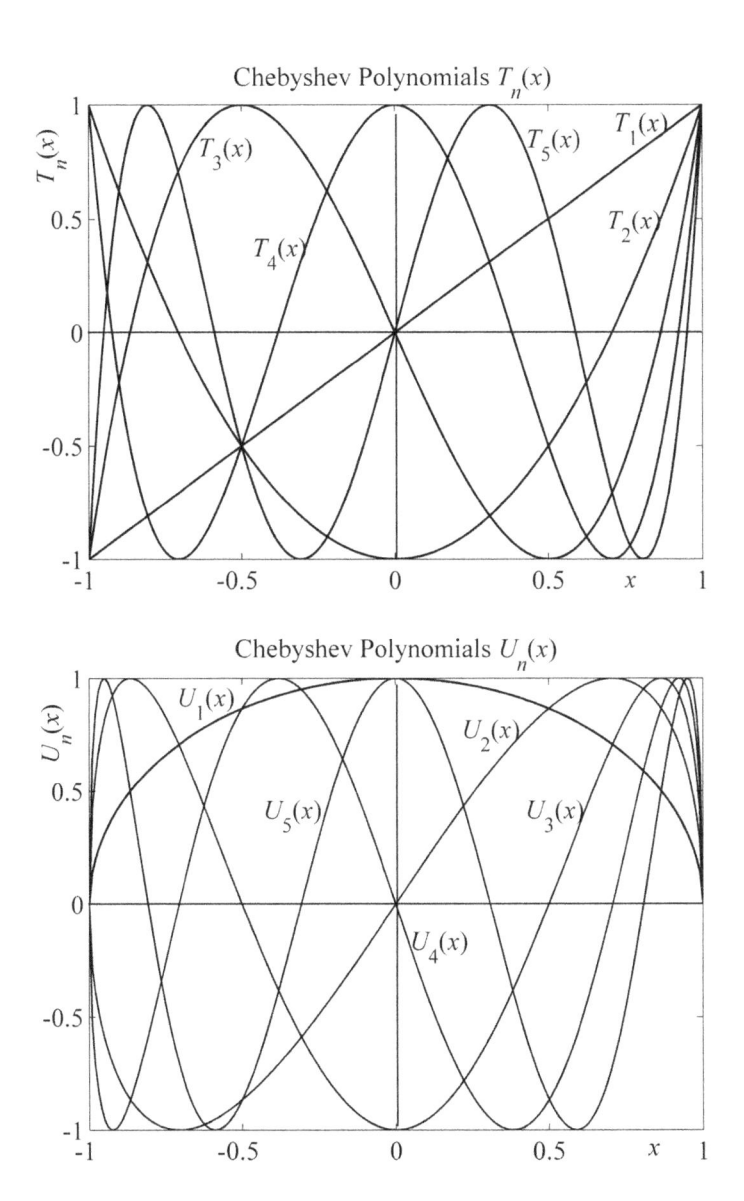

Chebyshev Polynomials $T_n(x)$

Chebyshev Polynomials $U_n(x)$

For various values of n, the polynomials $T_n(x)$ and $U_n(x)$ are

$T_0(x) = 1$ $\qquad\qquad U_0(x) = 1$

$T_1(x) = x$ $\qquad\qquad U_1(x) = 2x$

$T_2(x) = 2x^2 - 1$ $\qquad U_2(x) = 4x^2 - 1$

$T_3(x) = 4x^3 - 3x$ $\qquad U_3(x) = 8x^3 - 4x$

$T_4(x) = 8x^4 - 8x^2 + 1$ $\qquad U_4(x) = 16x^4 - 12x^2 + 1$

$T_5(x) = 16x^5 - 20x^3 + 5x$ $\qquad U_5(x) = 32x^5 - 32x^3 + 6x$

$T_6(x) = 32x^6 - 48x^4 + 18x^2 - 1$ $\qquad U_6(x) = 64x^6 - 80x^4 + 24x^2 - 1$

Example 2: Show that:

$$T_n(x) = \frac{1}{2}\left[\left(x + i\sqrt{1-x^2}\right)^n + \left(x - i\sqrt{1-x^2}\right)^n\right] \qquad (28)$$

Solution: We have $T_n(x) = \cos(n\cos^{-1} x)$.

Let $x = \cos\theta$, then

$$T_n(\cos\theta) = \cos(n\cos^{-1}\cos\theta) = \cos(n\theta) = \frac{1}{2}\left(e^{in\theta} + e^{-in\theta}\right)$$

$$= \frac{1}{2}\left[(\cos\theta + i\sin\theta)^n + (\cos\theta - i\sin\theta)^n\right]$$

Back-substituting, we obtain

$$T_n(x) = \frac{1}{2}\left[\left(x + i\sqrt{1-x^2}\right)^n + \left(x - i\sqrt{1-x^2}\right)^n\right]. \qquad \square$$

5.7.1. Generating Functions

Chebyshev Polynomials of the first kind satisfy the following two relations:

$$\left.\begin{array}{l} \dfrac{1-t^2}{1-2xt+t^2} = T_0(x) + \sum_{n=1}^{\infty} T_n(x) t^n \\[2mm] \dfrac{1-xt}{1-2xt+t^2} = \sum_{n=0}^{\infty} T_n(x) t^n \end{array}\right\}, |x| \leq 1 \text{ and } |t| < 1 \qquad (29)$$

And Chebyshev Polynomials of the second kind satisfy the following two relations:

$$\left.\begin{array}{l}\dfrac{1}{1-2xt+t^2} = \sum_{n=0}^{\infty} U_n(x)t^n \\[2mm] \dfrac{1-t^2}{(1-2xt+t^2)^2} = \sum_{n=0}^{\infty} (n+1)U_n(x)t^n \end{array}\right\}, |x|<1 \text{ and } |t|<1 \qquad (30)$$

5.7.2. Recurrence Relations

Chebyshev polynomials satisfy the following recurrence relations:

I. $T_{n+1}(x) - 2xT_n(x) + T_{n-1}(x) = 0$

II. $T_{n+1}(x) = xT_n(x) - \sqrt{(1-x^2)\{1-[T_n(x)]^2\}}$

III. $(1-x^2)T_n'(x) = nxT_n(x) + nT_{n-1}(x) = 0$

IV. $U_{n+1}(x) - 2xU_n(x) + U_{n-1}(x) = 0$

V. $(1-x^2)U_n'(x) = nxU_n(x) + nU_{n-1}(x) = 0$

5.7.3. Orthogonality Property

The orthogonality properties for Chebyshev polynomials are

$$\int_{-1}^{1} \frac{T_n(x)T_m(x)}{\sqrt{1-x^2}} dx = \begin{cases} 0 & m \neq n \\ \pi/2 & m = n \neq 0 \\ \pi & m = n = 0 \end{cases} \qquad (31)$$

and

$$\int_{-1}^{1} \frac{U_n(x)U_m(x)}{\sqrt{1-x^2}} dx = \begin{cases} 0 & m \neq n \\ \pi/2 & m = n \neq 0 \\ 0 & m = n = 0 \end{cases} \qquad (32)$$

5.7.4. Fourier-Chebyshev Series Expansion

Using the Orthogonality Property of Chebyshev Polynomials, a piecewise continuous function $f(x)$ in $-1 \leq x \leq 1$ can by expanded into a series of Chebyshev polynomials of the form

$$f(x) = \sum_{n=0}^{\infty} c_n T_n(x) \qquad (33)$$

where
$$c_n = \frac{2}{\pi}\int_{-1}^{1}\frac{f(x)T_n(x)}{\sqrt{1-x^2}}dx, \quad n \neq 0 \tag{34}$$

and
$$c_0 = \frac{1}{\pi}\int_{-1}^{1}\frac{f(x)T_0(x)}{\sqrt{1-x^2}}dx \tag{35}$$

at the point of finite discontinuity, we replace $f(x)$ by $\dfrac{f(x^-)+f(x^+)}{2}$.

5.7.5. Rodrigues' Formula

The differential form for Chebyshev polynomial is given by the Rodrigues representation:

$$T_n(x) = \frac{(-1)^n\sqrt{\pi}\sqrt{1-x^2}}{2n(n-1/2)!}\frac{d^n}{dx^n}\left[(1-x^2)^{n-1/2}\right] \tag{36}$$

5.7.6. Chebyshev Determinants

$T_n(x)$ and $U_n(x)$ satisfy the following determinant equations:

$$T_n(x) = \begin{vmatrix} x & 1 & 0 & 0 & \cdots & 0 & 0 \\ 1 & 2x & 1 & \cdots & \cdots & 0 & 0 \\ 0 & 1 & 2x & 1 & \cdots & 0 & 0 \\ 0 & 0 & 1 & 2x & \ddots & \vdots & 0 \\ 0 & 0 & 0 & 1 & \ddots & \ddots & \vdots \\ \vdots & \vdots & \vdots & \vdots & \ddots & \ddots & 1 \\ 0 & 0 & 0 & 0 & \cdots & 1 & 2x \end{vmatrix},$$

$$U_n(x) = \begin{vmatrix} 2x & 1 & 0 & 0 & \cdots & 0 & 0 \\ 1 & 2x & 1 & \cdots & \cdots & 0 & 0 \\ 0 & 1 & 2x & 1 & \cdots & 0 & 0 \\ 0 & 0 & 1 & 2x & \ddots & \vdots & 0 \\ 0 & 0 & 0 & 1 & \ddots & \ddots & \vdots \\ \vdots & \vdots & \vdots & \vdots & \ddots & \ddots & 1 \\ 0 & 0 & 0 & 0 & \cdots & 1 & 2x \end{vmatrix}.$$

The determinants are the tri-diagonal type.

5.8. Gegenbauer Polynomials

They are solutions of the differential equation

$$(1-x^2)y'' - (2\lambda+1)y' + n(n+2\lambda)y = 0 \qquad (37)$$

and are denoted by $C_n^\lambda(x)$. We note that if $\lambda = 1/2$, the *Gegenbauer Differential Equation* reduces to *Legendre Differential Equation*. In this case, *Gegenbauer Polynomials* $C_n^\lambda(x)$ reduce to *Legendre Polynomials* $P_n(x)$. Gegenbauer Polynomials are called sometimes *Ultraspherical Polynomials*. The series representation of these polynomials is

$$C_n^\lambda(x) = \sum_{k=0}^{N} \frac{(-1)^k \Gamma(n-k+\lambda)}{\Gamma(\lambda) k!(n-2k)!} (2x)^{n-2k} \qquad (38)$$

where the floor function $N = n/2$ if n is even, and $N = (n-1)/2$ if n is odd.

The first few polynomials are given by

$$C_0^\lambda(x) = 1,$$

$$C_1^\lambda(x) = 2\lambda x,$$

$$C_2^\lambda(x) = -\lambda + 2\lambda(1+\lambda)x^2,$$

$$C_3^\lambda(x) = -2\lambda(1+\lambda)x + \frac{4}{3}\lambda(1+\lambda)(2+\lambda)x^3.$$

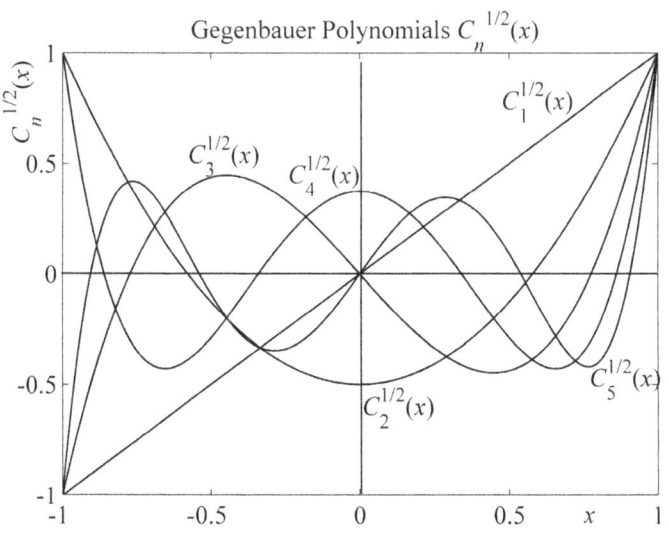

Special Functions and Orthogonal Polynomials

Gegenbauer Polynomials $C_n^{\,1}(x)$

Gegenbauer Polynomials $C_n^{\,2}(x)$

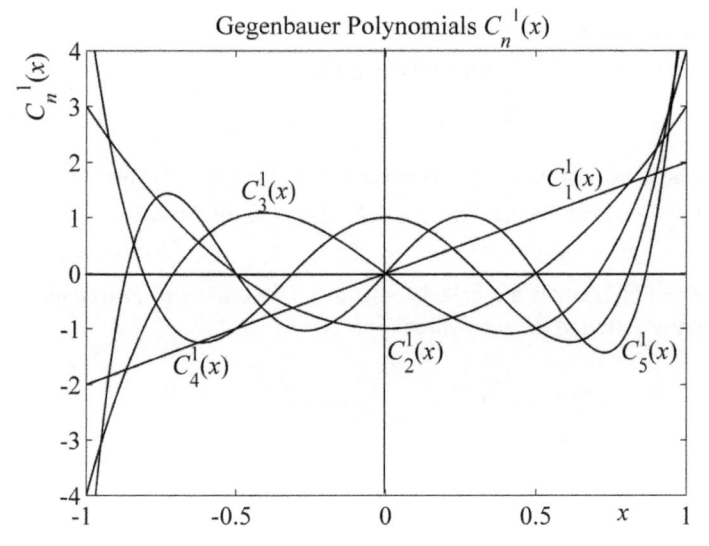

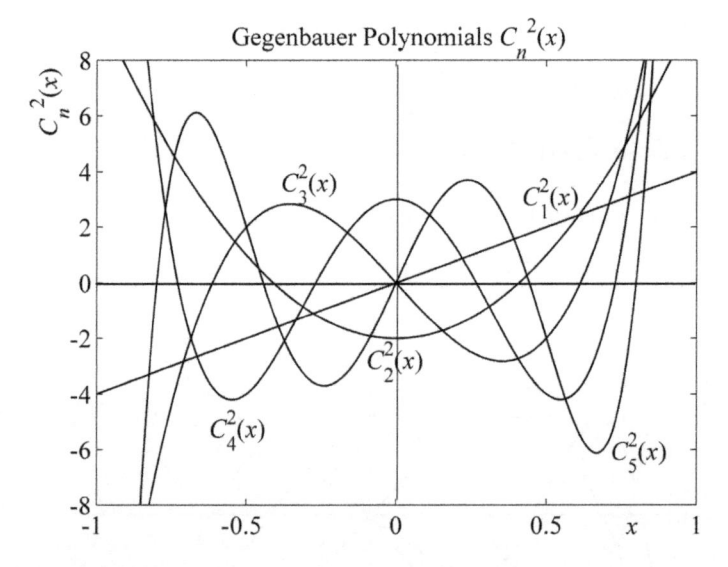

The generating function is given by

$$\frac{1}{(1-2xt+t^2)^\lambda} = \sum_{n=0}^{\infty} C_n^\lambda(x)\, t^n \qquad (39)$$

The recurrence relations of these polynomials are

I. $(n+2)C_{n+2}^{\lambda}(x) = 2(\lambda+n+1)C_{n+1}^{\lambda}(x) - (2\lambda+n)C_n^{\lambda}(x)$

II. $nC_n^{\lambda}(x) = 2\lambda\{x\,C_{n-1}^{\lambda+1}(x) - C_{n-2}^{\lambda+1}(x)\}$

III. $(n+2\lambda)C_n^{\lambda}(x) = 2\lambda\{x\,C_n^{\lambda+1}(x) - C_{n-1}^{\lambda+1}(x)\}$

IV. $nC_n^{\lambda}(x) = (2\lambda+n-1)x\,C_{n-1}^{\lambda}(x) - 2\lambda(1-x^2)C_{n-2}^{\lambda-1}(x)$

V. $\dfrac{d}{dx}C_n^{\lambda}(x) = 2\lambda C_{n+1}^{\lambda+1}(x)$

Finally, the orthogonality property is

$$\int_{-1}^{1}(1-x^2)^{\lambda-1/2}C_n^{\lambda}(x)C_m^{\lambda}(x)\,dx = \begin{cases} 0 & m \neq n \\[4pt] \dfrac{\Gamma(n+2\lambda)}{(n+\lambda)\Gamma^2(\lambda)\Gamma(n+1)} & m = n \end{cases} \quad (40)$$

5.9. Jacobi Polynomials

They satisfy the differential equation

$$(1-x^2)y'' + [\beta-\alpha-(\alpha+\beta+2)x]y' + n(n+\alpha+\beta+1)y = 0 \quad (41)$$

and are denoted by $P_n^{(\alpha,\beta)}(x)$. Their generating function is

$$\frac{2^{\alpha+\beta}}{\sqrt{1-2xt+t^2}\left(1-t+\sqrt{1-2xt+t^2}\right)^{\alpha}\left(1+t+\sqrt{1-2xt+t^2}\right)^{\beta}} = \sum_{n=0}^{\infty} P_n^{(\alpha,\beta)}(x)t^n \quad (42)$$

Jacobi polynomials are also given by the series expansions

$$P_n^{(\alpha,\beta)}(x) = \sum_{k=0}^{n} \frac{\Gamma(n+\alpha+1)\Gamma(n+\beta+1)}{\Gamma(\alpha+k+1)\Gamma(n+\beta-k+1)(n-k)!k!}\left(\frac{x-1}{2}\right)^k\left(\frac{x+1}{2}\right)^{n-k} \quad (43)$$

$$P_n^{(\alpha,\beta)}(x) = \sum_{k=0}^{n} \frac{\Gamma(n+\alpha+1)\Gamma(n+k+\alpha+\beta+1)}{\Gamma(\alpha+k+1)\Gamma(n+\alpha+\beta+1)(n-k)!k!}\left(\frac{x-1}{2}\right)^k \quad (44)$$

or

$$P_n^{(\alpha,\beta)}(x) = \sum_{k=0}^{n} \frac{(-1)^{n-k}\Gamma(n+\beta+1)\Gamma(n+k+\alpha+\beta+1)}{\Gamma(\beta+k+1)\Gamma(n+\alpha+\beta+1)(n-k)!k!}\left(\frac{x+1}{2}\right)^k \quad (45)$$

The orthogonality property for Jacobi Polynomials is

$$\int_{-1}^{1}(1-x)^{\alpha}(1+x)^{\beta}P_{n}^{(\alpha,\beta)}(x)P_{m}^{(\alpha,\beta)}(x)dx = \begin{cases} 0 & m \neq n \\ \dfrac{2^{\alpha+\beta+1}\Gamma(n+\alpha+1)\Gamma(n+\beta+1)}{(2n+\alpha+\beta+1)n!\Gamma(n+\alpha+\beta+1)} & m = n \end{cases}$$

We consider Jacobi and Gegenbauer polynomials as a generalization of Legendre, Hermite, Laguerre as well as Chebyshev polynomials. In fact, it can be shown that:

$$C_n^{\lambda}(x) = \frac{\Gamma(\lambda+1/2)\Gamma(n+2\lambda)}{\Gamma(2\lambda)\Gamma(n+\lambda+1/2)} P_n^{(\lambda-1/2,\lambda-1/2)}(x)$$

$$T_n(x) = \frac{n}{2}\lim_{\lambda \to 0}\frac{C_n^{\lambda}(x)}{\lambda}, \quad n \geq 1$$

$$U_n(x) = \sqrt{1-x^2}\, C_{n-1}^{1}(x)$$

$$L_n^{\alpha}(x) = \lim_{\beta \to \infty} P_n^{(\alpha,\beta)}(1-2x/\beta)$$

$$H_n(x) = n!\lim_{\lambda \to \infty}\lambda^{-n/2}C_n^{\lambda}(x/\sqrt{\lambda})$$

$$C_{n-m}^{m+1/2}(x) = \frac{1}{(2m-1)!!}\frac{d^m}{dx^m}P_n(x)$$

Exercise:

1. Show that:

i) $L_n^{\alpha+\beta+1}(x+y) = \sum_{k=0}^{\infty} L_k^{\alpha}(x)L_{n-k}^{\beta}(y)$

ii) $J_m(2\sqrt{xt}) = e^{-t}(xt)^{m/2}\sum_{k=0}^{\infty}\frac{L_n^m(x)}{(n+m)!}t^n$

iii) $\int_x^{\infty} e^{-t}L_n^m(t)dt = e^{-x}\left[L_n^m(x) - L_{n-1}^m(x)\right]$

iv) $n!\dfrac{d^m}{dx^m}\left[e^{-x}x^k L_n^k(x)\right] = (m+n)!e^{-x}x^{k-m}L_{m+n}^{k-m}(x)$

220

v) $\int_0^\infty e^{-x} x^{k+1} \left[L_n^k(x)\right]^2 dx = \dfrac{(n+k)!(2n+k+1)}{n!}$

vi) $x^n = \sum_{k=0}^{n} \dfrac{(-1)^k n!(m+1)_n L_k^m(x)}{(n-k)!(m+1)_k}$

vii) $\int_0^\infty e^{-st} L_n(t) dt = \dfrac{1}{s}\left(1-\dfrac{1}{s}\right)^n$

viii) $L_{2n}(2x) = n! \sum_{k=0}^{n} \dfrac{(-1)^k 2^{n-k} L_{n-1}(x)}{k!(n-k)!}$, $n > k$

2. Show that:

i) $\sqrt{1-x^2}\, T_n(x) = U_{n+1}(x) - xU_n(x)$

ii) $\sqrt{1-x^2}\, U_n(x) = xT_n(x) - T_{n+1}(x)$

iii) $T_{m+n}(x) + T_{m-n}(x) = 2T_m(x) T_n(x)$

iv) $T_n'(x) = nU_n(x)/\sqrt{1-x^2}$

v) $2T_n^2(x) = 1 + T_{2n}(x)$

vi) $T_n^2(x) - T_{n+1}(x) T_{n-1}(x) = 1 - x^2$

vii) $T_m[T_n(x)] = T_n[T_m(x)] = T_{nm}(x)$

3. Show that: $\dfrac{d^m}{dx^m} C_n^\lambda(x) = 2^m \dfrac{\Gamma(\lambda+m)}{\Gamma(\lambda)} C_{n-m}^{\lambda+m}(x)$.

4. Show that:

i) $P_n^{(\alpha,\beta)}(x) = \dfrac{(-1)^n}{2^n n!(1-x)^\alpha(1+x)^\beta} \dfrac{d^n}{dx^n}\left[(1-x)^{\alpha+n}(1+x)^{\beta+n}\right]$

ii) $P_n^{(\alpha,\beta)}(-x) = (-1)^n P_n^{(\beta,\alpha)}(x)$

iii) $P_n^{(\alpha,\beta)}(1) = \dfrac{\Gamma(\alpha+n+1)}{n!\Gamma(\alpha+1)}$

iv) $P_n^{(\alpha,\beta-1)}(x) - P_n^{(\alpha-1,\beta)}(x) = P_{n-1}^{(\alpha,\beta)}(x)$

221

Special Functions and Orthogonal Polynomials

Chapter Six

Bessel Functions

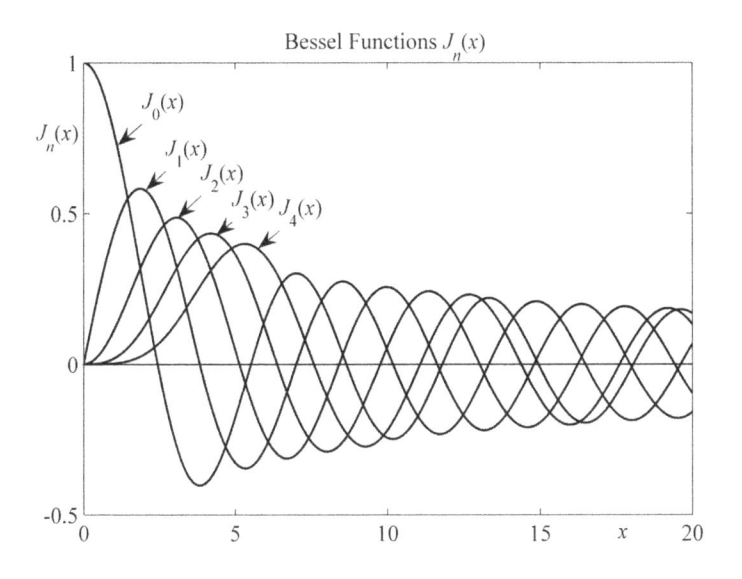

$$J_n(x) = \sum_{k=0}^{\infty} \frac{(-1)^k (x/2)^{n+2k}}{k!\,\Gamma(n+k+1)}$$

Special Functions and Orthogonal Polynomials

Chapter 6
Bessel Functions

6.1. Introduction

Bessel[29] *Functions* are solutions of the differential equation

$$x^2 \frac{d^2y}{dx^2} + x\frac{dy}{dx} + (x^2 - n^2)y = 0 \tag{1}$$

where n is a (possibly complex) constant. Equation (1) is known as *Bessel Differential equation* of order n. It arises in a variety of problems such as applications involving partial differential equations like the wave equation or the heat equation. It can be seen that the origin is a regular point of the equation, and all other values of x are ordinary points.

If we replace the independent variable x by λx, the resulting equation is

$$x^2 \frac{d^2y}{dx^2} + x\frac{dy}{dx} + (\lambda^2 x^2 - n^2)y = 0 \tag{2}$$

which is another form for *Bessel Differential equation*.

6.2. Solution of Bessel Differential equation

Using Frobenius method, we can obtain a series solutions for Equation (1) at the origin by assuming a solution of the form

$$y = \sum_{k=0}^{\infty} c_k x^{k+\alpha} \tag{3}$$

where the index α and the coefficients c_k, $k = 0, 1, 2, \cdots$ are to be determined. Arranging Equation (1) in the form

$$\left(x^2 \frac{d^2}{dx^2} + x\frac{dy}{dx} - n^2 \right) y + x^2 y = 0 \tag{4}$$

and substituting from (3), we get

---

29. Friedreich Wilhelm Bessel 1784-1846.

$$\sum_{k=0}^{\infty}\left[(\alpha+k)^2 - n^2\right]c_k x^{k+\alpha} + \sum_{k=0}^{\infty} c_k x^{k+\alpha+2} = 0 \qquad (5)$$

Equating the coefficients of all powers of x to zero, we get

$$\left.\begin{array}{l}(\alpha^2 - n^2)c_0 = 0 \\ [(\alpha+1)^2 - n^2]c_1 = 0 \\ [(\alpha+k)^2 - n^2]c_k + c_{k-2} = 0, k \geq 2\end{array}\right\} \qquad (6)$$

If c_0 is not zero, then from the first equation in (6) we have $\alpha = \pm n$. First take $\alpha = n$, the other Equations in (6) become

$$\left.\begin{array}{l}(2n+1)c_1 = 0 \\ k(2n+k)c_k + c_{k-2} = 0, k \geq 2\end{array}\right\} \qquad (7)$$

If <u>$2n$ is not a negative integer</u>, we get

$$c_1 = c_3 = \cdots = c_{2k-1} = \cdots = 0 \qquad (8)$$

$$c_{2k} = -\frac{c_{2k-2}}{2^2 k(n+k)} \qquad (9)$$

Hence,

$$c_2 = -\frac{c_0}{2^2 \cdot (n+1)},$$

$$c_4 = -\frac{c_2}{2^2 \cdot 2(n+2)} = \frac{c_0}{2^4 \cdot 2(n+1)(n+2)},$$

$$c_6 = -\frac{c_4}{2^2 \cdot 3(n+3)} = -\frac{c_0}{2^6 \cdot 3!(n+1)(n+2)(n+3)}, \cdots$$

And in general $c_{2k} = \frac{(-1)^k c_0}{2^{2k} \cdot k!(n+1)(n+2)\cdots(n+k)}.$

and one solution of Bessel differential equations is given by

$$y_1 = c_0 \sum_{k=0}^{\infty} \frac{(-1)^k x^{n+2k}}{2^{2k} \cdot k!(n+1)(n+2)\cdots(n+k)} \qquad (10)$$

Bessel Functions

Since c_0 is at our disposal, we may take $c_0 = \dfrac{1}{2^n \Gamma(n+1)}$. This will yield to the solution denoted by $J_n(x)$ and given by

$$J_n(x) = \frac{1}{2^n \Gamma(n+1)} \sum_{k=0}^{\infty} \frac{(-1)^k x^{n+2k}}{2^{2k} \cdot k!(n+1)(n+2)\cdots(n+k)} \tag{11}$$

or

$$J_n(x) = \sum_{k=0}^{\infty} \frac{(-1)^k (x/2)^{n+2k}}{k!\,\Gamma(n+k+1)} \tag{12}$$

where $J_n(x)$ is called the Bessel function of the first kind of order n and argument x.

Similarly, for $\alpha = -n$ in Equation (6) and <u>2n is not a positive integer</u>, we get

$$J_{-n}(x) = \sum_{k=0}^{\infty} \frac{(-1)^k (x/2)^{-n+2k}}{k!\,\Gamma(-n+k+1)} \tag{13}$$

For the values of n considered, (12) and (13) converge for all values of x except $x = 0$. Since $J_{-n}(x)$ contains negative powers of x while $J_n(x)$ does not, then near the origin $J_{-n}(x)$ is unbounded while $J_n(x)$ remains finite. Hence, when 2n is not an integer, $J_n(x)$ and $J_{-n}(x)$ are two linearly independent solutions of Bessel differential equation. Then a general solution of Bessel differential equation when <u>2n is not an integer</u> is

$$y(x) = A J_n(x) + B J_{-n}(x) \tag{14}$$

Where A and B are two arbitrary constants.

Next we use the definition of the Wronskian to prove that the two solutions are in fact linearly independent.

6.3. Independence of the Solutions $J_n(x)$ and $J_{-n}(x)$

Recall that two functions $f(x)$ and $g(x)$ are said to be linearly independent if the Wronskian does not vanish, i.e.,

$$W = \begin{vmatrix} f(x) & g(x) \\ f'(x) & g'(x) \end{vmatrix} \neq 0 \tag{15}$$

Now, since $J_n(x)$ and $J_{-n}(x)$ both satisfy Bessel Differential Equation (1),

then
$$x^2 J_n''(x) + x J_n'(x) + (x^2 - n^2) J_n(x) = 0$$
$$x^2 J_{-n}''(x) + x J_{-n}'(x) + (x^2 - n^2) J_{-n}(x) = 0$$

Multiplying the first equation by $J_{-n}(x)$ and the second equation by $J_n(x)$ and subtracting, we get

$$x \left[J_n(x) J_{-n}''(x) - J_{-n}(x) J_n''(x) \right] + \left[J_n(x) J_{-n}'(x) - J_{-n}(x) J_n'(x) \right] = 0 \quad (16)$$

or
$$\frac{d}{dx} \left\{ x \left[J_n(x) J_{-n}'(x) - J_{-n}(x) J_n'(x) \right] \right\} = 0 \quad (17)$$

i.e.,
$$\frac{d}{dx}(xW) = 0 \quad (18)$$

Hence, $W = c/x$, where c is a constant to be evaluated. To determine c, we have from Equations (12) and (13):

$$J_n(x) = \frac{(x/2)^n}{\Gamma(n+1)} \left[1 + O(x^2) \right] \quad (19)$$

and
$$J_n'(x) = \frac{(x/2)^{n-1}}{2\Gamma(n)} \left[1 + O(x^2) \right] \quad (20)$$

with similar expressions for $J_{-n}(x)$ and $J_{-n}'(x)$. $O(x^2)$ are terms in x of second degree and higher. From these equations and the definition of the Wronskian, we have,

$$W = \frac{1}{x} \left[\frac{1}{\Gamma(n+1)\Gamma(-n)} - \frac{1}{\Gamma(-n+1)\Gamma(n)} \right] + O(x) \quad (21)$$

But $\Gamma(1-n)\Gamma(n) = \dfrac{\pi}{\sin n\pi}$, then

$$W = -\frac{2 \sin n\pi}{\pi x} + O(x) \quad (22)$$

Comparing this result with $W = c/x$, it can observe that $O(x)$ must vanish and $c = -\dfrac{2 \sin n\pi}{\pi}$. Since <u>$2n$ is not an integer</u>, then $\sin n\pi \neq 0$ and $W \neq 0$.

The two functions $J_n(x)$ and $J_{-n}(x)$ are linearly independent provided that $2n$ in not an integer. We will show now that they are linearly dependent if n is an integer or zero.

Bessel Functions
Bessel Functions $J_n(x)$

Example 1: If n is an integer, show that $J_{-n}(x) = (-1)^n J_n(x)$.

Solution: From the properties of Gamma function, $\Gamma(-n+k+1)$ is infinite when $k = 0, 1, 2, ..., (n-1)$, then Equation (13) becomes

$$J_{-n}(x) = \sum_{k=n}^{\infty} \frac{(-1)^k (x/2)^{-n+2k}}{k!\,\Gamma(-n+k+1)}.$$

Let $k = n + s$, we get

$$J_{-n}(x) = \sum_{s=0}^{\infty} \frac{(-1)^{n+s}(x/2)^{n+2s}}{\Gamma(n+s+1)\Gamma(s+1)} = (-1)^n J_n(x) \qquad \square$$

Example 2: Show that: i) $J_{-1/2}(x) = \sqrt{\dfrac{2}{\pi x}} \cos x$ ii) $J_{1/2}(x) = \sqrt{\dfrac{2}{\pi x}} \sin x$

iii) $J_{1/2}^2(x) + J_{-1/2}^2(x) = \dfrac{2}{\pi x}$

Solution: We start from the expression of Bessel function

$$J_n(x) = \sum_{k=0}^{\infty} \frac{(-1)^k (x/2)^{n+2k}}{k!\,\Gamma(n+k+1)}.$$

229

Special Functions and Orthogonal Polynomials

i) Letting $n = -1/2$, we get

$$J_{-1/2}(x) = \sum_{k=0}^{\infty} \frac{(-1)^k (x/2)^{-1/2+2k}}{k! \Gamma(k+1/2)} = \sqrt{\frac{2}{x}} \sum_{k=0}^{\infty} \frac{(-1)^k x^{2k}}{2^{2k} k! \Gamma(k+1/2)}$$

From Legendre Duplication Formula of Gamma function,

$$\Gamma(k+1/2) = \frac{\sqrt{\pi}(2k-1)!}{2^{2k-1}(k-1)!}, \text{ we get}$$

$$J_{-1/2}(x) = \sqrt{\frac{2}{\pi x}} \sum_{k=0}^{\infty} \frac{(-1)^k 2^{2k-1}(k-1)! x^{2k}}{2^{2k} k!(2k-1)!}$$

$$= \sqrt{\frac{2}{\pi x}} \sum_{k=0}^{\infty} \frac{(-1)^k x^{2k}}{(2k)!}$$

And from Taylor's expansion, we have

$$\cos x = \sum_{k=0}^{\infty} \frac{(-1)^k x^{2k}}{(2k)!}, \text{ thus}$$

$$J_{-1/2}(x) = \sqrt{\frac{2}{\pi x}} \cos x.$$

ii) Letting $n = 1/2$, we get

$$J_{1/2}(x) = \sum_{k=0}^{\infty} \frac{(-1)^k (x/2)^{1/2+2k}}{k! \Gamma(k+3/2)} = \sqrt{\frac{2}{x}} \sum_{k=0}^{\infty} \frac{(-1)^k x^{2k+1}}{2^{2k+1} k! \Gamma(k+3/2)}$$

Again, from Legendre duplication formula of Gamma function,

$$\Gamma(k+3/2) = \frac{\sqrt{\pi}(2k+1)!}{2^{2k+1} k!}, \text{ we get}$$

$$J_{1/2}(x) = \sqrt{\frac{2}{\pi x}} \sum_{k=0}^{\infty} \frac{(-1)^k 2^{2k+1} k! x^{2k+1}}{2^{2k+1} k!(2k+1)!}$$

$$= \sqrt{\frac{2}{\pi x}} \sum_{k=0}^{\infty} \frac{(-1)^k x^{2k+1}}{(2k+1)!}$$

And from Taylor's expansion, we have

Bessel Functions

$$\sin x = \sum_{k=0}^{\infty} \frac{(-1)^k x^{2k+1}}{(2k+1)!}, \text{ thus}$$

$$J_{1/2}(x) = \sqrt{\frac{2}{\pi x}} \sin x .$$

iii) Squaring i) and ii) and adding, we get

$$J_{1/2}^2(x) + J_{-1/2}^2(x) = \frac{2}{\pi x}(\cos^2 x + \sin^2 x) = \frac{2}{\pi x} \qquad \square$$

Example 3: Show that:

$$J_n(x) = \frac{2(x/2)^{n-m}}{\Gamma(n-m)} \int_0^1 (1-t^2)^{n-m-1} t^{m+1} J_m(xt) dt, \quad n > m > -1$$

Solution: Let $I = \int_0^1 (1-t^2)^{n-m-1} t^{m+1} J_m(xt) dt$, then from the expression of Bessel function, we have

$$J_m(xt) = \sum_{k=0}^{\infty} \frac{(-1)^k (xt/2)^{m+2k}}{k! \Gamma(m+k+1)} .$$

The integral becomes

$$I = \int_0^1 (1-t^2)^{n-m-1} t^{m+1} \sum_{k=0}^{\infty} \frac{(-1)^k (xt/2)^{m+2k}}{k! \Gamma(m+k+1)} dt$$

$$= \sum_{k=0}^{\infty} \frac{(-1)^k (x/2)^{m+2k}}{k! \Gamma(m+k+1)} \int_0^1 (1-t^2)^{n-m-1} t^{2m+2k+1} dt$$

Letting $t^2 = y$ in the last integral, we obtain

$$I = \sum_{k=0}^{\infty} \frac{(-1)^k (x/2)^{m+2k}}{k! \Gamma(m+k+1)} \int_0^1 (1-y)^{n-m-1} y^{m+k} dy .$$

From the properties of Beta and Gamma functions, the last integral equals to

$$\int_0^1 (1-y)^{n-m-1} y^{m+k} dy = \frac{\Gamma(n-m)\Gamma(m+k+1)}{\Gamma(n+k+1)}, \text{ then}$$

231

Special Functions and Orthogonal Polynomials

$$I = \frac{\Gamma(n-m)(x/2)^{m-n}}{2} \sum_{k=0}^{\infty} \frac{(-1)^k (x/2)^{m+2k}}{k!\Gamma(m+k+1)}$$

$$= \frac{\Gamma(n-m)(x/2)^{m-n}}{2} J_n(x)$$

Thus,

$$J_n(x) = \frac{2(x/2)^{n-m}}{\Gamma(n-m)} \int_0^1 (1-t^2)^{n-m-1} t^{m+1} J_m(xt)dt, \quad n > m > -1 \;\square$$

Note: The product of two Bessel functions is given by (the proof is omitted)

$$J_n(x)J_m(x) = \sum_{k=0}^{\infty} \frac{(-1)^k \Gamma(n+m+2k+1)(x/2)^{n+m+2k}}{k!\Gamma(n+k+1)\Gamma(m+k+1)\Gamma(n+m+k+1)}$$

If $m = n$, then

$$J_n^2(x) = \sum_{k=0}^{\infty} \frac{(-1)^k \Gamma(2n+2k+1)(x/2)^{2n+2k}}{k!\Gamma^2(n+k+1)\Gamma(2n+k+1)}$$

Moreover, if n is a positive integer, we get

$$J_n^2(x) = \sum_{k=0}^{\infty} \frac{(-1)^k (2n+2k)!(x/2)^{2n+2k}}{k![(n+k)!]^2 (2n+k)!}$$

Example 4: Show that $\displaystyle\int_0^{\pi/2} J_1^2(x\sin\theta)\csc\theta\, d\theta = \frac{1}{2} - \frac{J_1(2x)}{2x}$

Solution: From the previous note, we have

$$J_1^2(x\sin\theta) = \sum_{k=0}^{\infty} \frac{(-1)^k (2k+2)!(x/2)^{2k+2} \sin^{2k+2}\theta}{k![(k+1)!]^2 (k+2)!}, \text{ then}$$

$$\int_0^{\pi/2} J_1^2(x\sin\theta)\csc\theta\, d\theta$$

$$= \sum_{k=0}^{\infty} \frac{(-1)^k (2k+2)!(x/2)^{2k+2}}{k![(k+1)!]^2 (k+2)!} \int_0^{\pi/2} \sin^{2k+1}\theta\, d\theta$$

232

Bessel Functions

But, from Wallis' Formula, we have

$$\int_0^{\pi/2} \sin^{2k+1}\theta\, d\theta = \frac{2k}{2k+1}\cdot\frac{2k-2}{2k-1}\cdots\frac{4}{5}\cdot\frac{2}{3}\cdot 1 = \frac{2^{2k}(k!)^2}{(2k+1)!}, \text{ then}$$

Then

$$\int_0^{\pi/2} J_1^2(x\sin\theta)\operatorname{cosec}\theta\, d\theta$$

$$= \sum_{k=0}^{\infty} \frac{(-1)^k (2k+2)!(x/2)^{2k+2}}{k![(k+1)!]^2(k+2)!}\cdot\frac{2^{2k}(k!)^2}{(2k+1)!}$$

$$= \sum_{k=0}^{\infty} \frac{(-1)^k(x/2)^{2k+2}}{(k+1)!(k+2)!} = \frac{1}{2} - \frac{1}{2}\sum_{m=0}^{\infty}\frac{(-1)^m x^{2m}}{m!(m+1)!}$$

From the expression for Bessel function, the last summation is nothing but $\dfrac{J_1(2x)}{2x}$, then we have

$$\int_0^{\pi/2} J_1^2(x\sin\theta)\operatorname{cosec}\theta\, d\theta = \frac{1}{2} - \frac{J_1(2x)}{2x}. \qquad \square$$

Example 5: Show that $J_n(x) = \dfrac{(x/2)^n}{\sqrt{\pi}\,\Gamma(n+1/2)}\displaystyle\int_0^{\pi}\cos(x\sin\theta)\cos^{2n}\theta\, d\theta$.

Solution: We have

$$I = \int_0^{\pi}\cos(x\sin\theta)\cos^{2n}\theta\, d\theta = 2\int_0^{\pi/2}\cos(x\sin\theta)\cos^{2n}\theta\, d\theta$$

$$= 2\int_0^{\pi/2}\cos^{2n}\theta\left(1 - \frac{x^2\sin^2\theta}{2!} + \frac{x^4\sin^4\theta}{4!} - \cdots\right)d\theta$$

From the properties of Beta function, we have

$$2\int_0^{\pi/2}\cos^{2p-1}\theta\sin^{2q-1}\theta\, d\theta = \beta(p,q) = \frac{\Gamma(p)\Gamma(q)}{\Gamma(p+q)}$$

Using this property, the integral becomes

233

Special Functions and Orthogonal Polynomials

$$I = \frac{\Gamma(n+1/2)\Gamma(1/2)}{\Gamma(n+1)} - \frac{x^2}{2!}\frac{\Gamma(n+1/2)\Gamma(3/2)}{\Gamma(n+2)} + \frac{x^4}{4!}\frac{\Gamma(n+1/2)\Gamma(5/2)}{\Gamma(n+3)} - \cdots$$

$$= \frac{\Gamma(n+1/2)\sqrt{\pi}}{\Gamma(n+1)}\left[1 - \frac{x^2}{4(n+1)} + \frac{x^4}{4\cdot 8(n+1)(n+2)} - \cdots\right]$$

Multiplying both sides by $\dfrac{x^n}{2^n\,\Gamma(n+1/2)\sqrt{\pi}}$, we get

$$\frac{x^n}{2^n\,\Gamma(n+1/2)\sqrt{\pi}} I = \frac{x^n}{2^n\,\Gamma(n+1)}\left[1 - \frac{x^2}{4(n+1)} + \frac{x^4}{4\cdot 8(n+1)(n+2)} - \cdots\right]$$

But $J_n(x) = \dfrac{x^n}{2^n\,\Gamma(n+1)}\left[1 - \dfrac{x^2}{4(n+1)} + \dfrac{x^4}{4\cdot 8(n+1)(n+2)} - \cdots\right]$, then

$$J_n(x) = \frac{(x/2)^n}{\sqrt{\pi}\,\Gamma(n+1/2)}\int_0^{\pi}\cos(x\sin\theta)\cos^{2n}\theta\,d\theta. \qquad \square$$

We can also show by the same approach that:

$$J_n(x) = \frac{(x/2)^n}{\sqrt{\pi}\,\Gamma(n+1/2)}\int_0^{\pi}\cos(x\cos\theta)\sin^{2n}\theta\,d\theta.$$

Example 6: Show that $x^n J_n(x)$ is a solution of the differential equation:

$$x^2\frac{d^2y}{dx^2} + (1-2n)\frac{dy}{dx} + xy = 0.$$

Solution: Let $y = x^n J_n(x)$, then $y' = x^n J_n'(x) + nx^{n-1}J_n(x)$ and

$$y'' = x^n J_n''(x) + 2nx^{n-1}J_n'(x) + n(n-1)x^{n-2}J_n(x)$$

Substitution in the left hand side of the differential equation, we obtain

$$x^{n+1}J_n''(x) + 2nx^n J_n'(x) + n(n-1)x^{n-1}J_n(x) + x^n J_n'(x)$$

$$+ nx^{n-1}J_n(x) - 2nx^n J_n'(x) - 2n^2 x^{n-1}J_n(x) + x^{n+1}J_n(x)$$

$$= \underbrace{x^2 J_n''(x) + xJ_n'(x) + (x^2 - n^2)J_n(x)}_{J_n(x)\text{ is a solution of Bessel equation}} \triangleq 0 \qquad \square$$

234

Bessel Functions

Example 7: Show that $\int_0^{\pi/2} J_1(x\cos\theta)d\theta = \dfrac{1-\cos x}{x}$.

Solution: From the summation expression of Bessel Function

$$J_n(x) = \sum_{k=0}^{\infty} \frac{(-1)^k (x/2)^{n+2k}}{k!\,\Gamma(n+k+1)}.$$

Let $n=1$, we have $J_1(x) = \sum_{k=0}^{\infty} \dfrac{(-1)^k (x/2)^{2k+1}}{k!(k+1)!}$.

Replacing x by $x\cos\theta$, we get

$$J_1(x\cos\theta) = \sum_{k=0}^{\infty} \frac{(-1)^k (x/2)^{2k+1}}{k!(k+1)!}\cos^{2k+1}\theta.$$

Integrating with respect to θ from 0 to $\pi/2$, we obtain

$$\int_0^{\pi/2} J_1(x\cos\theta)d\theta = \sum_{k=0}^{\infty} \frac{(-1)^k (x/2)^{2k+1}}{k!(k+1)!}\int_0^{\pi/2}\cos^{2k+1}\theta\, d\theta.$$

But, from Walli's Formula, we have

$$\int_0^{\pi/2}\cos^{2k+1}\theta\, d\theta = \frac{2k}{2k+1}\cdot\frac{2k-2}{2k-1}\cdots\frac{4}{5}\cdot\frac{2}{3}\cdot 1 = \frac{2^{2k}(k!)^2}{(2k+1)!},$$

We obtain

$$\int_0^{\pi/2} J_1(x\cos\theta)d\theta = \sum_{k=0}^{\infty} \frac{(-1)^k (x/2)^{2k+1}}{k!(k+1)!}\cdot\frac{2^{2k}(k!)^2}{(2k+1)!}$$

$$= \sum_{k=0}^{\infty}\frac{(-1)^k x^{2k+1}}{2(k+1)(2k+1)!} = \frac{1}{x}\sum_{k=0}^{\infty}\frac{(-1)^k x^{2k+2}}{(2k+2)!}$$

$$= \frac{1}{x}\left[\frac{x^2}{2!} - \frac{x^4}{4!} + \frac{x^6}{6!} - \cdots\right]$$

$$= \frac{1}{x}\left[1 - \left\{1 - \frac{x^2}{2!} + \frac{x^4}{4!} - \frac{x^6}{6!} - \cdots\right\}\right] = \frac{1-\cos x}{x}.\;\square$$

Special Functions and Orthogonal Polynomials

Example 8: Show that:

i. $J_n(x) = \dfrac{(x/2)^n}{\sqrt{\pi}\Gamma(n+1/2)} \displaystyle\int_{-1}^{1}(1-t^2)^{n-1/2}e^{ixt}\,dt$, $n \geq -1/2$.

ii. $J_n(x) = \dfrac{x^n}{2^{n-1}\sqrt{\pi}\Gamma(n+1/2)} \displaystyle\int_{0}^{1}(1-t^2)^{n-1/2}\cos(xt)\,dt$

Solution: i. Let $I = \displaystyle\int_{-1}^{1}(1-t^2)^{n-1/2}e^{ixt}\,dt$, also, we have

$$e^{ixt} = 1 + ixt + \frac{(ixt)^2}{2!} + \cdots + \frac{(ixt)^k}{k!} + \cdots = \sum_{k=0}^{\infty}\frac{(ixt)^k}{k!}.$$ Then

$$I = \int_{-1}^{1}(1-t^2)^{n-1/2}\sum_{k=0}^{\infty}\frac{(ixt)^k}{k!}dt = \sum_{k=0}^{\infty}\frac{(ix)^k}{k!}\underbrace{\int_{-1}^{1}(1-t^2)^{n-1/2}t^k\,dt}_{I_1}.$$

If k is odd, the integrand is an odd function and the integral vanishes. On the other hand, If k is even, say $2m$, we have

$$I_1 = \int_{-1}^{1}(1-t^2)^{n-1/2}t^{2m}\,dt = 2\int_{0}^{1}(1-t^2)^{n-1/2}t^{2m}\,dt.$$

Let $t^2 = y$, we can write

$$I_1 = 2\int_{0}^{1}(1-y)^{n-1/2}y^m \cdot \frac{dy}{2\sqrt{y}} = \int_{0}^{1}(1-y)^{n-1/2}y^{m-1/2}dy$$
$$= \beta(n+1/2, m+1/2) = \frac{\Gamma(n+1/2)\Gamma(m+1/2)}{\Gamma(n+m+1)}$$

Thus, $I = \displaystyle\sum_{m=0}^{\infty}\frac{(-1)^m x^{2m}}{(2m)!}\cdot\frac{\Gamma(n+1/2)\Gamma(m+1/2)}{\Gamma(n+m+1)}$.

From Legendre Duplication Formula: $\Gamma(m+1/2) = \dfrac{(2m)!\sqrt{\pi}}{2^{2m}m!}$, we obtain

Bessel Functions

$$I = \Gamma(n+1/2) \sum_{m=0}^{\infty} \frac{(-1)^m (2m)! \sqrt{\pi} x^{2m}}{(2m)! 2^{2m} m! \Gamma(n+m+1)}$$

$$= \sqrt{\pi}\Gamma(n+1/2) \sum_{m=0}^{\infty} \frac{(-1)^m (x/2)^{2m}}{m! \Gamma(n+m+1)}$$

$$= \sqrt{\pi}\Gamma(n+1/2)(x/2)^{-n} \underbrace{\sum_{m=0}^{\infty} \frac{(-1)^m (x/2)^{n+2m}}{m! \Gamma(n+m+1)}}_{J_n(x)}$$

Therefore

$$J_n(x) = \frac{(x/2)^n}{\sqrt{\pi}\Gamma(n+1/2)} \int_{-1}^{1} (1-t^2)^{n-1/2} e^{ixt} dt, \quad n \geq -1/2.$$

ii. From Euler's formula, we have $e^{ixt} = \cos(xt) + i \sin(xt)$. Then

$$J_n(x) = \frac{(x/2)^n}{\sqrt{\pi}\Gamma(n+1/2)} \int_{-1}^{1} (1-t^2)^{n-1/2} \left[\cos(xt) + i \sin(xt)\right] dt.$$

The second term in the integrand is an odd function and vanishes on integration, while the first term is an even function, then we have

$$J_n(x) = \frac{x^n}{2^{n-1}\sqrt{\pi}\Gamma(n+1/2)} \int_{0}^{1} (1-t^2)^{n-1/2} \cos(xt) dt. \quad \square$$

Example 9: Show that $\lim_{x \to 0} \frac{J_n(x)}{x^n} = \frac{1}{2^n \Gamma(n+1)}$.

Solution: From the expression of Bessel function, we have

$$J_n(x) = \frac{x^n}{2^n \Gamma(n+1)} \left[1 - \frac{x^2}{4(n+1)} + \frac{x^4}{4 \cdot 8(n+1)(n+2)} - \cdots \right].$$

Therefore $\lim_{x \to 0} \frac{J_n(x)}{x^n} = \frac{1}{2^n \Gamma(n+1)}$. $\quad \square$

6.4. Solution of Bessel's equation when n is an integer

Consider the function $Y_n(x)$ defined by

$$Y_n(x) = \frac{(\cos n\pi)J_n(x) - J_{-n}(x)}{\sin n\pi} \tag{23}$$

It is clear that the expression on the right hand side of Equation (23) is a linear combination of $J_n(x)$ and $J_{-n}(x)$. Then, the general solution of Bessel's equation can take the form

$$y(x) = AJ_n(x) + BY_n(x) \tag{24}$$

whenever n is not an integer or zero.

When n is an integer or zero, the expression on the right of (23) takes the form (0/0), but

$$\lim_{k \to n} \frac{(\cos k\pi)J_k(x) - J_{-k}(x)}{\sin k\pi} \tag{25}$$

exists. Using l'Hôpital rule, we get

$$Y_n(x) = \lim_{k \to n} \frac{(-\pi \sin k\pi)J_k(x) + (\cos k\pi)\frac{\partial}{\partial k}J_k(x) - \frac{\partial}{\partial k}J_{-k}(x)}{\pi \cos k\pi} \tag{26}$$

or

$$Y_n(x) = \frac{1}{\pi}\left[\frac{\partial}{\partial k}J_k(x) - (-1)^n \frac{\partial}{\partial k}J_{-k}(x)\right]_{k=n} \tag{27}$$

Now, to show that $Y_n(x)$ satisfies Bessel Differential Equation when n is an integer or zero we proceed as follows. Since $J_{\pm n}(x)$ both satisfy Bessel's equation (1),

$$\left(x^2 \frac{d^2}{dx^2} + x\frac{dy}{dx} + x^2 - n^2\right)J_{\pm n}(x) = 0 \tag{28}$$

then, differentiating with respect to n, we obtain

$$\left(x^2 \frac{d^2}{dx^2} + x\frac{dy}{dx} + x^2 - n^2\right)\frac{\partial}{\partial n}J_{\pm n}(x) = 2nJ_{\pm n}(x) \tag{29}$$

Hence, equation (1.27) gives

$$\left(x^2 \frac{d^2}{dx^2} + x\frac{dy}{dx} + x^2 - n^2\right)Y_n(x) = \left[2kJ_k(x) - (-1)^n 2kJ_{-k}(x)\right]_{k=n}$$

$$= 2n\left[J_n(x) - (-1)^n J_{-n}(x)\right] = 0. \tag{30}$$

Hence $Y_n(x)$ satisfies Bessel's differential equation.

Bessel Functions

The function $Y_n(x)$ is called the *Bessel function of the second kind* of order n and argument x.

If n is integer, we can give an explicit expression of $Y_n(x)$. The proof is omitted.

$$Y_n(x) = \frac{2}{\pi}\left(\ln\frac{x}{2} + \gamma - \frac{1}{2}\sum_{k=1}^{n-1}\frac{1}{k}\right)J_n(x) - \frac{1}{\pi}\sum_{m=0}^{n-1}\frac{(n-m-1)!}{m!}\left(\frac{x}{2}\right)^{-n+2m}$$

$$-\frac{1}{\pi}\sum_{m=0}^{\infty}\frac{(-1)^m}{m!(n+m)!}\left(\frac{x}{2}\right)^{n+2m}\sum_{k=1}^{m}\left(\frac{1}{k}+\frac{1}{k+n}\right)$$

$Y_n(x)$ is sometimes called the *Neumann function of order n and argument x*, and is denoted by $N_n(x)$. One last word, $J_n(x)$ is always finite at $x=0$, while $Y_n(x)$ is always infinite at $x=0$.

Now, If n is an integer, the general solution of Bessel differential equation takes the form

$$y(x) = AJ_n(x) + BY_n(x)$$

To summarize, we can say that

$$Y_n(x) = \frac{(\cos n\pi)J_n(x) - J_{-n}(x)}{\sin n\pi} \quad \text{if } n \text{ is not an integer}$$

and

$$Y_n(x) = \lim_{k \to n} \frac{(\cos k\pi)J_k(x) - J_{-k}(x)}{\sin k\pi} \quad \text{if } n \text{ is an integer.}$$

Example 10: If n is a positive integer or zero, show that:

i. $Y_{n+1/2}(x) = (-1)^{n+1}J_{-n-1/2}(x)$;

ii. $Y_{-n-1/2}(x) = (-1)^{n+1}J_{n+1/2}(x)$.

Solution: i. We have $Y_n(x) = \frac{(\cos n\pi)J_n(x) - J_{-n}(x)}{\sin n\pi}$

Replace n by $(n+1/2)$, we get

$$Y_{n+1/2}(x) = \frac{\cos(n\pi + \pi/2)J_{n+1/2}(x) - J_{-n-1/2}(x)}{\sin(n\pi + \pi/2)}$$

$$= (-1)^{n+1}J_{-n-1/2}(x)$$

Special Functions and Orthogonal Polynomials

ii. Replace n by $-(n+1/2)$, we get

$$Y_{-n-1/2}(x) = \frac{\cos(n\pi + \pi/2)J_{-n-1/2}(x) - J_{n+1/2}(x)}{-\sin(n\pi + \pi/2)}$$

$$= (-1)^{n+1}J_{n+1/2}(x) \qquad \square$$

Bessel Functions $Y_n(x)$

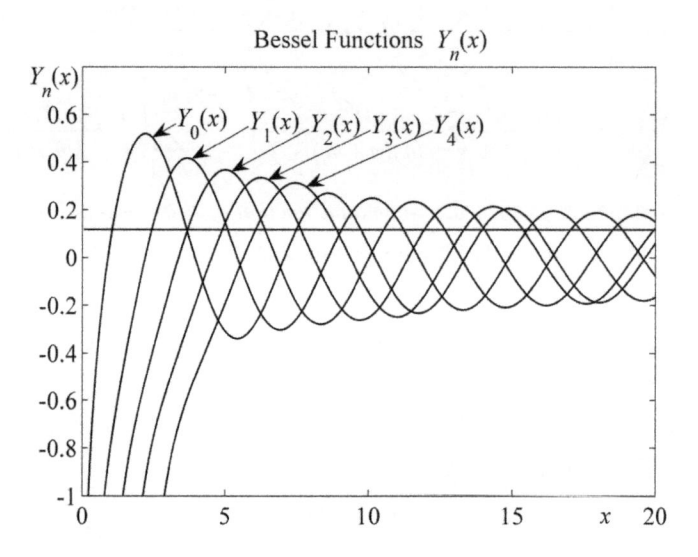

[Graph showing $Y_0(x)$, $Y_1(x)$, $Y_2(x)$, $Y_3(x)$, $Y_4(x)$ on axes with $Y_n(x)$ ranging from -1 to 0.6 and x from 0 to 20.]

***Example* 10**: If n is a positive integer or zero, show that:

i. $Y_{n+1/2}(x) = (-1)^{n+1}J_{-n-1/2}(x)$;

ii. $Y_{-n-1/2}(x) = (-1)^{n+1}J_{n+1/2}(x)$.

Solution: i. We have $Y_n(x) = \dfrac{(\cos n\pi)J_n(x) - J_{-n}(x)}{\sin n\pi}$

Replace n by $(n+1/2)$, we get

$$Y_{n+1/2}(x) = \frac{\cos(n\pi + \pi/2)J_{n+1/2}(x) - J_{-n-1/2}(x)}{\sin(n\pi + \pi/2)}$$

$$= (-1)^{n+1}J_{-n-1/2}(x)$$

ii. Replace n by $-(n+1/2)$, we get

$$Y_{-n-1/2}(x) = \frac{\cos(n\pi + \pi/2)J_{-n-1/2}(x) - J_{n+1/2}(x)}{-\sin(n\pi + \pi/2)}$$

$$= (-1)^{n+1}J_{n+1/2}(x) \qquad \square$$

Bessel Functions

Example 11: Find a series representation for $Y_0(x)$.

Solution: From Equation (27), let $n = 0$, we get

$$\pi Y_0(x) = \left[\frac{\partial}{\partial k}J_k(x) - (-1)^n \frac{\partial}{\partial k}J_{-k}(x)\right]_{k=0} = 2\left[\frac{\partial}{\partial k}J_k(x)\right]_{k=0}$$

From Equation (12), differentiating with respect to n, we get

$$\frac{\partial}{\partial n}J_n(x) = \sum_{k=0}^{\infty} \frac{(-1)^k (x/2)^{n+2k}}{k!\Gamma(n+k+1)}\left[\ln(x/2) - \frac{\Gamma'(n+k+1)}{\Gamma(n+k+1)}\right]$$

Hence

$$\pi Y_0(x) = 2\sum_{k=0}^{\infty} \frac{(-1)^k (x/2)^{2k}}{k!k!}\left[\ln(x/2) - \frac{\Gamma'(k+1)}{\Gamma(k+1)}\right]$$

It is known that

$$\frac{\Gamma'(k+1)}{\Gamma(k+1)} = -\gamma + 1 + \frac{1}{2} + \frac{1}{3} + \cdots + \frac{1}{k}, \quad \frac{\Gamma'(1)}{\Gamma(1)} = -\gamma$$

where $\gamma = 0.5772...$ is Euler's constant, then

$$\frac{\pi}{2}Y_0(x) = (\gamma + \ln(x/2))J_0(x) - \sum_{k=1}^{\infty} \frac{(-1)^k (x/2)^{2k}}{k!k!}\left[1 + \frac{1}{2} + \frac{1}{3} + \cdots + \frac{1}{k}\right] \square$$

Example 12: If n is an integer, show that $Y_{-n}(x) = (-1)^n Y_n(x)$.

Solution: From Equation (27), we have

$$Y_{-n}(x) = \frac{1}{\pi}\left[\frac{\partial}{\partial k}J_k(x) - (-1)^n \frac{\partial}{\partial k}J_{-k}(x)\right]_{k=-n}$$

Replacing k by $-k$

$$Y_{-n}(x) = \frac{1}{\pi}\left[\frac{\partial}{\partial(-k)}J_{-k}(x) - (-1)^n \frac{\partial}{\partial(-k)}J_k(x)\right]_{k=n}$$

$$= \frac{1}{\pi}\left[-\frac{\partial}{\partial k}J_{-k}(x) + (-1)^n \frac{\partial}{\partial k}J_k(x)\right]_{k=n}$$

$$= (-1)^n \frac{1}{\pi}\left[\frac{\partial}{\partial k}J_k(x) - (-1)^n \frac{\partial}{\partial k}J_{-k}(x)\right]_{k=n}$$

$$= (-1)^n Y_n(x) \quad \square$$

6.5. Generating Function for Bessel Functions

Bessel functions of the first kind $J_n(x)$ satisfy the generating expression

$$\boxed{e^{\frac{x}{2}(t-1/t)} = \sum_{n=-\infty}^{\infty} t^n J_n(x)} \qquad (31)$$

To prove this expression, we proceed as follows. Provided that t is not zero, the exponential functions $e^{xt/2}$ and $e^{-x/2t}$ can be expanded in powers of t using MacLaurin's expansion to give

$$e^{\frac{x}{2}(t-1/t)} = e^{xt/2} \cdot e^{-x/2t} = \sum_{k=0}^{\infty} \frac{(xt/2)^k}{k!} \cdot \sum_{m=0}^{\infty} \frac{(-x/2t)^m}{m!}$$

$$= \sum_{k=0}^{\infty} \sum_{m=0}^{\infty} (-1)^m (1/2)^{k+m} \frac{x^{k+m} t^{k-m}}{k!\, m!}$$

It will be sufficient to show that the coefficient of t^n in this double summation is $J_n(x)$. First, if n is a positive integer or zero, the coefficient of t^n can be found by taking $k = n+m$ and letting m vary from 0 to ∞, then the coefficient of t^n is

$$\sum_{m=0}^{\infty} \frac{(-1)^m (x/2)^{n+2m}}{(n+m)!\, m!} = J_n(x)$$

The coefficient of t^{-n} is found by taking $k = -n+m$ and letting m vary from n to ∞, then the coefficient of t^{-n} is

$$\sum_{m=0}^{\infty} \frac{(-1)^m (x/2)^{-n+2m}}{(-n+m)!\, m!} = J_{-n}(x) = (-1)^n J_n(x)$$

Then, the total coefficient of t^n for n varying from $-\infty$ to ∞ is $J_n(x)$ and

$$e^{\frac{x}{2}(t-1/t)} = \sum_{n=-\infty}^{\infty} t^n J_n(x) \qquad \square$$

Bessel Functions

Example 13: If n is an integer, show that $J_n(x+y) = \sum_{k=-\infty}^{\infty} J_k(x) J_{n-k}(y)$.

Solution: From the generating function, we have

$$e^{\frac{(x+y)}{2}(t-1/t)} = \sum_{n=-\infty}^{\infty} t^n J_n(x+y)$$

Then $J_n(x+y)$ is the coefficient of t^n in the expression $e^{\frac{(x+y)}{2}(t-1/t)}$. Now,

$$e^{\frac{(x+y)}{2}(t-1/t)} = e^{\frac{x}{2}(t-1/t)} e^{\frac{y}{2}(t-1/t)}$$

$$= \sum_{k=-\infty}^{\infty} t^k J_k(x) \sum_{m=-\infty}^{\infty} t^m J_m(y)$$

$$= \sum_{k=-\infty}^{\infty} \sum_{m=-\infty}^{\infty} J_k(x) J_m(y) t^{k+m}$$

For a particular value of k, the coefficient of t^n is found by taking $m = n - k$, then the total coefficient of t^n is obtained by letting k vary from $-\infty$ to ∞ to get:

coefficient of $t^n = \sum_{k=-\infty}^{\infty} J_k(x) J_{n-k}(y)$. Then

$$J_n(x+y) = \sum_{k=-\infty}^{\infty} J_k(x) J_{n-k}(y) \qquad \square$$

6.6. Integral Form of Bessel Functions

If n is integer, then from the generating function, we have

$$e^{\frac{x}{2}(t-1/t)} = J_0(x) + \sum_{n=1}^{\infty}\left[t^n + (-1)^n t^{-n}\right] J_n(x) \qquad (32)$$

Letting $t = e^{i\theta}$, then

$$t - \frac{1}{t} = e^{i\theta} - e^{-i\theta} = 2i\sin\theta,$$

and Equation (32) becomes

$$e^{ix\sin\theta} = J_0(x) + \sum_{n=1}^{\infty}\left[e^{in\theta} + (-1)^n e^{-in\theta}\right] J_n(x).$$

But $\quad e^{in\theta} + (-1)^n e^{-in\theta} = \begin{cases} e^{in\theta} + e^{-in\theta} = 2\cos n\theta & \text{if } n \text{ is even} \\ e^{in\theta} - e^{-in\theta} = 2i\sin n\theta & \text{if } n \text{ is odd} \end{cases}$

Then, we have

$$e^{ix\sin\theta} = J_0(x) + \sum_{k=1}^{\infty} 2\cos(2k\theta)J_{2k}(x) + i\,2\sin[(2k-1)\theta]J_{2k-1}(x)$$

Equating the real and imaginary parts in both sides of this equation, we obtain

$$\cos(x\sin\theta) = J_0(x) + 2\sum_{k=1}^{\infty}\cos(2k\theta)J_{2k}(x) \qquad (33)$$

$$\sin(x\sin\theta) = 2\sum_{k=1}^{\infty}\sin[(2k-1)\theta]J_{2k-1}(x) \qquad (34)$$

These two expressions are known as *Jacobi Series*. Multiplying both sides of Equation (33) by $\cos n\theta$, $n \geq 0$, and both sides of Equation (34) by $\sin n\theta$, $n \geq 1$, and integrating with respect to θ from 0 to π, noting that

$$\int_0^\pi \cos m\theta \cos n\theta \, d\theta = \begin{cases} 0 & m \neq n \\ \pi/2 & m = n \neq 0 \\ \pi & m = n = 0 \end{cases}$$

Bessel Functions

and
$$\int_0^\pi \sin m\theta \sin n\theta \, d\theta = \begin{cases} 0 & m \neq n \\ \pi/2 & m = n \neq 0 \end{cases}$$

we obtain

$$\int_0^\pi \cos n\theta \cos(x \sin \theta) \, d\theta = \begin{cases} \pi J_n(x) & n \text{ even} \\ 0 & n \text{ odd} \end{cases}$$

and
$$\int_0^\pi \sin n\theta \sin(x \sin \theta) \, d\theta = \begin{cases} 0 & n \text{ even} \\ \pi J_n(x) & n \text{ odd} \end{cases}$$

Adding these last two equations, we get

$$\int_0^\pi [\cos n\theta \cos(x \sin \theta) + \sin n\theta \sin(x \sin \theta)] \, d\theta = \pi J_n(x)$$

or upon using trigonometric identity, we get

$$J_n(x) = \frac{1}{\pi} \int_0^\pi \cos(n\theta - x \sin \theta) \, d\theta \quad (35)$$

It can also be shown that

$$J_{-n}(x) = \frac{(-1)^n}{\pi} \int_0^\pi \cos(n\theta - x \sin \theta) \, d\theta \quad (36)$$

Example 14: Show that:

i. $\cos(x \cos \theta) = J_0(x) + 2 \sum_{k=1}^{\infty} (-1)^k \cos(2k\theta) J_{2k}(x)$;

ii. $\sin(x \cos \theta) = 2 \sum_{k=1}^{\infty} (-1)^{k+1} \cos[(2k-1)\theta] J_{2k-1}(x)$;

iii. $\cos x = J_0(x) + 2 \sum_{k=1}^{\infty} (-1)^k J_{2k}(x)$;

iv. $\sin x = 2 \sum_{k=1}^{\infty} (-1)^{k+1} J_{2k-1}(x)$

Solution: i. Replacing θ by $\pi/2-\theta$ in Equation (33), we obtain

$$\cos[x\,\sin(\pi/2-\theta)] = J_0(x) + 2\sum_{k=1}^{\infty} \cos[2k\,(\pi/2-\theta)]J_{2k}(x)$$

But $\sin(\pi/2-\theta) = \cos\theta$,

and $\cos[2k\,(\pi/2-\theta)] = (-1)^k \cos(2k\,\theta)$, then

$$\cos(x\,\cos\theta) = J_0(x) + 2\sum_{k=1}^{\infty} (-1)^k \cos(2k\,\theta)J_{2k}(x)$$

ii. Replacing θ by $\pi/2-\theta$ in Equation (34), we obtain

$$\sin[x\,\sin(\pi/2-\theta)] = 2\sum_{k=1}^{\infty} \sin[(2k-1)(\pi/2-\theta)]J_{2k-1}(x)$$

But $\sin(\pi/2-\theta) = \cos\theta$,

and $\sin[(2k-1)(\pi/2-\theta)] = (-1)^{k+1} \cos[(2k-1)\theta]$, then

$$\sin(x\,\cos\theta) = 2\sum_{k=1}^{\infty} (-1)^{k+1} \cos[(2k-1)\theta]J_{2k-1}(x)$$

iii. Replacing θ by 0 in i., we obtain

$$\cos x = J_0(x) + 2\sum_{k=1}^{\infty} (-1)^k J_{2k}(x)$$

iv. Replacing θ by 0 in ii., we obtain

$$\sin x = 2\sum_{k=1}^{\infty} (-1)^{k+1} J_{2k-1}(x). \qquad \square$$

Example 15: Show that $\displaystyle\int_0^{\infty} e^{-\alpha x} J_0(\beta x)\,dx = \frac{1}{\sqrt{\alpha^2+\beta^2}}$, $\alpha > 0$.

Solution: Let $I = \displaystyle\int_0^{\infty} e^{-\alpha x} J_0(\beta x)\,dx$. From the integral form of Bessel function (Equation(35)), letting $n = 0$ and replacing x by βx, we obtain

Bessel Functions

$$J_0(\beta x) = \frac{1}{\pi} \int_0^\pi \cos(\beta x \sin\theta) d\theta.$$

Then the integral becomes

$$I = \frac{1}{\pi} \int_0^\infty e^{-\alpha x} \left\{ \int_0^\pi \cos(\beta x \sin\theta) d\theta \right\} dx$$

Interchanging the order of integration (allowed), we obtain

$$I = \frac{1}{\pi} \int_0^\pi \left\{ \int_0^\infty e^{-\alpha x} \cos(\beta x \sin\theta) dx \right\} d\theta$$

$$= \frac{1}{2\pi} \int_0^\pi \left\{ \int_0^\infty e^{-\alpha x} \left(e^{i\beta x \sin\theta} - e^{-i\beta x \sin\theta} \right) dx \right\} d\theta$$

$$= \frac{1}{2\pi} \int_0^\pi \left[\frac{e^{-(\alpha - i\beta\sin\theta)x}}{-(\alpha - i\beta\sin\theta)} - \frac{e^{-(\alpha + i\beta\sin\theta)x}}{-(\alpha + i\beta\sin\theta)} \right]_0^\infty d\theta$$

$$= \frac{1}{2\pi} \int_0^\pi \frac{2\alpha}{\alpha^2 + \beta^2 \sin^2\theta} d\theta = \cdots = \frac{1}{\sqrt{\alpha^2 + \beta^2}}$$

Therefore

$$\int_0^\infty e^{-\alpha x} J_0(\beta x) dx = \frac{1}{\sqrt{\alpha^2 + \beta^2}}, \quad \alpha > 0.$$

Note that, as $\alpha \to 0$, we have

$$\int_0^\infty J_0(\beta x) dx = \frac{1}{\beta}. \qquad \square$$

247

6.7. Recurrence Relations for Bessel Functions

Bessel function of the first kind $J_n(x)$ satisfies the following recurrence relations:

I. $\boxed{\dfrac{d}{dx}\{x^n J_n(x)\} = x^n J_{n-1}(x)}$

Proof: From the expression of Bessel function

$$J_n(x) = \sum_{k=0}^{\infty} \frac{(-1)^k (x/2)^{n+2k}}{k!\,\Gamma(n+k+1)}, \text{ we have}$$

$$\frac{d}{dx}\{x^n J_n(x)\} = \frac{d}{dx}\left\{ x^n \sum_{k=0}^{\infty} \frac{(-1)^k (x/2)^{n+2k}}{k!\,\Gamma(n+k+1)} \right\}$$

$$= \sum_{k=0}^{\infty} \frac{(-1)^k}{2^{n+2k}\, k!\,\Gamma(n+k+1)} \frac{d}{dx} x^{2n+2k}$$

$$= \sum_{k=0}^{\infty} \frac{(-1)^k\, 2(n+k) x^{2n+2k-1}}{2^{n+2k}\, k!\,\Gamma(n+k+1)}$$

$$= x^n \sum_{k=0}^{\infty} \frac{(-1)^k (x/2)^{n+2k-1}}{k!\,\Gamma(n+k)} = x^n J_{n-1}(x) \qquad \square$$

II. $\boxed{\dfrac{d}{dx}\{x^{-n} J_n(x)\} = -x^{-n} J_{n+1}(x)}$

Proof: From the expression of Bessel function

$$J_n(x) = \sum_{k=0}^{\infty} \frac{(-1)^k (x/2)^{n+2k}}{k!\,\Gamma(n+k+1)}, \text{ we have}$$

$$\frac{d}{dx}\{x^{-n} J_n(x)\} = \frac{d}{dx}\left\{ x^{-n} \sum_{k=0}^{\infty} \frac{(-1)^k (x/2)^{n+2k}}{k!\,\Gamma(n+k+1)} \right\}$$

$$= \sum_{k=0}^{\infty} \frac{(-1)^k}{2^{n+2k}\, k!\,\Gamma(n+k+1)} \frac{d}{dx} x^{2k} = \sum_{k=1}^{\infty} \frac{(-1)^k\, x^{2k-1}}{2^{n+2k-1}(k-1)!\,\Gamma(n+k+1)}$$

$$= \sum_{k=0}^{\infty} \frac{(-1)^{k+1} x^{2k+1}}{2^{n+2k+1} k!\,\Gamma(n+k+2)} = -x^{-n} \sum_{k=0}^{\infty} \frac{(-1)^k (x/2)^{n+2k+1}}{k!\,\Gamma(n+k+2)}$$

$$= -x^{-n} J_{n+1}(x) \qquad \square$$

Bessel Functions

III. $\boxed{J_n'(x) = J_{n-1}(x) - \dfrac{n}{x} J_n(x)}$

Proof: Carrying out the differentiation in RR **I**, we obtain

$$nx^{n-1} J_n(x) + x^n J_n'(x) = x^n J_{n-1}(x)$$

Dividing by x^n throughout and rearranging, we get

$$J_n'(x) = J_{n-1}(x) - \dfrac{n}{x} J_n(x). \qquad \square$$

IV. $\boxed{J_n'(x) = \dfrac{n}{x} J_n(x) - J_{n+1}(x)}$

Proof: Carrying out the differentiation in RR **II**, we obtain

$$-nx^{-n-1} J_n(x) + x^{-n} J_n'(x) = -x^{-n} J_{n+1}(x)$$

Multiplying by x^n and rearranging, we get

$$J_n'(x) = \dfrac{n}{x} J_n(x) - J_{n+1}(x). \qquad \square$$

V. $\boxed{J_n'(x) = \dfrac{1}{2}[J_{n-1}(x) - J_{n+1}(x)]}$

Proof: Adding RRs **III** and **IV**, the result follows immediately. $\qquad \square$

VI. $\boxed{J_{n-1}(x) + J_{n+1}(x) = \dfrac{2n}{x} J_n(x)}$

Proof: Subtracting RR **IV** from RR **III**, the result follows immediately. $\qquad \square$

Note: The Bessel function of the second kind $Y_n(x)$ 2 satisfies the same recurrence relations as that of the first kind, namely,

I. $\dfrac{d}{dx}\{x^n Y_n(x)\} = x^n Y_{n-1}(x)$

II. $\dfrac{d}{dx}\{x^{-n} Y_n(x)\} = -x^{-n} Y_{n+1}(x)$

III. $Y_n'(x) = Y_{n-1}(x) - \dfrac{n}{x} Y_n(x)$

IV. $Y_n'(x) = \dfrac{1}{2}[Y_{n-1}(x) - Y_{n+1}(x)]$

Special Functions and Orthogonal Polynomials

V. $Y_{n-1}(x) + Y_{n+1}(x) = \dfrac{2n}{x} Y_n(x)$

VI. $Y_n'(x) = \dfrac{n}{x} Y_n(x) - Y_{n+1}(x)$

Example 16: Show that:

$$\left(\dfrac{1}{x}\dfrac{d}{dx}\right)^m \{x^n J_n(x)\} = x^{n-m} J_{n-m}(x), \quad \text{for positive } m < n.$$

And

$$\left(\dfrac{1}{x}\dfrac{d}{dx}\right)^m \{x^{-n} J_n(x)\} = (-1)^m x^{-n-m} J_{n+m}(x).$$

Solution: From RR **I**, $\dfrac{d}{dx}\{x^n J_n(x)\} = x^n J_{n-1}(x)$, we may write

$$\left(\dfrac{1}{x}\dfrac{d}{dx}\right)\{x^n J_n(x)\} = x^{n-1} J_{n-1}(x), \text{ and}$$

$$\left(\dfrac{1}{x}\dfrac{d}{dx}\right)^2 \{x^n J_n(x)\} = x^{n-2} J_{n-2}(x).$$

Repeating this process $m-2$ times and $m < n$, we get

$$\left(\dfrac{1}{x}\dfrac{d}{dx}\right)^m \{x^n J_n(x)\} = x^{n-m} J_{n-m}(x).$$

In a similar manner, and from RR **II**, it is easy to obtain

$$\left(\dfrac{1}{x}\dfrac{d}{dx}\right)^m \{x^{-n} J_n(x)\} = (-1)^m x^{-n-m} J_{n+m}(x). \qquad \square$$

Note that, for $n = 0$,

$$\left(\dfrac{1}{x}\dfrac{d}{dx}\right)^m J_0(x) = x^{-m} J_{-m}(x) = (-1)^m x^{-m} J_m(x), \text{ or}$$

$$J_m(x) = (-1)^m x^m \left(\dfrac{1}{x}\dfrac{d}{dx}\right)^m J_0(x), \text{ also } J_0(x) = -J_0'(x)$$

Example 17: Show that $J_n(x) = 0$ has no repeated roots except at $x = 0$.

Solution: We will prove this by contradiction. Suppose λ is a repeated root of

Bessel Functions

$J_n(x) = 0$, then

$J_n(\lambda) = 0$ and $J'_n(\lambda) = 0$. (*)

From RR IV: $J_{n+1}(x) = \dfrac{n}{x} J_n(x) - J'_n(x)$.

From RR III: $J_{n-1}(x) = \dfrac{n}{x} J_n(x) + J'_n(x)$.

Then from (*), we will have

$J_{n+1}(\lambda) = 0$ and $J_{n-1}(\lambda) = 0$.

This cannot be true, then our supposition is not true and $J_n(x) = 0$ has no repeated roots except at $x = 0$. □

Example 18: Show that:

i. $J_{-3/2}(x) = -\sqrt{\dfrac{2}{\pi x}} \left(\dfrac{\cos x}{x} + \sin x \right)$

ii. $J_{3/2}(x) = \sqrt{\dfrac{2}{\pi x}} \left(\dfrac{\sin x}{x} - \cos x \right)$

Solution: We know that $J_{-1/2}(x) = \sqrt{\dfrac{2}{\pi x}} \cos x$ and $J_{1/2}(x) = \sqrt{\dfrac{2}{\pi x}} \sin x$.

Also from RR VI: $J_{n-1}(x) + J_{n+1}(x) = \dfrac{2n}{x} J_n(x)$

i. Letting $n = -1/2$, we get

$J_{-3/2}(x) = -J_{-1/2}(x) - \dfrac{1}{x} J_{1/2}(x)$.

Substituting for $J_{-1/2}(x)$ and $J_{1/2}(x)$, we get

$J_{-3/2}(x) = -\sqrt{\dfrac{2}{\pi x}} \left(\dfrac{\cos x}{x} + \sin x \right)$.

ii. Letting $n = 1/2$, we get

$J_{3/2}(x) = \dfrac{1}{x} J_{1/2}(x) - J_{-1/2}(x)$.

Substituting for $J_{-1/2}(x)$ and $J_{1/2}(x)$, we get

$J_{3/2}(x) = \sqrt{\dfrac{2}{\pi x}} \left(\dfrac{\sin x}{x} - \cos x \right)$. □

Special Functions and Orthogonal Polynomials

Example 19: Express $J_3(x)$ in terms of $J_0(x)$ and $J_1(x)$.

Solution: From RR **VI**: $J_{n-1}(x) + J_{n+1}(x) = \dfrac{2n}{x} J_n(x)$, letting $n = 1$, we get

$$J_2(x) = \frac{2}{x} J_1(x) - J_0(x).$$

Letting $n = 2$, we get $J_3(x) = \dfrac{4}{x} J_2(x) - J_1(x)$.

Substituting for $J_2(x)$ and rearranging, we obtain

$$J_3(x) = \frac{8 - x^2}{x^2} J_1(x) - \frac{4}{x} J_0(x). \qquad \square$$

Example 20: Show that: $xJ_n(x) = 2 \sum_{k=0}^{\infty} (-1)^k (2k + n + 1) J_{2k+n+1}(x)$

Solution: From RR **VI**, we have

$$2nJ_n(x) = x \left[J_{n-1}(x) + J_{n+1}(x) \right].$$

Replacing n by $n+1$, we get

$$2(n+1)J_{n+1}(x) = x \left[J_n(x) + J_{n+2}(x) \right].$$

Then $xJ_n(x) = 2(n+1)J_{n+1}(x) - xJ_{n+2}(x)$

Replacing n by $n+2$, we get

$$xJ_{n+2}(x) = 2(n+3)J_{n+3}(x) - xJ_{n+4}(x)$$

Substituting this into the preceding equation, we get

$$xJ_n(x) = 2(n+1)J_{n+1}(x) - 2(n+3)J_{n+3}(x) - xJ_{n+4}(x)$$

If we proceed likewise for $xJ_{n+4}(x)$, $xJ_{n+6}(x)$, ... we get

$$xJ_n(x) = 2\{(n+1)J_{n+1}(x) - (n+3)J_{n+3}(x) + (n+5)J_{n+5}(x) + \cdots\}$$

And in compact form

$$xJ_n(x) = 2 \sum_{k=0}^{\infty} (-1)^k (2k + n + 1) J_{2k+n+1}(x). \qquad \square$$

Example 21: Show that:

$$\frac{d}{dx}\left[J_n^2(x) + J_{n+1}^2(x) \right] = \frac{2}{x}\left[n J_n^2(x) - (n+1) J_{n+1}^2(x) \right].$$

Solution: We have

$$\frac{d}{dx}\left[J_n^2(x) + J_{n+1}^2(x) \right] = 2J_n(x) J_n'(x) + 2J_{n+1}(x) J_{n+1}'(x). \quad (*)$$

Bessel Functions

But from RR **IV**, we have $J'_n(x) = \frac{n}{x} J_n(x) - J_{n+1}(x)$

Also, from RR **III**, upon replacing n by $n+1$, we have

$$J'_{n+1}(x) = J_n(x) - \frac{n+1}{x} J_{n+1}(x).$$

Substituting for $J'_n(x)$ and $J'_{n+1}(x)$ in (*), we obtain

$$\frac{d}{dx}\left[J_n^2(x) + J_{n+1}^2(x)\right] = 2J_n(x)\left[\frac{n}{x}J_n(x) - J_{n+1}(x)\right]$$

$$+ 2J_{n+1}(x)\left[J_n(x) - \frac{n+1}{x}J_{n+1}(x)\right]$$

Rearranging, we obtain

$$\frac{d}{dx}\left[J_n^2(x) + J_{n+1}^2(x)\right] = \frac{2}{x}\left[nJ_n^2(x) - (n+1)J_{n+1}^2(x)\right]. \qquad \square$$

***Example* 22:** Show that:

i. $J_0^2(x) + 2\sum_{k=1}^{\infty} J_k^2(x) = 1$,   ii. $|J_0(x)| \leq 1$,

iii. $|J_n(x)| \leq 1/\sqrt{2}$, $n \geq 1$.

***Solution*:**   i. From the previous illustration, we have

$$\frac{d}{dx}\left[J_n^2(x) + J_{n+1}^2(x)\right] = \frac{2}{x}\left[nJ_n^2(x) - (n+1)J_{n+1}^2(x)\right].$$

Replacing n by 0, 1, 2, ... successively, we get

$$\frac{d}{dx}\left[J_0^2(x) + J_1^2(x)\right] = \frac{2}{x}\left[-J_1^2(x)\right]$$

$$\frac{d}{dx}\left[J_1^2(x) + J_2^2(x)\right] = \frac{2}{x}\left[J_1^2(x) - 2J_2^2(x)\right]$$

$$\frac{d}{dx}\left[J_2^2(x) + J_3^2(x)\right] = \frac{2}{x}\left[2J_2^2(x) - 3J_3^2(x)\right]$$

$$\frac{d}{dx}\left[J_3^2(x) + J_4^2(x)\right] = \frac{2}{x}\left[3J_3^2(x) - 4J_4^2(x)\right]$$

...

Adding all these equations, we get

$$\frac{d}{dx}\left\{J_0^2(x)+2\sum_{k=1}^{\infty}J_k^2(x)\right\}=0$$

Integrating with respect to x, we obtain

$$J_0^2(x)+2\sum_{k=1}^{\infty}J_k^2(x)=c$$

where c is the constant of integration. To obtain c, let $x=0$ and noting that $J_0(0)=1$ and $J_n(0)=0$, $n\geq 1$ and, we found that $c=1$. Thus $J_0^2(x)+2\sum_{k=1}^{\infty}J_k^2(x)=1$.

ii. From i., since $J_1^2(x), J_2^2(x), J_3^2(x), \cdots$ are all positive or zero, then $J_0^2(x)\leq 1$ or $|J_0(x)|\leq 1$.

iii. From i., solving for $J_n^2(x)$, we get

$$J_n^2(x)=\frac{1}{2}\left[1-J_0^2(x)\right]-\sum_{\substack{k=1\\k\neq n}}^{\infty}J_k^2(x).$$

And since $J_k^2(x)$, $k=1,2,3,\cdots$, $k\neq n$ are all positive or zero, then

$$J_n^2(x)\leq 1/2 \quad \text{or} \quad |J_n(x)|\leq 1/\sqrt{2},\quad n\geq 1. \qquad \Box$$

Example 23: Show that $\int_0^x tJ_n^2(t)dt = \frac{x^2}{2}\left[J_n^2(x)-J_{n-1}(x)J_{n+1}(x)\right]$

Solution: We start as follows:

Bessel Functions

$$\frac{d}{dt}\left\{\frac{t^2}{2}\left[J_n^2(t)-J_{n-1}(t)J_{n+1}(t)\right]\right\}=t\left[J_n^2(t)-J_{n-1}(t)J_{n+1}(t)\right]$$

$$+\frac{t^2}{2}\left[2J_n(t)J_n'(t)-J_{n-1}(t)J_{n+1}'(t)-J_{n+1}(t)J_{n-1}'(t)\right]$$

Now, from recurrence relation **V**, we have

$$J_n'(t)=\frac{1}{2}\left[J_{n-1}(t)-J_{n+1}(t)\right].$$

From recurrence relation **III**, upon replacing n by $n+1$,

$$J_{n+1}'(t)=J_n(t)-\frac{n+1}{t}J_{n+1}(t)$$

And from recurrence relation **IV**, upon replacing n by $n-1$,

$$J_{n-1}'(t)=\frac{n-1}{t}J_{n-1}(t)-J_n(t).$$

Substituting for all these derivatives in the preceding equation, we obtain upon simplification

$$\frac{d}{dt}\left\{\frac{t^2}{2}\left[J_n^2(t)-J_{n-1}(t)J_{n+1}(t)\right]\right\}=tJ_n^2(t).$$

Integrating with respect to t from 0 to x, and noting that $J_n(0)=J_{n-1}(0)=J_{n+1}(0)=0$, we get

$$\int_0^x tJ_n^2(t)dt=\left\{\frac{t^2}{2}\left[J_n^2(t)-J_{n-1}(t)J_{n+1}(t)\right]\right\}_0^x$$

$$=\frac{x^2}{2}\left[J_n^2(x)-J_{n-1}(x)J_{n+1}(x)\right] \qquad \square$$

6.8. Integrals Involving Bessel Functions

Some integrals involving $J_n(x)$ are obtained directly from recurrence relations **I** and **II**:

$$\frac{d}{dx}\{x^n J_n(x)\} = x^n J_{n-1}(x)$$

$$\frac{d}{dx}\{x^{-n} J_n(x)\} = -x^{-n} J_{n+1}(x)$$

Integrating with respect to x, we obtain

$$\int x^n J_{n-1}(x)\,dx = x^n J_n(x) + c \tag{37}$$

$$\int x^{-n} J_{n+1}(x)\,dx = -x^{-n} J_n(x) + c \tag{38}$$

These two integrals enable us to obtain a reduction formula for integrals of the form

$$u_{m,n} = \int x^{m+1} J_n(x)\,dx$$

Integrating by parts and using relation (37), we get

$$u_{m,n} = \int x^{m+1} J_n(x)\,dx = \int x^{m-n}\{x^{n+1} J_n(x)\}\,dx$$

$$= \int x^{m-n} d\{x^{n+1} J_{n+1}(x)\} = x^{m+1} J_{n+1}(x) - \int x^{n+1} J_{n+1}(x)\,dx^{m-n}$$

$$= x^{m+1} J_{n+1}(x) - (m-n) \int x^m J_{n+1}(x)\,dx \tag{39}$$

Evaluating the last integral by parts and using relation (38), we obtain

$$\int x^m J_{n+1}(x)\,dx = \int x^{m+n}\{x^{-n} J_{n+1}(x)\}\,dx = -\int x^{m+n} d\{x^{-n} J_n(x)\}$$

$$= -x^m J_n(x) + \int x^{-n} J_{n+1}(x)\,dx^{m+n}$$

$$= -x^m J_n(x) + (m+n) \int x^{m-1} J_n(x)\,dx$$

and equation (39) becomes

$$u_{m,n} = x^{m+1} J_{n+1}(x) - (m-n) x^m J_n(x) - (m^2 - n^2) \int x^{m-1} J_n(x)\,dx \tag{40}$$

From recurrence relation **V**, replacing n by $n+1$, we have

Bessel Functions

$$J_n(x) - J_{n+2}(x) = 2J'_{n+1}(x)$$

Integrating with respect to x, we obtain

$$\int J_n(x)\,dx - \int J_{n+2}(x)\,dx = 2J_{n+1}(x) \tag{41}$$

Similarly

$$\int J_{n+2}(x)\,dx - \int J_{n+4}(x)\,dx = 2J_{n+3}(x) \tag{42}$$

and so on. If we add $m-1$ relations of the type (41), we obtain

$$\int J_n(x)\,dx - \int J_{n+2m}(x)\,dx = 2\sum_{k=0}^{m-1} J_{n+2k+1}(x) \tag{43}$$

or

$$\boxed{\int J_{n+2m}(x)\,dx = \int J_n(x)\,dx - 2\sum_{k=0}^{m-1} J_{n+2k+1}(x)} \tag{44}$$

Relations (37), (38), (40) and (44) remain valid if we replace $J_n(x)$ by $Y_n(x)$:

$$\int x^n Y_{n-1}(x)\,dx = x^n Y_n(x) + c \tag{45}$$

$$\int x^{-n} Y_{n+1}(x)\,dx = -x^{-n} Y_n(x) + c \tag{46}$$

$$u_{m,n} = x^{m+1} Y_{n+1}(x) - (m-n)x^m Y_n(x) - (m^2 - n^2)\int x^{m-1} Y_n(x)\,dx \tag{47}$$

$$\int Y_{n+2m}(x)\,dx = \int Y_n(x)\,dx - 2\sum_{k=0}^{m-1} Y_{n+2k+1}(x) \tag{48}$$

Example 24: Evaluate the following integrals:

    i. $\int J_3(x)\,dx$ ii. $\int x^3 J_3(x)\,dx$ iii. $\int x^4 J_1(x)\,dx$

Solution: i. $\int J_3(x)\,dx = \int x^2 \{x^{-2} J_3(x)\}\,dx$

Integrating by parts using relation (38), we obtain

$$\int J_3(x)\,dx = -\int x^2\,d\{x^{-2} J_2(x)\} = -J_2(x) + 2\int x^{-1} J_2(x)\,dx .$$

Using relation (38) again, we obtain

$$\int J_3(x)\,dx = -J_2(x) - 2x^{-1} J_1(x) + c .$$

257

ii. $\int x^3 J_3(x)\,dx = \int x^5 \{x^{-2} J_3(x)\}\,dx$

Integrating by parts using relation (38), we obtain

$\int x^3 J_3(x)\,dx = -\int x^5 d\{x^{-2} J_2(x)\}$

$= -x^3 J_2(x) + 5\int x^2 J_2(x)\,dx$

$= -x^3 J_2(x) + 5\int x^3 \{x^{-1} J_2(x)\}\,dx$

$= -x^3 J_2(x) - 5\int x^3 d\{x^{-1} J_1(x)\}$

$= -x^3 J_2(x) - 5x^2 J_1(x) + 15\int x J_1(x)\,dx$

$= -x^3 J_2(x) - 5x^2 J_1(x) + 15\int x\, d\{J_0(x)\}$

$= -x^3 J_2(x) - 5x^2 J_1(x) + 15x J_0(x) + 15\int J_0(x)\,dx$ □

iii. $\int x^4 J_1(x)\,dx = \int x^2 \{x^2 J_1(x)\}\,dx$

Integrating by parts using relation (37), we obtain

$\int x^3 J_3(x)\,dx = \int x^5 d\{x^2 J_2(x)\}$

$= x^4 J_2(x) - 2\int x^3 J_2(x)\,dx = x^4 J_2(x) - 2x^3 J_3(x) + c.$ □

Note: For an integral of the form $\int x^m J_n(x)\,dx$, $m + n \geq 0$, we have

1. If $m + n$ is odd, the integral can be completely evaluated.

2. If $m + n$ is even, the integral will be given in terms of $\int J_0(x)\,dx$ which cannot be evaluated in closed form.

Example 25: Show that: $\int_0^x t^3 J_0(t)\,dt = 2x^2 J_0(x) + x(x^2 - 4) J_1(x)$.

Solution: Using the reduction relation (40), we get

Bessel Functions

$$\int_0^x t^3 J_0(t)\,dt = \left[t^3 J_1(t) + 2t^2 J_0(t)\right]_0^x - 4\int_0^x t J_0(t)\,dt$$

$$= x^3 J_1(x) + 2x^2 J_0(x) - 4\int_0^x t J_0(t)\,dt.$$

Using relation (37) for the last integral, we obtain

$$\int_0^x t^3 J_0(t)\,dt = x^3 J_1(x) + 2x^2 J_0(x) - 4\left[t J_1(t)\right]_0^x$$

$$= 2x^2 J_0(x) + x(x^2 - 4)J_1(x) \qquad \square$$

Example 26: Show that: $J_{n+1}(x) = x \int_0^1 J_n(xy) y^{n+1}\,dy$

Solution: Letting $xy = t$ so that $x\,dy = dt$ and the integral on the right hand side becomes

$$\int_0^1 J_n(xy) y^{n+1}\,dy = \int_0^x J_n(t)\left(\frac{t}{x}\right)^{n+1} \frac{dt}{x} = \frac{1}{x^{n+2}} \int_0^x t^{n+1} J_n(t)\,dt.$$

Using relation (37), we get

$$\int_0^1 J_n(xy) y^{n+1}\,dy = \frac{1}{x^{n+2}}\left[t^{n+1} J_{n+1}(t)\right]_0^x = \frac{1}{x} J_{n+1}(x).$$

Therefore

$$J_{n+1}(x) = x \int_0^1 J_n(xy) y^{n+1}\,dy. \qquad \square$$

Example 27: Show that:

i. $\displaystyle\int_0^x t^{-n} J_{n+1}(t)\,dt = \frac{1}{2^n \Gamma(n+1)} - x^{-n} J_n(x), \quad n > 1$

ii. $\displaystyle\int_0^\infty t^{-n} J_{n+1}(t)\,dt = \frac{1}{2^n \Gamma(n+1)}, \quad n > -1/2.$

259

Solution: i. From recurrence relation **II**, we have

$$\frac{d}{dx}\{t^{-n}J_n(t)\} = -t^{-n}J_{n+1}(t)$$

Integrating with respect to t from 0 to x, we obtain

$$\left[t^{-n}J_n(t)\right]_0^x = -\int_0^x t^{-n}J_{n+1}(t)\,dt.$$

Therefore

$$\int_0^x t^{-n}J_{n+1}(t)\,dt = -x^{-n}J_n(x) + \lim_{t \to 0}\frac{J_n(t)}{t^n}.$$

But

$$\lim_{t \to 0}\frac{J_n(t)}{t^n} = \lim_{t \to 0}\frac{1}{t^n}\sum_{k=0}^{\infty}\frac{(-1)^k(t/2)^{n+2k}}{k!\,\Gamma(n+k+1)}$$

$$= \lim_{t \to 0}\sum_{k=0}^{\infty}\frac{(-1)^k\,t^{2k}}{k!\,2^{n+2k}\,\Gamma(n+k+1)} = \frac{1}{2^n\,\Gamma(n+1)}.$$

Therefore

$$\int_0^x t^{-n}J_{n+1}(t)\,dt = \frac{1}{2^n\,\Gamma(n+1)} - x^{-n}J_n(x), \quad n > 1.$$

ii. From recurrence relation **II**, we have

$$\frac{d}{dx}\{t^{-n}J_n(t)\} = -t^{-n}J_{n+1}(t)$$

Integrating with respect to t from 0 to ∞, we obtain

$$\left[t^{-n}J_n(t)\right]_0^{\infty} = -\int_0^{\infty} t^{-n}J_{n+1}(t)\,dt.$$

Therefore

$$\int_0^{\infty} t^{-n}J_{n+1}(t)\,dt = \lim_{t \to 0}\frac{J_n(t)}{t^n} - \lim_{t \to \infty}\frac{J_n(t)}{t^n}.$$

But

$$\lim_{t \to 0} \frac{J_n(t)}{t^n} = \frac{1}{2^n \Gamma(n+1)} \quad \text{and} \quad \lim_{t \to \infty} \frac{J_n(t)}{t^n} = 0$$

Therefore

$$\int_0^\infty t^{-n} J_{n+1}(t)\, dt = \frac{1}{2^n \Gamma(n+1)}, \quad n > -1/2. \qquad \square$$

6.9. Generalized Bessel Differential Equation

Many differential equations can be reduced to Bessel differential equation by suitable transformation. We consider here a generalized form of Bessel differential equation whose solutions are found to be some variation of Bessel functions. We show that this generalized equation can be transformed to the original Bessel equation and hence solved.

Consider the differential equation

$$x^2 \frac{d^2y}{dx^2} + (1-2\alpha)x \frac{dy}{dx} + \left[\beta^2\gamma^2 x^{2\gamma} + (\alpha^2 - n^2\gamma^2)\right] y = 0 \qquad (49)$$

We first change the dependent variable y using the transformation $y = x^{\alpha} z$, then

$$\frac{dy}{dx} = x^{\alpha} \frac{dz}{dx} + \alpha x^{\alpha-1} z$$

and

$$\frac{d^2y}{dx^2} = x^{\alpha} \frac{d^2z}{dx^2} + 2\alpha x^{\alpha-1} \frac{dz}{dx} + \alpha(\alpha-1) z$$

Substituting in equation (49) and simplifying, we obtain

$$x^2 \frac{d^2z}{dx^2} + x \frac{dz}{dx} + \left[\beta^2\gamma^2 x^{2\gamma} - n^2\gamma^2\right] z = 0 \qquad (50)$$

Now, we change the independent variable x using the transformation $t = x^{\gamma}$, then

$$\frac{dz}{dx} = \gamma x^{\gamma-1} \frac{dz}{dt}$$

and

$$\frac{d^2z}{dx^2} = \gamma^2 x^{2\gamma-2} \frac{d^2z}{dt^2} + \gamma(\gamma-1) x^{\gamma-2} \frac{dz}{dt}$$

Substituting in equation (50) and simplifying, we get

$$t^2 \frac{d^2z}{dt^2} + t \frac{dz}{dt} + \left[\beta^2 t^2 - n^2\right] z = 0 \qquad (51)$$

Equation (51) is Bessel differential equation whose solution is $z = AJ_n(\beta t) + BY_n(\beta t)$. Hence the solution of the generalized equation is found by back substituting the relations $y = x^{\alpha} z$ and $t = x^{\gamma}$ to give

$$\boxed{y = Ax^{\alpha} J_n(\beta x^{\gamma}) + Bx^{\alpha} Y_n(\beta x^{\gamma})} \qquad (52)$$

Bessel Functions

Example 28: Show that the general solution of the differential equation

$$\frac{d^2y}{dx^2} + \left[e^{2x} - n^2\right]y = 0 \text{ is } y = AJ_n(e^x) + BY_n(e^x).$$

Solution: Using the transformation $e^x = t$, then

$$\frac{dy}{dx} = \frac{dy}{dt}\frac{dt}{dx} = t\frac{dy}{dt} \text{ and } \frac{d^2y}{dx^2} = \frac{d}{dt}\left(t\frac{dy}{dt}\right)\frac{dt}{dx} = t^2\frac{d^2y}{dt^2} + t\frac{dy}{dt}.$$

Substituting in the differential equation, and rearranging we get

$$t^2\frac{d^2y}{dt^2} + t\frac{dy}{dt} + \left[t^2 - n^2\right]y = 0,$$

which is Bessel differential equation whose solution is

$$y = AJ_n(t) + BY_n(t)$$

Then the solution of the original equation is

$$y = AJ_n(e^x) + BY_n(e^x). \qquad \square$$

Example 29: Find the general solution of the differential equations:

i. $x\dfrac{d^2y}{dx^2} + y = 0$ ii. $x^2\dfrac{d^2y}{dx^2} + x\dfrac{dy}{dx} + 4(x^4 - m^2)y = 0$.

Solution: i. The equation can be put in the form $x^2\dfrac{d^2y}{dx^2} + xy = 0$.

Comparing this equation with the generalized Bessel differential equation (equation (49)), we have

$1 - 2\alpha = 0$ and $\beta^2\gamma^2 x^{2\gamma} + (\alpha^2 - n^2\gamma^2) = x$.

From which we obtain $\alpha = 1/2$, $\gamma = 1/2$, $\beta = 2$ and $n = 1$.

Then, the general solution will be (equation (52)),

$$y = A\sqrt{x}\,J_1(\sqrt{x}) + B\sqrt{x}\,Y_1(\sqrt{x}).$$

ii. Comparing this equation with the generalized Bessel differential equation (equation (49)), we have

$1 - 2\alpha = 1$ and $\beta^2\gamma^2 x^{2\gamma} + (\alpha^2 - n^2\gamma^2) = 4(x^4 - m^2)$.

From which we obtain $\alpha = 0$, $\gamma = 2$, $\beta = 1$ and $n = m$.

Then, the general solution will be (equation (52)),

$$y = AJ_m(x^2) + BY_m(x^2). \qquad \square$$

6.10. Hankel Functions

Sometimes it is more convenient to take the fundamental solutions in a slightly different forms. *Hankel Functions*, often called *Bessel Functions of the Third Kind and of Order n*, are defined as

$$H_n^{(1)}(x) = J_n(x) + iY_n(x) \tag{53}$$

$$H_n^{(2)}(x) = J_n(x) - iY_n(x) \tag{54}$$

They are linearly independent solutions of Bessel's differential equation, and the general solution can be put in the form

$$y = A H_n^{(1)}(x) + B H_n^{(2)}(x) \tag{55}$$

Both Hankel functions are infinite at $x = 0$; and their are especially useful for their behavior for large values of x. They satisfy recurrence relations identical to $J_n(x)$, i.e.

I. $\dfrac{d}{dx}\{x^n H_n^{(1)}(x)\} = x^n H_{n-1}^{(1)}(x)$ $\qquad \dfrac{d}{dx}\{x^n H_n^{(2)}(x)\} = x^n H_{n-1}^{(2)}(x)$

II. $\dfrac{d}{dx}\{x^{-n} H_n^{(1)}(x)\} = -x^{-n} H_{n+1}^{(1)}(x)$

$\qquad\qquad\qquad\qquad\qquad \dfrac{d}{dx}\{x^{-n} H_n^{(2)}(x)\} = -x^{-n} H_{n+1}^{(2)}(x)$

III. $\dfrac{d}{dx} H_n^{(1)}(x) = H_{n-1}^{(1)}(x) - \dfrac{n}{x} H_n^{(1)}(x)$

$\qquad\qquad\qquad\qquad\qquad \dfrac{d}{dx} H_n^{(2)}(x) = H_{n-1}^{(2)}(x) - \dfrac{n}{x} H_n^{(2)}(x)$

IV. $\dfrac{d}{dx} H_n^{(1)}(x) = \dfrac{1}{2}\left[H_{n-1}^{(1)}(x) - H_{n+1}^{(1)}(x)\right]$

$\qquad\qquad\qquad\qquad\qquad \dfrac{d}{dx} H_n^{(2)}(x) = \dfrac{1}{2}\left[H_{n-1}^{(2)}(x) - H_{n+1}^{(2)}(x)\right]$

V. $H_{n-1}^{(1)}(x) + H_{n+1}^{(1)}(x) = \dfrac{2n}{x} H_n^{(1)}(x)$

$\qquad\qquad\qquad\qquad\qquad H_{n-1}^{(2)}(x) + H_{n+1}^{(2)}(x) = \dfrac{2n}{x} H_n^{(2)}(x)$

VI. $\dfrac{d}{dx} H_n^{(1)}(x) = \dfrac{n}{x} H_n^{(1)} - H_{n+1}^{(1)}(x)$ $\qquad \dfrac{d}{dx} H_n^{(2)}(x) = \dfrac{n}{x} H_n^{(2)} - H_{n+1}^{(2)}(x)$

6.11. Modified Bessel Functions

We start from Bessel differential equation

$$x^2 \frac{d^2y}{dx^2} + x \frac{dy}{dx} + (x^2 - n^2)y = 0 \qquad (56)$$

whose solution is

$$y(x) = A J_n(x) + B Y_n(x) \qquad (57)$$

If we replace the independent variable x by λx, the resulting equation is

$$x^2 \frac{d^2y}{dx^2} + x \frac{dy}{dx} + (\lambda^2 x^2 - n^2)y = 0 \qquad (58)$$

with solution of the form

$$y(x) = A J_n(\lambda x) + B Y_n(\lambda x) \qquad (59)$$

Now, if $\lambda^2 = -1$, equation (58) reduces to

$$\boxed{x^2 \frac{d^2y}{dx^2} + x \frac{dy}{dx} - (x^2 + n^2)y = 0} \qquad (60)$$

This equation is called the *Modified Bessel Differential Equation* whose solution is

$$y(x) = A J_n(ix) + B Y_n(ix) \qquad (61)$$

The solutions $J_n(ix)$ and $Y_n(ix)$ are not necessarily real when x is real. To express the solution $J_n(ix)$ in real terms when x is real, consider the function

$$I_n(x) = i^{-n} J_n(ix) \qquad (62)$$

The factor i^{-n} is constant, then $I_n(x)$ is a solution of the modified Bessel differential equation. Moreover, we will show that $I_n(x)$ is a real function of x.

$$I_n(x) = i^{-n} J_n(ix) = i^{-n} \sum_{k=0}^{\infty} \frac{(-1)^k (ix/2)^{n+2k}}{k!\,\Gamma(n+k+1)}$$

$$= i^{-n} \sum_{k=0}^{\infty} \frac{(-1)^k i^{2k} i^n (x/2)^{n+2k}}{k!\,\Gamma(n+k+1)}.$$

But $i^{2k} = (-1)^k$, then $(-1)^k i^{2k} = (-1)^k (-1)^k = 1$, and

$$\boxed{I_n(x) = \sum_{k=0}^{\infty} \frac{(x/2)^{n+2k}}{k!\,\Gamma(n+k+1)}} \qquad (63)$$

$I_n(x)$ is called the *Modified Bessel Function of the First Kind*. It is easy to show that if *n* is an integer, then

$$I_{-n}(x) = I_n(x) \tag{64}$$

i.e. $I_n(x)$ and $I_{-n}(x)$ are linearly dependent. But if *n* is not an integer then $I_n(x)$ and $I_{-n}(x)$ are linearly independent.

In other words, if *n* in not an integer, the solution of the modified Bessel differential equations can be put in the form

$$y(x) = AI_n(x) + BI_{-n}(x) \tag{65}$$

When n is an integer, we obtain a second linearly independent solution in a way similar to what we did for $Y_n(x)$. In fact this second solution is

$$K_n(x) = \frac{\pi}{2} \frac{I_{-n}(x) - I_n(x)}{\sin n\pi} \tag{66}$$

This is the *Modified Bessel Function of the Second Kind*. It can be shown that $K_n(x)$ satisfies the modified Bessel differential equation whose solution is now written as

$$y(x) = AI_n(x) + BK_n(x) \tag{67}$$

$I_n(x)$ and $K_n(x)$ satisfy recurrence relations similar to, but not identical to those of $J_n(x)$ and $Y_n(x)$. It can be shown that for $I_n(x)$, we have

I. $\dfrac{d}{dx}\{x^n I_n(x)\} = x^n I_{n-1}(x)$

II. $\dfrac{d}{dx}\{x^{-n} I_n(x)\} = x^{-n} I_{n+1}(x)$

III. $\dfrac{d}{dx} I_n(x) = I_{n-1}(x) - \dfrac{n}{x} I_n(x)$

IV. $\dfrac{d}{dx} I_n(x) = \dfrac{1}{2}[I_{n-1}(x) + I_{n+1}(x)]$

V. $I_{n-1}(x) - I_{n+1}(x) = \dfrac{2n}{x} I_n(x)$

VI. $\dfrac{d}{dx} I_n(x) = \dfrac{n}{x} I_n(x) + I_{n+1}(x)$

And for $K_n(x)$, we have

I. $\dfrac{d}{dx}\{x^n K_n(x)\} = -x^n K_{n-1}(x)$

II. $\dfrac{d}{dx}\{x^{-n}K_n(x)\} = -x^{-n}K_{n+1}(x)$

III. $\dfrac{d}{dx}K_n(x) = -K_{n-1}(x) - \dfrac{n}{x}K_n(x)$

IV. $\dfrac{d}{dx}K_n(x) = -\dfrac{1}{2}[K_{n-1}(x) + K_{n+1}(x)]$

V. $K_{n-1}(x) - K_{n+1}(x) = -\dfrac{2n}{x}K_n(x)$

VI. $\dfrac{d}{dx}K_n(x) = \dfrac{n}{x}K_n(x) - K_{n+1}(x)$

It can also be shown that $I_n(x)$ and $K_n(x)$ have the integral representations

$$I_n(x) = \dfrac{(x/2)^n}{\sqrt{\pi}\,\Gamma(n+1/2)} \int_{-1}^{1} e^{-xt}(1-t^2)^{n-1/2}\,dt, \quad n > -1/2.$$

$$K_n(x) = \dfrac{\sqrt{\pi}\,(x/2)^n}{\Gamma(n+1/2)} \int_{1}^{\infty} e^{-xt}(t^2-1)^{n-1/2}\,dt, \quad n > -1/2,\; x > 0.$$

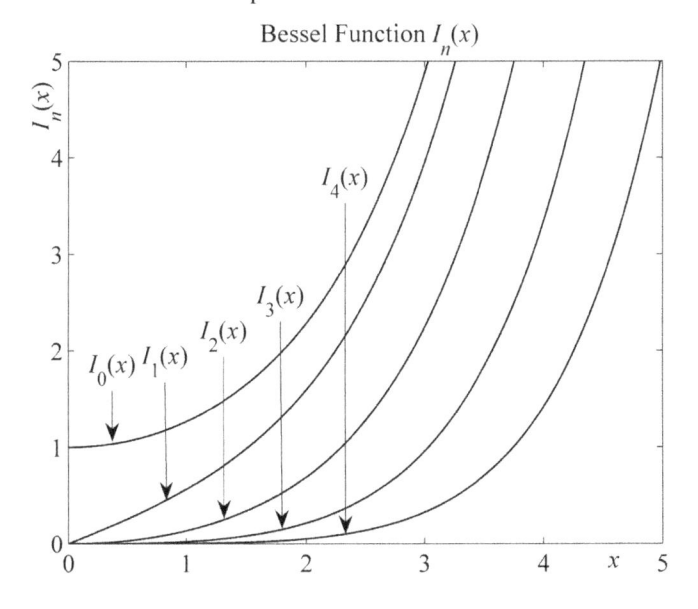

Bessel Function $I_n(x)$

Bessel Function $K_n(x)$

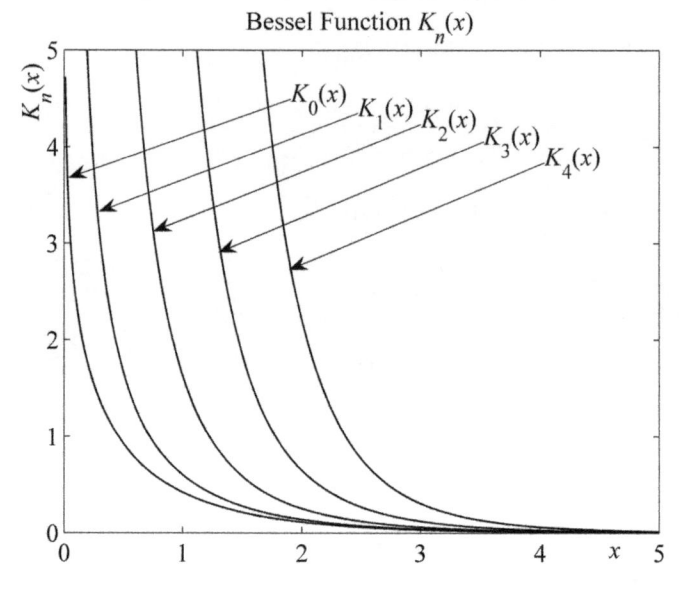

Graph showing $K_0(x), K_1(x), K_2(x), K_3(x), K_4(x)$

Example 30: Show that $xI_1(x) = 4\sum_{k=1}^{\infty} k\, I_{2k}(x)$.

Solution: From recurrence relation **V** of $I_n(x)$, we have

$$x\left[I_{n-1}(x) - I_{n+1}(x)\right] = 2nI_n(x).$$

Letting $n = 2, 4, 6, 8, \cdots$ successively, we obtain

$$x\left[I_1(x) - I_3(x)\right] = 4I_2(x)$$

$$x\left[I_3(x) - I_5(x)\right] = 8I_4(x)$$

$$x\left[I_5(x) - I_7(x)\right] = 12I_6(x)$$

$$x\left[I_7(x) - I_9(x)\right] = 16I_8(x)$$

…, and so on. Adding all these equations, we get

$$xI_1(x) = 4\{I_2(x) + 2I_4(x) + 3I_6(x) + 4I_8(x) + \cdots\}$$

$$= 4\sum_{k=1}^{\infty} k\, I_{2k}(x). \qquad \square$$

6.12. Kelvin Functions

From the modified Bessel differential equation

$$x^2 \frac{d^2 y}{dx^2} + x \frac{dy}{dx} - (x^2 + n^2) y = 0 \qquad (68)$$

whose solution is $y(x) = A I_n(x) + B K_n(x)$. If we replace x^2 by $im^2 x^2$, we obtain the equation

$$x^2 \frac{d^2 y}{dx^2} + x \frac{dy}{dx} - (im^2 x^2 + n^2) y = 0 \qquad (69)$$

with general solution $y(x) = A I_n(\sqrt{i}\, mx) + B K_n(\sqrt{i}\, mx)$. But $\sqrt{i}$ is a two-valued function with values $e^{i\pi/4}$ and $e^{i 5\pi/4}$. we take $\sqrt{i} = e^{i\pi/4}$ to remove the ambiguity, we get

$$y(x) = A I_n(e^{i\pi/4} mx) + B K_n(e^{i\pi/4} mx) \qquad (70)$$

Now, since $I_n(x) = i^{-n} J_n(ix)$, we may take the two independent solutions of equation (69) as $J_n(i^{3/2} mx)$ and $K_n(i^{1/2} mx)$. When x is real $J_n(i^{3/2} mx)$ and $K_n(i^{1/2} mx)$ are not necessarily real. Then we define the Kelvin functions, that are real functions, as

$$\text{ber}_n(x) = \text{Re}\{J_n(i^{3/2} x)\} \quad \text{and}$$

$$\text{bei}_n(x) = \text{Im}\{J_n(i^{3/2} x)\},$$

$$J_n(i^{3/2} x) = \text{ber}_n(x) + i\, \text{bei}_n(x)$$

$$\text{ker}_n(x) = \text{Re}\{i^{-n} K_n(i^{1/2} x)\} \quad \text{and}$$

$$\text{kei}_n(x) = \text{Im}\{i^{-n} K_n(i^{1/2} x)\},$$

$$i^{-n} K_n(i^{1/2} x) = \text{ker}_n(x) + i\, \text{kei}_n(x)$$

Thus, the general solution of equation (69) is

$$y(x) = A_1 \{\text{ber}_n(mx) + i\, \text{bei}_n(mx)\} + B_1 \{\text{ker}_n(mx) + i\, \text{kei}_n(mx)\} \qquad (71)$$

6.13. Orthogonality Property and Bessel Series Expansion

Bessel functions $J_n(x)$ exhibits some kind of orthogonality. This is given by the following theorem

Theorem: If α_i and α_j are two roots of the equation $J_n(\alpha x) = 0$, then

$$\int_0^a x J_n(\alpha_i x) J_n(\alpha_j x) dx = \begin{cases} 0 & \text{if } i \neq j \ (\alpha_i \neq \alpha_j) \\ \dfrac{a^2}{2} J_{n+1}^2(\alpha_i a) & \text{if } i = j \end{cases}$$

Proof: Let $i \neq j$, then α_i and α_j are distinct roots of $J_n(\alpha x) = 0$. Since $J_n(\alpha_i x)$ and $J_n(\alpha_j x)$ are solutions of the Bessel differential equation, then

$$x^2 J_n''(\alpha_i x) + x J_n'(\alpha_i x) + (\alpha_i^2 x^2 - n^2) J_n(\alpha_i x) = 0, \tag{72}$$

and $x^2 J_n''(\alpha_j x) + x J_n'(\alpha_j x) + (\alpha_j^2 x^2 - n^2) J_n(\alpha_j x) = 0.$ (73)

Equations (72) and (73) may be written in the form

$$x \frac{d}{dx}\left\{x \frac{d}{dx} J_n(\alpha_i x)\right\} + (\alpha_i^2 x^2 - n^2) J_n(\alpha_i x) = 0, \tag{74}$$

and $x \dfrac{d}{dx}\left\{x \dfrac{d}{dx} J_n(\alpha_j x)\right\} + (\alpha_j^2 x^2 - n^2) J_n(\alpha_j x) = 0.$ (75)

Multiplying equation (74) by $J_n(\alpha_j x)/x$ and equation (75) by $J_n(\alpha_i x)/x$ and subtracting, we obtain

$$J_n(\alpha_j x) \frac{d}{dx}\left\{x \frac{d}{dx} J_n(\alpha_i x)\right\} - J_n(\alpha_i x) \frac{d}{dx}\left\{x \frac{d}{dx} J_n(\alpha_j x)\right\}$$
$$+ (\alpha_i^2 - \alpha_j^2) x J_n(\alpha_i x) J_n(\alpha_j x) = 0. \tag{76}$$

This equation may be put in the form

$$\frac{d}{dx}\left\{J_n(\alpha_j x) x \frac{d}{dx} J_n(\alpha_i x)\right\} - \left\{\frac{d}{dx} J_n(\alpha_j x)\right\} x \frac{d}{dx} J_n(\alpha_i x)$$
$$- \frac{d}{dx}\left\{J_n(\alpha_i x) x \frac{d}{dx} J_n(\alpha_j x)\right\} + \left\{\frac{d}{dx} J_n(\alpha_i x)\right\} x \frac{d}{dx} J_n(\alpha_j x)$$
$$+ (\alpha_i^2 - \alpha_j^2) x J_n(\alpha_i x) J_n(\alpha_j x) = 0. \tag{77}$$

Bessel Functions

Thus

$$\frac{d}{dx}\left\{J_n(\alpha_j x)x\frac{d}{dx}J_n(\alpha_i x)\right\} - \frac{d}{dx}\left\{J_n(\alpha_i x)x\frac{d}{dx}J_n(\alpha_j x)\right\}$$
$$+(\alpha_i^2 - \alpha_j^2)x\, J_n(\alpha_i x)J_n(\alpha_j x) = 0. \quad (78)$$

Integrating equation (78) with respect to x from 0 to a, noting that $J_n(\alpha_i a) = J_n(\alpha_j a) = 0$, we get

$$(\alpha_i^2 - \alpha_j^2)\int_0^a x\, J_n(\alpha_i x)J_n(\alpha_j x)\, dx = 0. \quad (79)$$

And since $\alpha_i \neq \alpha_j$, then

$$\int_0^a x\, J_n(\alpha_i x)J_n(\alpha_j x)\, dx = 0. \quad (80)$$

To prove the second part, let $u = J_n(\alpha x)$, then

$$x^2 u'' + xu' + (\alpha^2 x^2 - n^2)u = 0. \quad (81)$$

Multiplying by $2u'$, we get

$$2x^2 u''u' + x\, u'^2 + (\alpha^2 x^2 - n^2)u u' = 0. \quad (82)$$

This equation is readily seen as

$$\frac{d}{dx}\{x^2 u'^2 - n^2 u^2 + \alpha^2 x^2 u^2\} - 2\alpha^2 x\, u^2 = 0. \quad (83)$$

Integrating with respect to x from 0 to a, we get

$$\left[x^2 u'^2 - n^2 u^2 + \alpha^2 x^2 u^2\right]_0^a - 2\alpha^2\int_0^a x\, u^2 dx = 0. \quad (84)$$

Or

$$\left[x^2\left\{\frac{d}{dx}J_n(\alpha x)\right\}^2 - n^2 J_n^2(\alpha x) + \alpha^2 x^2 J_n^2(\alpha x)\right]_0^a$$
$$-2\alpha^2\int_0^a x\, J_n^2(\alpha x)\, dx = 0.$$

Special Functions and Orthogonal Polynomials

But $J_n(\alpha a) = 0$ and $nJ_n(0) = 0$, then

$$\left[a^2\left\{\frac{d}{dx}J_n(\alpha x)\right\}^2\right]\bigg|_{x=a} - 2\alpha^2 \int_0^a x J_n^2(\alpha x)\,dx = 0.$$

Hence

$$\int_0^a x J_n^2(\alpha x)\,dx = \frac{1}{2\alpha^2}\left[a^2\left\{\frac{d}{dx}J_n(\alpha x)\right\}^2\right]\bigg|_{x=a} \tag{85}$$

And from recurrence relation **IV** of Bessel function, we have

$$\frac{d}{d(\alpha x)}J_n(\alpha x) = \frac{n}{\alpha x}J_n(\alpha x) - J_{n+1}(\alpha x), \text{ or}$$

$$\frac{d}{dx}J_n(\alpha x) = \frac{n}{x}J_n(\alpha x) - \alpha J_{n+1}(\alpha x).$$

Then equation (85) becomes

$$\int_0^a x J_n^2(\alpha x)\,dx = \frac{1}{2\alpha^2}\left[a^2\left\{\frac{n}{x}J_n(\alpha x) - \alpha J_{n+1}(\alpha x)\right\}^2\right]\bigg|_{x=a}$$

$$= \frac{a^2}{2}J_{n+1}^2(\alpha a). \qquad \square$$

Theorem: If the function $f(x)$, defined in the interval $0 \le x \le a$, has an expansion of the form

$$f(x) = \sum_{i=1}^{\infty} c_i J_n(\alpha_i x) \tag{86}$$

where α_i are the roots of the equation $J_n(\alpha_i x) = 0$, then the coefficients c_i are given by

$$c_i = \frac{2}{a^2 J_{n+1}^2(\alpha_i a)} \int_0^a x f(x) J_n(\alpha_i x)\,dx \tag{87}$$

Proof: Multiplying both sides of equation (86) by $x J_n(\alpha_j x)$, we get

$$x f(x) J_n(\alpha_j x) = \sum_{i=1}^{\infty} c_i x J_n(\alpha_i x) J_n(\alpha_j x). \tag{88}$$

Bessel Functions

Integrating both sides of equation (88) with respect to x from 0 to a, we obtain

$$\int_0^a x f(x) J_n(\alpha_j x) dx = \sum_{i=1}^{\infty} c_i \int_0^a x J_n(\alpha_i x) J_n(\alpha_j x) dx. \qquad (89)$$

From the orthogonality property of Bessel function,

$$\int_0^a x J_n(\alpha_i x) J_n(\alpha_j x) dx = \begin{cases} 0 & \text{if } i \neq j \ (\alpha_i \neq \alpha_j) \\ \dfrac{a^2}{2} J_{n+1}^2(\alpha_i a) & \text{if } i = j \end{cases}$$

equation (89) becomes

$$\int_0^a x f(x) J_n(\alpha_j x) dx = c_j \frac{a^2}{2} J_{n+1}^2(\alpha_j a). \qquad (90)$$

Replacing j by i, the coefficients c_i are

$$c_i = \frac{2}{a^2 J_{n+1}^2(\alpha_i a)} \int_0^a x f(x) J_n(\alpha_i x) dx. \qquad \square$$

Example 30: Expand the function $f(x) = 1$, $0 \le x \le 1$ into a Bessel series of the form $f(x) = \sum_{i=1}^{\infty} c_i J_0(\alpha_i x)$, where α_i are the roots of the equation $J_0(\alpha) = 0$.

Solution: The coefficients c_i are $c_i = \dfrac{2}{J_1^2(\alpha_i)} \int_0^1 x J_0(\alpha_i x) dx$.

Let $\alpha_i x = t$, then $dx = dt / \alpha_i$ and the integral in last equation becomes $\int_0^1 x J_0(\alpha_i x) dx = \dfrac{1}{\alpha_i^2} \int_0^{\alpha_i} t J_0(t) dt$.

But from recurrence relation **I**, we have $\boxed{\dfrac{d}{dx} \{x J_1(x)\} = x J_0(x)}$, then

$$\int_0^1 x\, J_0(\alpha_i x)\, dx = \frac{1}{\alpha_i^2} \int_0^{\alpha_i} \frac{d}{dt}\{t\, J_1(t)\}\, dt$$
$$= \frac{1}{\alpha_i^2}[t\, J_1(t)]_0^{\alpha_i} = \frac{1}{\alpha_i} J_1(\alpha_i).$$

The coefficients c_i becomes

$$c_i = \frac{2}{J_1^2(\alpha_i)} \frac{1}{\alpha_i} J_1(\alpha_i) = \frac{2}{\alpha_i J_1(\alpha_i)}.$$

And the required expansion is $1 = 2 \sum_{i=1}^{\infty} \frac{J_0(\alpha_i x)}{\alpha_i J_1(\alpha_i)}$. □

6.14. Spherical Bessel Functions

Recall the generalized Bessel differential equation

$$x^2 \frac{d^2 y}{dx^2} + (1-2\alpha)x \frac{dy}{dx} + \left[\beta^2 \gamma^2 x^{2\gamma} + (\alpha^2 - n^2\gamma^2)\right] y = 0 \qquad (91)$$

whose solution is $y = Ax^\alpha J_n(\beta x^\gamma) + Bx^\alpha Y_n(\beta x^\gamma)$. Consider now the equation

$$x^2 \frac{d^2 y}{dx^2} + 2x \frac{dy}{dx} + \left[k^2 x^2 - m(m+1)\right] y = 0 \qquad (92)$$

Comparing equation (92) with equation (91), we can see that

$$1-2\alpha = 2, \quad \gamma = 1, \quad \beta^2 \gamma^2 = k^2 \text{ and } \alpha^2 - n^2\gamma^2 = -m(m+1).$$

Solving for α, β, γ and n, we get

$$\alpha = -1/2, \quad \beta = k, \quad \gamma = 1 \text{ and } n = m + 1/2.$$

And the solution of equation (92) will be

$$y = Ax^{-1/2} J_{m+1/2}(kx) + Bx^{-1/2} Y_{m+1/2}(kx)$$
$$= A_1 j_{m+1/2}(kx) + B_1 y_{m+1/2}(kx)$$

where $j_m(x)$ and $y_m(x)$ are defined as the spherical Bessel functions and are given by

$$j_m(x) = \sqrt{\frac{\pi}{2x}} J_{m+1/2}(x) \qquad (93)$$

$$y_m(x) = \sqrt{\frac{\pi}{2x}} Y_{m+1/2}(x) \, 3 \qquad (94)$$

The most interesting fact about spherical Bessel functions is that they can be written in closed form in terms of elementary functions.

Spherical Hankel functions are defined in the same way as the original Hankel functions, namely

$$h_n^{(1)}(x) = j_n(x) + i y_n(x) \qquad (95)$$

$$h_n^{(2)}(x) = j_n(x) - i y_n(x) \qquad (96)$$

Spherical Bessel functions and spherical Hankel functions satisfy analogous recurrence relations to those of Bessel functions, for example, we have

I. $\dfrac{d}{dx}\{x^n j_n(x)\} = x^n j_{n-1}(x)$

II. $\dfrac{d}{dx}\{x^{-n} j_n(x)\} = -x^{-n} j_{n+1}(x)$

III. $\dfrac{d}{dx} j_n(x) = j_{n-1}(x) - \dfrac{n+1}{x} j_n(x)$

IV. $j_{n-1}(x) + j_{n+1}(x) = \dfrac{2n+1}{x} j_n(x)$

V. $(2n+1)\dfrac{d}{dx} j_n(x) = n\, j_{n-1}(x) - (n+1) j_{n+1}(x)$

VI. $\dfrac{d}{dx} j_n(x) = \dfrac{n}{x} j_n(x) + j_{n+1}(x)$

with identical relations if we replace $j_n(x)$ by $y_n(x)$, $h_n^{(1)}(x)$ or $h_n^{(2)}(x)$.

If n is a non-negative integer, the spherical Bessel functions can be put in differential forms (called Rayleigh's Formulae). These formulae are given by

I. $j_n(x) = (-1)^n x^n \left(\dfrac{1}{x}\dfrac{d}{dx}\right)^n \left(\dfrac{\sin x}{x}\right)$

II. $y_n(x) = -(-1)^n x^n \left(\dfrac{1}{x}\dfrac{d}{dx}\right)^n \left(\dfrac{\cos x}{x}\right)$

III. $h_n^{(1)}(x) = -i(-1)^n x^n \left(\dfrac{1}{x}\dfrac{d}{dx}\right)^n \left(\dfrac{e^{ix}}{x}\right)$

IV. $h_n^{(2)}(x) = i(-1)^n x^n \left(\dfrac{1}{x}\dfrac{d}{dx}\right)^n \left(\dfrac{e^{-ix}}{x}\right)$

We can use Rayleigh's formulae to find explicit expressions for few spherical Bessel functions of integer order, for example

$$j_1(x) = \dfrac{\sin x}{x^2} - \dfrac{\cos x}{x},$$

$$j_3(x) = \left(\dfrac{3}{x^3} - \dfrac{1}{x}\right)\sin x - \dfrac{3}{x^2}\cos x$$

Bessel Functions

$$y_1(x) = -\frac{\cos x}{x^2} - \frac{\sin x}{x},$$

$$y_3(x) = -\left(\frac{3}{x^3} - \frac{1}{x}\right)\cos x - \frac{3}{x^2}\sin x.$$

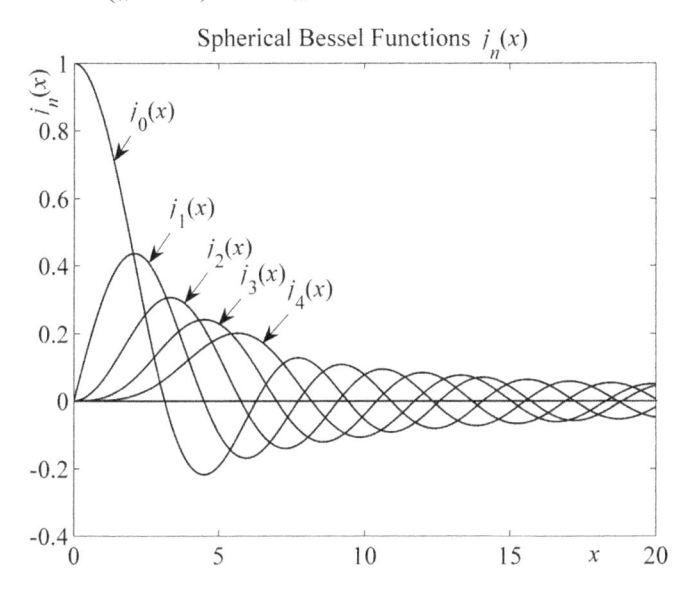

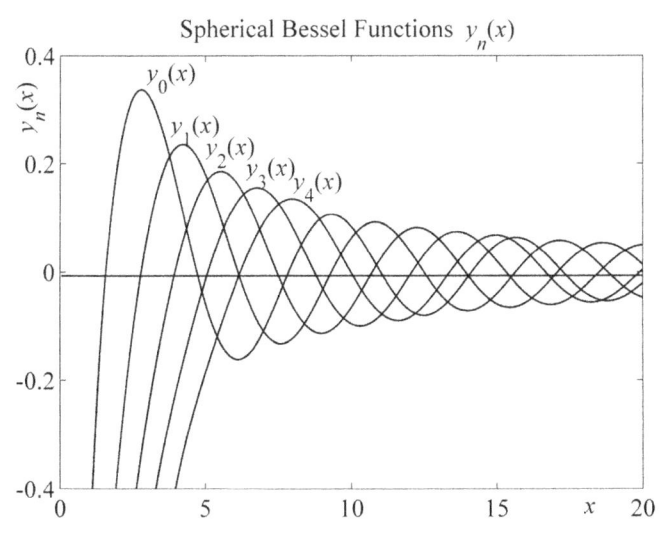

Spherical Bessel Functions $j_n(x)$

Spherical Bessel Functions $y_n(x)$

Special Functions and Orthogonal Polynomials

Example 31: Show that: i. $j_0(x) = \dfrac{\sin x}{x}$ ii. $y_0(x) = -\dfrac{\cos x}{x}$.

Solution: i. From equation (93), we have

$$j_0(x) = \sqrt{\dfrac{\pi}{2x}} J_{1/2}(x) = \sqrt{\dfrac{\pi}{2x}} \sum_{k=0}^{\infty} \dfrac{(-1)^k (x/2)^{1/2+2k}}{k!\,\Gamma(k+3/2)}.$$

But from Legendre duplication formula, and since k is an integer, we have

$$\Gamma(k+3/2) = \dfrac{\sqrt{\pi}(2k+2)!}{2^{2k+2}(k+1)!}, \text{ then}$$

$$j_0(x) = \sqrt{\dfrac{\pi}{2x}} \sum_{k=0}^{\infty} \dfrac{(-1)^k \, 2^{2k+2}(k+1)!(x/2)^{1/2+2k}}{k!\,2^{2k+1/2}\sqrt{\pi}(2k+2)!}$$

$$= \sum_{k=0}^{\infty} \dfrac{(-1)^k \, 2(k+1)x^{2k}}{(2k+2)!} = \dfrac{1}{x}\sum_{k=0}^{\infty} \dfrac{(-1)^k x^{2k+1}}{(2k+1)!} = \dfrac{\sin x}{x}.$$

ii. From equation (94), we have

$$y_0(x) = \sqrt{\dfrac{\pi}{2x}} Y_{1/2}(x) = \sqrt{\dfrac{\pi}{2x}} \dfrac{\cos(\pi/2)J_{1/2}(x) - J_{-1/2}(x)}{\sin(\pi/2)}$$

$$= -\sqrt{\dfrac{\pi}{2x}} J_{-1/2}(x) = -\sqrt{\dfrac{\pi}{2x}} \sum_{k=0}^{\infty} \dfrac{(-1)^k (x/2)^{2k-1/2}}{k!\,\Gamma(k+1/2)}.$$

But from Legendre duplication formula, and since k is an integer, we have

$$\Gamma(k+1/2) = \dfrac{\sqrt{\pi}(2k)!}{2^{2k}\,k!}, \text{ then}$$

$$y_0(x) = -\sqrt{\dfrac{\pi}{2x}} \sum_{k=0}^{\infty} \dfrac{(-1)^k \, 2^{2k} \, k!\, x^{2k-1/2}}{k!\,2^{2k-1/2}\sqrt{\pi}(2k)!}$$

$$= -\sum_{k=0}^{\infty} \dfrac{(-1)^k x^{2k-1}}{(2k)!} = -\dfrac{1}{x}\sum_{k=0}^{\infty} \dfrac{(-1)^k x^{2k}}{(2k)!} = -\dfrac{\cos x}{x}. \quad \square$$

6.15. Modified Spherical Bessel Functions

The *Modified Spherical Bessel Functions* of the first and second kinds are defined as

$$i_n(x) = \sqrt{\frac{\pi}{2x}} I_{n+1/2}(x) \tag{97}$$

$$k_n(x) = \sqrt{\frac{2}{\pi x}} K_{n+1/2}(x) \tag{98}$$

For $x > 0$, the first few Modified Spherical Bessel Functions are

$$i_0(x) = \frac{\sinh x}{x}, \quad i_1(x) = \frac{x \cosh x - \sinh x}{x^2},$$

$$i_2(x) = \frac{(x^2+3)\sinh x - 3x \cosh x}{x^3}$$

$$i_3(x) = \frac{(x^3+15x)\cosh x - (3x^2+15)\sinh x}{x^4}$$

$$i_4(x) = \frac{(x^4+45x^2+105)\sinh x - (10x^3+105x)\cosh x}{x^5}$$

$$k_0(x) = \frac{e^{-x}}{x}, \quad k_1(x) = \frac{e^{-x}(x+1)}{x^2}, \quad k_2(x) = \frac{e^{-x}(x^2+3x+3)}{x^3}$$

$$k_3(x) = \frac{e^{-x}(x^3+6x^2+15x+15)}{x^4}$$

$$k_4(x) = \frac{e^{-x}(x^4+10x^3+45x^2+105x+105)}{x^5}$$

The Modified Spherical Bessel functions satisfy the following relations.

$$i_{n+1}(x) = x^n \frac{d}{dx}\{x^{-n} i_n(x)\}, \quad i_n(x) = x^n \left(\frac{d}{x\,dx}\right)^n \frac{\sinh x}{x}$$

$$i_{n-1}(x) - i_{n+1}(x) = \frac{2n+1}{x} i_n(x), \quad i_n(x) = i^{-n} j_n(ix),$$

$$(2n+1)\frac{d}{dx} i_n(x) = n i_{n-1}(x) + (n+1) i_{n+1}(x).$$

$$k_{n+1}(x) = -x^n \frac{d}{dx}\{x^{-n} k_n(x)\}, \quad k_n(x) = (-1)^n x^n \left(\frac{d}{x\,dx}\right)^n \frac{e^{-x}}{x},$$

$$k_{n-1}(x) - k_{n+1}(x) = -\frac{2n+1}{x} k_n(x), \quad k_n(x) = -i^n h_n^{(1)}(ix),$$

Special Functions and Orthogonal Polynomials

$$(2n+1)\frac{d}{dx}k_n(x) = -nk_{n-1}(x) - (n+1)k_{n+1}(x).$$

Spherical Modified Bessel Functions $i_n(x)$

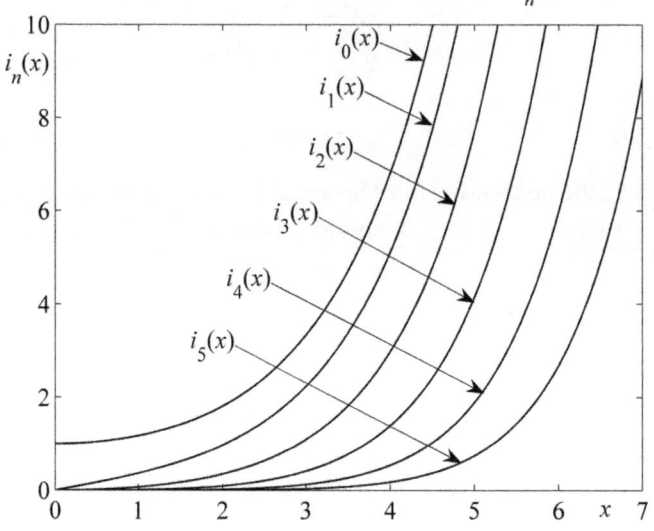

Spherical Modified Bessel Functions $k_n(x)$

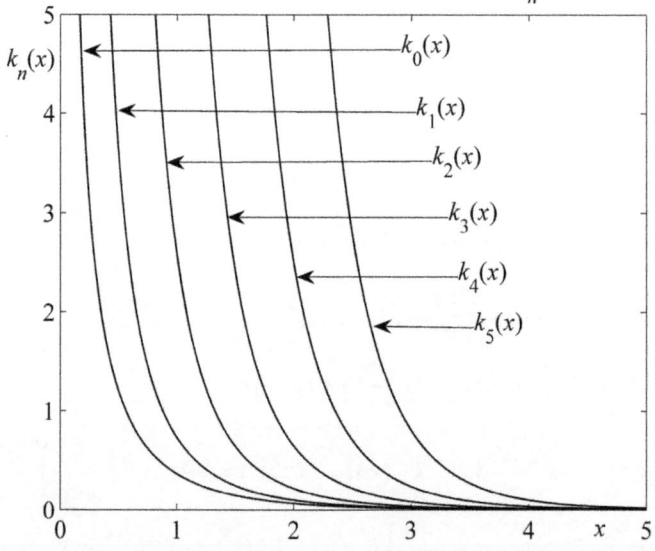

6.16. Airy Functions

The *Airy Functions* are defined as

$$\text{Ai}(x) = \frac{1}{\pi}\sqrt{\frac{x}{3}}\, K_{1/3}(2x^{3/2}/3) \tag{99}$$

$$\text{Bi}(x) = \sqrt{\frac{x}{3}}\left[I_{1/3}(2x^{3/2}/3) + I_{-1/3}(2x^{3/2}/3) \right] \tag{100}$$

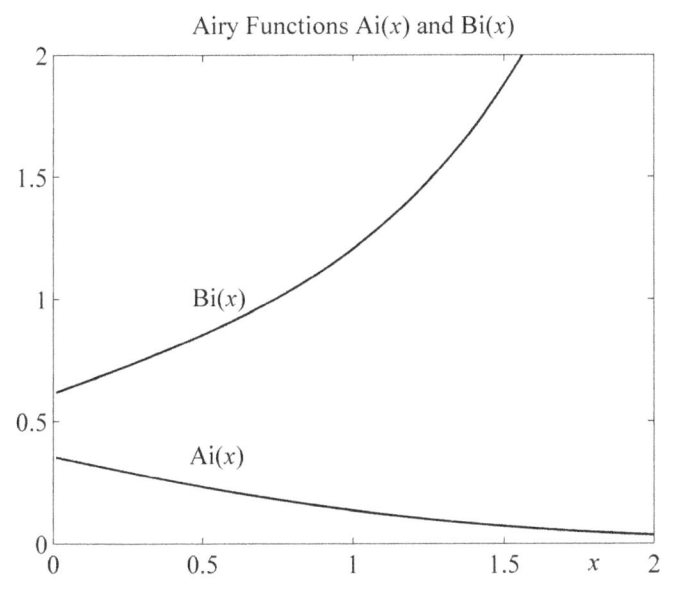

Airy Functions Ai(x) and Bi(x)

6.17. Summary of Bessel Functions

Bessel Differential Equation:

$$x^2 \frac{d^2 y}{dx^2} + x \frac{dy}{dx} + (x^2 - n^2) y = 0$$

Bessel Functions of the First Kind:

$$J_n(x) = \sum_{k=0}^{\infty} \frac{(-1)^k (x/2)^{n+2k}}{k!\, \Gamma(n+k+1)}$$

Bessel Functions of the Second Kind:

$$Y_n(x) = \frac{(\cos n\pi) J_n(x) - J_{-n}(x)}{\sin n\pi}$$

Generating Function:

$$e^{\frac{x}{2}(t - 1/t)} = \sum_{n=-\infty}^{\infty} t^n J_n(x)$$

Integral Form of Bessel Functions:

$$J_n(x) = \frac{1}{\pi} \int_0^{\pi} \cos(n\theta - x \sin\theta)\, d\theta$$

$$J_{-n}(x) = \frac{(-1)^n}{\pi} \int_0^{\pi} \cos(n\theta - x \sin\theta)\, d\theta$$

Recurrence Relations for Bessel Functions of the First Kind:

I. $\dfrac{d}{dx} \{x^n J_n(x)\} = x^n J_{n-1}(x)$

II. $\dfrac{d}{dx} \{x^{-n} J_n(x)\} = -x^{-n} J_{n+1}(x)$

III. $J'_n(x) = J_{n-1}(x) - \dfrac{n}{x} J_n(x)$

IV. $J'_n(x) = \dfrac{n}{x} J_n(x) - J_{n+1}(x)$

Bessel Functions

V. $J'_n(x) = \dfrac{1}{2}[J_{n-1}(x) - J_{n+1}(x)]$

VI. $J_{n-1}(x) + J_{n+1}(x) = \dfrac{2n}{x} J_n(x)$

Recurrence Relations for Bessel Functions of the Second Kind:

I. $\dfrac{d}{dx}\{x^n Y_n(x)\} = x^n Y_{n-1}(x)$

II. $\dfrac{d}{dx}\{x^{-n} Y_n(x)\} = -x^{-n} Y_{n+1}(x)$

III. $Y'_n(x) = Y_{n-1}(x) - \dfrac{n}{x} Y_n(x)$

IV. $Y'_n(x) = \dfrac{1}{2}[Y_{n-1}(x) - Y_{n+1}(x)]$

V. $Y_{n-1}(x) + Y_{n+1}(x) = \dfrac{2n}{x} Y_n(x)$

VI. $Y'_n(x) = \dfrac{n}{x} Y_n(x) - Y_{n+1}(x)$

Generalized Bessel Differential Equation and its Solution:

$$x^2 \dfrac{d^2 y}{dx^2} + (1 - 2\alpha)x \dfrac{dy}{dx} + \left[\beta^2 \gamma^2 x^{2\gamma} + (\alpha^2 - n^2 \gamma^2)\right] y = 0,$$

$$y = A x^\alpha J_n(\beta x^\gamma) + B x^\alpha Y_n(\beta x^\gamma)$$

Hankel Functions:

$$H_n^{(1)}(x) = J_n(x) + i Y_n(x)$$

$$H_n^{(2)}(x) = J_n(x) - i Y_n(x)$$

Recurrence Relations for Hankel Functions:

I. $\dfrac{d}{dx}\{x^n H_n^{(1)}(x)\} = x^n H_{n-1}^{(1)}(x)$ $\dfrac{d}{dx}\{x^n H_n^{(2)}(x)\} = x^n H_{n-1}^{(2)}(x)$

II. $\dfrac{d}{dx}\{x^{-n} H_n^{(1)}(x)\} = -x^{-n} H_{n+1}^{(1)}(x)$

Special Functions and Orthogonal Polynomials

$$\frac{d}{dx}\left\{x^{-n}H_n^{(2)}(x)\right\} = -x^{-n}H_{n+1}^{(2)}(x)$$

III. $\dfrac{d}{dx}H_n^{(1)}(x) = H_{n-1}^{(1)}(x) - \dfrac{n}{x}H_n^{(1)}(x)$

$$\frac{d}{dx}H_n^{(2)}(x) = H_{n-1}^{(2)}(x) - \frac{n}{x}H_n^{(2)}(x)$$

IV. $\dfrac{d}{dx}H_n^{(1)}(x) = \dfrac{1}{2}\left[H_{n-1}^{(1)}(x) - H_{n+1}^{(1)}(x)\right]$

$$\frac{d}{dx}H_n^{(2)}(x) = \frac{1}{2}\left[H_{n-1}^{(2)}(x) - H_{n+1}^{(2)}(x)\right]$$

V. $H_{n-1}^{(1)}(x) + H_{n+1}^{(1)}(x) = \dfrac{2n}{x}H_n^{(1)}(x) \qquad H_{n-1}^{(2)}(x) + H_{n=1}^{(2)}(x) = \dfrac{2n}{x}H_n^{(2)}(x)$

VI. $\dfrac{d}{dx}H_n^{(1)}(x) = \dfrac{n}{x}H_n^{(1)} - H_{n+1}^{(1)}(x) \qquad \dfrac{d}{dx}H_n^{(2)}(x) = \dfrac{n}{x}H_n^{(2)} - H_{n+1}^{(2)}(x)$

Modified Bessel Differential Equation:

$$x^2\frac{d^2y}{dx^2} + x\frac{dy}{dx} + (x^2 - n^2)y = 0$$

Modified Bessel Functions of the First and Second Kinds:

$$I_n(x) = \sum_{k=0}^{\infty} \frac{(x/2)^{n+2k}}{k!\,\Gamma(n+k+1)}$$

$$K_n(x) = \frac{\pi}{2}\frac{I_{-n}(x) - I_n(x)}{\sin n\pi}$$

Recurrence Relations for the Modified Bessel Functions $I_n(x)$ **and** $K_n(x)$:

For $I_n(x)$:

I. $\dfrac{d}{dx}\left\{x^n I_n(x)\right\} = x^n I_{n-1}(x)$

II. $\dfrac{d}{dx}\left\{x^{-n} I_n(x)\right\} = x^{-n} I_{n+1}(x)$

III. $\dfrac{d}{dx}I_n(x) = I_{n-1}(x) - \dfrac{n}{x}I_n(x)$

Bessel Functions

IV. $\dfrac{d}{dx} I_n(x) = \dfrac{1}{2}[I_{n-1}(x) + I_{n+1}(x)]$

V. $I_{n-1}(x) - I_{n+1}(x) = \dfrac{2n}{x} I_n(x)$

VI. $\dfrac{d}{dx} I_n(x) = \dfrac{n}{x} I_n(x) + I_{n+1}(x)$

And for $K_n(x)$

I. $\dfrac{d}{dx}\{x^n K_n(x)\} = -x^n K_{n-1}(x)$

II. $\dfrac{d}{dx}\{x^{-n} K_n(x)\} = -x^{-n} K_{n+1}(x)$

III. $\dfrac{d}{dx} K_n(x) = -K_{n-1}(x) - \dfrac{n}{x} K_n(x)$

IV. $\dfrac{d}{dx} K_n(x) = -\dfrac{1}{2}[K_{n-1}(x) + K_{n+1}(x)]$

V. $K_{n-1}(x) - K_{n+1}(x) = -\dfrac{2n}{x} K_n(x)$

VI. $\dfrac{d}{dx} K_n(x) = \dfrac{n}{x} K_n(x) - K_{n+1}(x)$

Kelvin Functions:

$$\text{ber}_n(x) = \text{Re}\{J_n(i^{3/2} x)\}$$

$$\text{bei}_n(x) = \text{Im}\{J_n(i^{3/2} x)\}$$

Orthogonality Property of Bessel Functions:

$$\int_0^a x J_n(\alpha_i x) J_n(\alpha_j x) dx = \begin{cases} 0 & \text{if } i \neq j \; (\alpha_i \neq \alpha_j) \\ \dfrac{a^2}{2} J_{n+1}^2(\alpha_i a) & \text{if } i = j \end{cases}$$

Spherical Bessel Functions:

$$j_m(x) = \sqrt{\dfrac{\pi}{2x}} J_{m+1/2}(x)$$

$$y_m(x) = \sqrt{\dfrac{\pi}{2x}} y_{m+1/2}(x)$$

Spherical Hankel functions:

$$h_n^{(1)}(x) = j_n(x) + i\, y_n(x)$$

$$h_n^{(2)}(x) = j_n(x) - i\, y_n(x)$$

Recurrence Relations For Spherical Bessel Functions $j_n(x)$:

I. $\dfrac{d}{dx}\{x^n j_n(x)\} = x^n j_{n-1}(x)$

II. $\dfrac{d}{dx}\{x^{-n} j_n(x)\} = -x^{-n} j_{n+1}(x)$

III. $\dfrac{d}{dx} j_n(x) = j_{n-1}(x) - \dfrac{n+1}{x} j_n(x)$

IV. $j_{n-1}(x) + j_{n+1}(x) = \dfrac{2n+1}{x} j_n(x)$

V. $(2n+1)\dfrac{d}{dx} j_n(x) = n j_{n-1}(x) - (n+1) j_{n+1}(x)$

VI. $\dfrac{d}{dx} j_n(x) = \dfrac{n}{x} j_n(x) + j_{n+1}(x)$

Modified Spherical Bessel Functions:

$$i_n(x) = \sqrt{\dfrac{\pi}{2x}}\, I_{n+1/2}(x)$$

$$k_n(x) = \sqrt{\dfrac{2}{\pi x}}\, K_{n+1/2}(x)$$

Airy Functions:

$$\text{Ai}(x) = \dfrac{1}{\pi}\sqrt{\dfrac{x}{3}}\, K_{1/3}(2x^{3/2}/3)$$

$$\text{Bi}(x) = \sqrt{\dfrac{x}{3}}\left[I_{1/3}(2x^{3/2}/3) + I_{-1/3}(2x^{3/2}/3)\right]$$

Exercises:

1. Show that:
   i. $J_n(x) = \dfrac{(x/2)^n}{\sqrt{\pi}\,\Gamma(n+1/2)} \displaystyle\int_{-1}^{1}(1-t^2)^{n-1/2} e^{ixt}\,dt, \quad n > -1/2$

   ii. $J_n(x) = \dfrac{x^n}{2^{n-1}\sqrt{\pi}\,\Gamma(n+1/2)} \displaystyle\int_{0}^{1}(1-t^2)^{n-1/2}\cos xt\,dt$

   iii. $J_n(x) = \dfrac{(x/2)^n}{\sqrt{\pi}\,\Gamma(n+1/2)} \displaystyle\int_{0}^{\pi}\cos(x\sin\theta)\cos^{2n}\theta\,d\theta$

   iv. $J_n(x) = \dfrac{(x/2)^n}{\sqrt{\pi}\,\Gamma(n+1/2)} \displaystyle\int_{0}^{\pi}\cos(x\sin\theta)\sin^{2n}\theta\,d\theta$

2. Show that $\displaystyle\lim_{x\to 0}\dfrac{J_n(x)}{x^n} = \dfrac{1}{2^n\,\Gamma(n+1)}$.

3. Show that $J_n(x) = (-1)^n\, 2^n\, x^n\, \dfrac{d^n}{d(x^2)^n} J_0(x)$.

4. Show that $y = x^{-n/2} J_n(2\sqrt{x})$ satisfies: $xy'' + (n+1)y' + y = 0$.

5. Show that $\displaystyle\int_{0}^{\pi/2} \sqrt{\pi x}\, J_{1/2}(2x)\,dx = 1$.

6. Show that:
   i. $J_0(x) = \dfrac{2}{\pi}\displaystyle\int_{0}^{1}\dfrac{\cos xt\,dt}{\sqrt{1-t^2}}$

   ii. $J_0(x) = \dfrac{1}{\pi}\displaystyle\int_{0}^{\pi}\cos(x\cos\theta)\,d\theta = \dfrac{1}{2\pi}\displaystyle\int_{0}^{2\pi} e^{\pm ix\cos\theta}\,d\theta$

7. Show that:

   i. $J_n(x)Y_n'(x) - Y_n(x)J_n'(x) = \dfrac{2}{\pi x}$,
   ii. $\displaystyle\int\dfrac{dx}{xJ_n^2(x)} = \dfrac{\pi}{2}\dfrac{Y_n(x)}{J_n(x)}$,

   iii. $\displaystyle\int\dfrac{dx}{xY_n^2(x)} = \dfrac{\pi}{2}\dfrac{J_n(x)}{Y_n(x)}$,
   iv. $\displaystyle\int\dfrac{dx}{xJ_n(x)Y_n(x)} = \dfrac{\pi}{2}\ln\left\{\dfrac{Y_n(x)}{J_n(x)}\right\}$

287

8. Show that $J_n''(x) = \left\{\dfrac{n(n+1)}{x^2} - 1\right\} J_n(x) - \dfrac{J_{n-1}(x)}{x}$.

9. Show that $\displaystyle\sum_{n=0}^{\infty} \dfrac{x^n J_n(1)}{n!} = J_0(\sqrt{1-2x})$.

10. Show that $J_n(x)J_{-n-1}(x) + J_{n+1}(x)J_{-n}(x) = -\dfrac{2\sin n\pi}{\pi x}$.

11. Show that:
    i. $\displaystyle\int x\, J_n^2(x)\, dx = \dfrac{x^2}{2}\{J_n^2(x) - J_{n-1}(x)J_{n+1}(x)\} + c$,

    ii. $\displaystyle\int \{J_{n-1}^2(x) - J_{n+1}^2(x)\} x\, dx = 2n J_n^2(x) + c$.

12. Expand x in a series of the form $\displaystyle\sum_{k=1}^{\infty} c_k J_1(\alpha_k x)$ valid in the interval $0 \leq x \leq 1$, where α_k are the roots of the equation $J_1(\alpha) = 0$.

    Ans: $x = 2\displaystyle\sum_{k=1}^{\infty} \dfrac{J_1(\alpha_k x)}{\alpha_k J_2(\alpha_k)}$, $0 \leq x \leq 1$

13. Expand x^2 in a series of the form $\displaystyle\sum_{k=1}^{\infty} c_k J_1(\alpha_k x)$ valid in the interval $0 \leq x \leq a$, where α_k are the roots of the equation $J_1(\alpha a) = 0$.

    Ans: $x^2 = \dfrac{2}{a}\displaystyle\sum_{k=1}^{\infty} \dfrac{\{(\alpha_k a)^2 - 4\} J_0(\alpha_k x)}{\alpha_k^3 J_0(\alpha_k a)}$, $0 \leq x \leq a$

14. Use Bessel series expansion to show that:

    $\ln x = -2\displaystyle\sum_{k=1}^{\infty} \dfrac{J_0(\alpha_k x)}{\alpha_k^2 J_1^2(\alpha_k)}$, $0 \leq x \leq 1$, where α_k are the roots of the equation $J_0(\alpha) = 0$.

15. Show that: $\displaystyle\int_0^1 x f^2(x)\, dx = \sum_{k=1}^{\infty} \alpha_k^2 J_k^2(\alpha_k)$, where α_k are such that $J_0(\alpha_k) = 0$, $k = 1, 2, 3, \cdots$.

Bessel Functions

16. If α_k are the positive roots of $J_0(\alpha) = 0$, show that:

$$\frac{1-x^2}{8} = \sum_{k=1}^{\infty} \frac{J_0(\alpha_k x)}{\alpha_k^3 J_1(\alpha_k)}, \quad -1 \leq x \leq 1.$$

17. If α_k are the positive roots of $J_1(\alpha) = 0$, show that:

$$x = 2 \sum_{k=1}^{\infty} \frac{J_1(\alpha_k x)}{\alpha_k J_2(\alpha_k)}, \quad -1 \leq x \leq 1.$$

18. If α_k are the positive roots of $J_1(\alpha) = 0$, show that:

$$x^3 = 2 \sum_{k=1}^{\infty} \frac{(8-\alpha_k^2) J_1(\alpha_k x)}{\alpha_k^3 J_1'(\alpha_k)}, \quad -1 \leq x \leq 1.$$

19. Find the general solution of the following differential equations:

i. $xy'' - y' + xy = 0$ **Ans:** $y = A x J_1(x) + B x Y_1(x)$

ii. $\sqrt{x} y'' + y = 0$ **Ans:** $y = A\sqrt{x} J_{2/3}(4x^{3/4}/3) + B\sqrt{x} Y_{2/3}(4x^{3/4}/3)$

iii. $x^2 y'' - 2xy' + 4(x^4 - 1) y = 0$

Ans: $y = A x^{3/2} J_{5/4}(x^2) + B x^{3/2} Y_{5/4}(x^2)$

iv. $x^2 y'' + 3xy' + 4x^2 y = 0$ **Ans:** $y = A \dfrac{J_1(2x)}{x} + B \dfrac{Y_1(2x)}{x}$

20. Show that: i. $\dfrac{d}{dx}\{x J_n(x) J_{n+1}(x)\} = x \{J_n^2(x) - J_{n+1}^2(x)\}$

ii. $x = 2 \sum_{k=0}^{\infty} (2k+1) J_k(x) J_{k+1}(x)$

21. Show that: i. $\left\{\dfrac{1}{x}\dfrac{d}{dx}\right\}^m \{x^n J_n(x)\} = x^{n-m} J_{n-m}(x), \quad m < n$

ii. $\left\{\dfrac{1}{x}\dfrac{d}{dx}\right\}^m \{x^{-n} J_n(x)\} = (-1)^m x^{-n-m} J_{n+m}(x)$

iii. $x^{-n} J_n(x) = (-1)^n \left\{\dfrac{1}{x}\dfrac{d}{dx}\right\}^n J_0(x)$

Special Functions and Orthogonal Polynomials

22. Show that: i. $J_{-5/2}(x) = \sqrt{\dfrac{2}{\pi x}} \left\{ \dfrac{3-x^2}{x^2} \cos x + \dfrac{3}{x} \sin x \right\}$

    ii. $J_{5/2}(x) = \sqrt{\dfrac{2}{\pi x}} \left\{ \dfrac{3-x^2}{x^2} \sin x + \dfrac{3}{x} \cos x \right\}$

23. Express i. $\int x^{-3} J_4(x)\,dx$ in terms of $J_0(x)$ and $J_1(x)$.

24. Show that: i. $\int x^{-1} J_4(x)\,dx = -x^{-1} J_3(x) - 2x^{-2} J_2(x) + c$

    ii. $\displaystyle\int_0^x t^3 J_0(t)\,dt = x^3 J_0(x) - 2x^2 J_2(x)$

    iii. $\displaystyle\int_0^x t^4 J_0(t) J_1(t)\,dt = \dfrac{x^2}{2} J_1^2(x)$

    iv. $\displaystyle\int \dfrac{J_2(x)}{x^2}\,dx = -\dfrac{J_2(x)}{3x} - \dfrac{J_1(x)}{3} + \dfrac{1}{3}\int J_0(x)\,dx$

    v. $\displaystyle\int J_0(x) \sin x\,dx = x J_0(x) \sin x - x J_1(x) \cos x + c$

25. Show that: $J_0(x) = \dfrac{2}{\pi} \displaystyle\int_0^1 \dfrac{\cos(xt)\,dt}{\sqrt{1-t^2}}$.

26. If $n \geq 0$, show that $\displaystyle\int_0^\infty J_n(\beta x)\,dx = \dfrac{1}{\beta}$.

27. Show that: $\displaystyle\int_0^\infty \dfrac{J_n(x)}{x}\,dx = \dfrac{1}{n}$.

28. Show that: i. $\displaystyle\int_0^\infty \sin \alpha x\, J_0(\beta x)\,dx = \begin{cases} 0 & \beta > \alpha \\ \dfrac{1}{\sqrt{\alpha^2 + \beta^2}} & \beta < \alpha \end{cases}$

    ii. $\displaystyle\int_0^\infty \cos \alpha x\, J_0(\beta x)\,dx = \begin{cases} \dfrac{1}{\sqrt{\alpha^2 + \beta^2}} & \beta > \alpha \\ 0 & \beta < \alpha \end{cases}$

Bessel Functions

29. Show that $\int_0^\infty e^{-\alpha x} J_0(\beta x)dx = \dfrac{1}{\sqrt{\alpha^2 + \beta^2}}$,

   then evaluate $\int_0^\infty x\, e^{-\alpha x} J_1(\beta x)dx$.

30. Show that: i. $\int_0^{\pi/2} J_0(x\cos\theta)\cos\theta\, d\theta = \dfrac{\sin x}{x}$

   ii. $\int_0^{\pi/2} J_1(x\cos\theta)d\theta = \dfrac{1-\cos x}{x}$

31. Show that $x^{n+1}J_{n+1}(x) = \dfrac{1}{2^n n!}\int_0^x t(x^2-t^2)^n J_0(t)dt$ then, using the

   substitution $t = x\sin\theta$, deduce that:

   $J_{n+1}(x) = \dfrac{x^{n+1}}{2^n n!}\int_0^{\pi/2}\sin\theta\cos^{2n+1}\theta J_0(x\sin\theta)d\theta$.

32. Show that: i. $I_n(x)I'_{-n}(x) - I_n(x)I'_{-n}(x) = -\dfrac{2\sin n\pi}{\pi x}$

   ii. $I_n(x)K'_n(x) - K_n(x)I'_n(x) = -\dfrac{1}{x}$

33. Show that $I_0(x) = \dfrac{1}{\pi}\int_0^\pi e^{x\cos\theta}d\theta$.

34. Prove that $\sum_{n=-\infty}^\infty J_n(kx)t^n = e^{\frac{x}{2t}\left(k-\frac{1}{k}\right)} \sum_{n=-\infty}^\infty k^n t^n J_n(x)$ and hence

   deduce that $I_n(x) = \sum_{k=0}^\infty \dfrac{x^k J_{n+k}(x)}{k!}$.

35. Show that the modified Bessel function $I_n(x)$ satisfy the generating

   relation $e^{\frac{x}{2}\left(t+\frac{1}{t}\right)} = \sum_{n=-\infty}^\infty t^n I_n(x)$.

36. Show that $J_0(\sqrt{i}\,x) = \text{ber}_0(x) - i\,\text{bei}_0(x)$.

37. Verify that: i. $\text{ber}_n(x) = \sum_{k=0}^{\infty} \dfrac{(-1)^k \cos[3(n+2k)\pi/4]x^{n+2k}}{2^{n+2k}\,k!\,\Gamma(n+k+1)}$

   ii. $\text{bei}_n(x) = \sum_{k=0}^{\infty} \dfrac{(-1)^k \sin[3(n+2k)\pi/4]x^{n+2k}}{2^{n+2k}\,k!\,\Gamma(n+k+1)}$.

38. Show that: i. $\int_0^x t\,\text{ber}_0(t)\,dt = x\,\text{bei}_0'(x)$ ii. $\int_0^x t\,\text{bei}_0(t)\,dt = -x\,\text{ber}_0'(x)$.

39. Show that: i. $h_0^{(1)}(x) = -\dfrac{i e^{ix}}{x}$ ii. $h_0^{(2)}(x) = \dfrac{i e^{-ix}}{x}$.

40. Show that:

   i. $j_1(x) = \dfrac{\sin x}{x^2} - \dfrac{\cos x}{x}$ ii. $j_3(x) = \left(\dfrac{3}{x^3} - \dfrac{1}{x}\right)\sin x - \dfrac{3}{x^2}\cos x$

   iii. $y_1(x) = -\dfrac{\cos x}{x^2} - \dfrac{\sin x}{x}$ iv. $y_3(x) = -\left(\dfrac{3}{x^3} - \dfrac{1}{x}\right)\cos x - \dfrac{3}{x^2}\sin x$

41. Show that the spherical Bessel functions satisfy the property:

$$\int_0^{\infty} j_m(x)\,j_n(x)\,dx = \begin{cases} \dfrac{\sin\{(n-m)\pi/2\}}{n(n+1)-m(m+1)} & \text{if } m \neq n \\[2mm] \dfrac{\pi}{2(2n+1)} & \text{if } m = n \end{cases}$$

42. Show that $J_0(t\sqrt{1-x^2}) = e^{-xt} \sum_{n=0}^{\infty} \dfrac{P_n(x)}{n!} t^n$, where $P_n(x)$ is Legendre polynomial of order n.

References

[1] Abramowitz, M. and Stegun, I.A. (Eds.), *Handbook of Mathematical Functions with Formulas, Graphs, and Mathematical Tables, 9th printing*, Dover Publications, New York, 1972, also John Wiley & Sons Inc; Reprint edition, **ISBN:** 0471800074, September 1993, 1060 pages.

[2] Aizenshtadt, V.S., Krylov, V.S. and Metel'skii, A.S., *Tables of Laguerre Polynomials and Functions*, Mathematical Table Series Volume 39, Pergamon, Oxford, England, 1966, 149 pages.

[3] Andrews, George E., Askey, Richard and Roy, Ranjan, *Special Functions*, Cambridge University Press, 1st ed., ISBN: 0521789885, February 2001, 620 pages. Also ISBN: 0521623219.

[4] Andrews, George E., *Special Functions*, Cambridge University Press, ISBN: 0521623219, January 1999, 664 pages.

[5] Andrews, Larry C., *Special Functions of Mathematics for Engineers*, Second Edition, SPIE-International Society for Optical Engine, November, 1997, ISBN: 0819426164, 504 pages.

[6] Andrews, Larry C., *Special Functions for Engineers and Applied Mathematics*, New York, Macmillan, ISBN: 0029486505, January 1985.

[7] Andrews, Larry C., *Special Functions of Mathematics for Engineers*, McGraw-Hill, ISBN: 0070018480, October 1991.

[8] Arfken, George B. and Weber, Hans, *Mathematical Methods for Physicists*, Academic Press, 5th edition, October, 2000, ISBN: 0120598256, 1112 pages.

[9] Artin, Emil, *The Gamma Function*, Holt, Rinehart and Winston, New York, 1964, 39 pages.

[10] Askey, Richard A. (Editor), *Theory and Application of Special Functions*, Proceedings of an Advanced Seminar Sponsored by the Mathematics Research Center, the University of Wisconsin-Madison, March 21-April 2, 1975, New York, Academic Press, ISNB: 0120648504, June 1975, 560 pages.

[11] Askey, Richard A., *Orthogonal Polynomials and Special Functions*, Regional Conference Series in Applied Mathematics 21, SIAM, Philadelphia, ISBN: 0898710189, June 1975, 110 pages.

[12] Askey, Richard A., Koornwinder, T.H. and Schempp, W. (Editors), "Special Functions: Group Theoretical Aspects and Applications." *Mathematics and Its Applications*, Vol. 18. Reidel, Dordrecht-Boston-Lancaster, ISBN: 9027718229, 1984, 352 pages.

[13] Askey, Richard and Ismail, Mourad, *Recurrence Relations, Continued Fractions, and Orthogonal Polynomials*, American Mathematical Society, ISBN: 0821823019, August 1984, 108 pages.

[14] Askey, Richard A., Schempp, W. and Koornwinder, T.H, *Special Functions: Group Theoretical Aspects and Applications,* (Mathematics and Its Applications), Kluwer Academic Publishers, January, 2002, ISBN: 1402003196, 348 pages.

[15] Askey, R.A. and Wilson, J., *Some basic hypergeometric orthogonal polynomials that generalize Jacobi polynomials*, Memoirs Amer. Math. Soc. 319, Amer Mathematical Society, Providence, Rhode Island, ISBN: 0821823213, April 1985, 55 pages.

[16] Babister, A.W., *Transcendental Functions Satisfying Nonhomogeneous Linear Differential Equations*, New York, Macmillan, **ASIN:** B0006BOKMW, 1967, 414 pages.

[17] Bailey, Wilfrid Norman, *Generalised Hypergeometric Series*, Hafner Pub. Co., 1972, 108 pages.

[18] Bailey, W.N., "On the Product of Two Legendre Polynomials." *Proc. Cambridge Philos. Soc.* 29, 173-177, 1933.

[19] Barbeau, E.J., *Polynomials*. New York, Springer-Verlag, ISBN: 0387969195, June 1989. 441 pages. Also Springer Verlag, October, 2003, ISBN: 0387406271, 455 pages.

[20] Barnerji, P.K. (Editor), *Special Functions: Selected Articles*, ISBN: 817233267X.

[21] Barnes, E.W., *The theory of the gamma function*, Messenger Math. (2),1900, vol. 29, p. 64-128.

[22] Bell, William Wallace, *Special Functions for Scientists and Engineers*, London, van Nostrand Comp. Ltd., ISBN: 0442006829, 1968. 247 pages; also Dover Publications, July, 2004, ISBN: 0486435210, 272 pages.

[23] Belousov, S. L., *Tables of normalized Associated Legendre Polynomials*, New York, MacMillan Company, 1962, 379 pages.

[24] Berlyand, Gavrilova, Prudnikov , *Tables of Integral Error Functions and Hermite Polynomials*, Pergamon Press, Oxford, NY, 1962, 163 pages.

[25] Berndt, B.C., *Ramanujan's Notebooks, Part IV*, Springer-Verlag, New York, 1994.

[26] Beyer, W.H., *CRC Standard Mathematical Tables, 28th ed*, CRC Press, Boca Raton, FL, 1987.

[27] Bickley, William G., *Bessel Functions and Formulae*, Cambridge, England, Cambridge University Press, ISBN: 052106158X, January 1953.

[28] Boas, Mary L., *Mathematical Methods in the Physical Sciences*, John Wiley & Sons; 3 edition, ISBN: 0471198269, 2005, 864 pages.

[29] Borwein, J. and Bailey, D., *Mathematics by Experiment: Plausible Reasoning in the 21st Century*, A. K. Peters, Natick, MA, 2003.

[30] Borwein, J. and Borwein, P. B., *Pi & the AGM: A Study in Analytic Number Theory and Computational Complexity*, Wiley, New York, 1987.

[31] Bourget, J.P.L., "Sur les intégrales Eulériennes et quelques autres fonctions uniformes", *Acta Mathematica*, v. 2, 1883, pp. 261-295.

[32] Bowman, Frank, *An Introduction to Bessel Functions*, New York, Dover Publications, ISBN: 0486604624, March 1968. 135 pages.

[33] Briggs, Lyman J. and Lowan, Arnold N., *Tables of Associated Legendre*

References

Functions, Columbia University Press, New York, NY, 1945, 303 pages.

[34] Brychkov, Yu.A. and Prudnikov, P.A., ***Integrals and Series: More Special Functions***, T&F STM; 3rd edition, ISBN: 2881246826, January 1990, 800 pages.

[35] Bustoz, Joaquin, Ismail, Mourad E.H. and Suslov, S.K. (Editors), ***Special Functions 2000: Current Perspective and Future Directions***, Proceedings of the NATO Advanced Study Institute on Special Functions 2000, held in Tempe, Arizona, USA, NATO Science Series II: Mathematics, Physics and Chemistry , Vol. 30, Kluwer Academic Pub., ISBN: 0-7923-7120-8, August, 2001, 536 pages.

[36] Byerly, William Elwood, ***An Elementary Treatise on Fourier's Series, and Spherical, Cylindrical, and Ellipsoidal Harmonics, with Applications to Problems in Mathematical Physics***, Dover Publications, December, 2003, ISBN: 0486495469, 304 pages.

[37] Cambi, Enzo, ***Eleven and fifteen-place tables of Bessel functions of the first kind to all significant orders***, New York, Dover Publications, Inc., 1948, 154 pages.

[38] Campbell, R., ***Les intégrales eulériennes et leurs applications***, Dunod, Paris, 1966.

[39] Carlson, Bille Chandler, ***Special Functions of Applied Mathematics***, New York, Academic Press, ISBN: 0121601501, June 1977, 335 pages.

[40] Chakrabarti, A., ***Elements of Ordinary Differential Equations and Special Functions***, Wiley, ISBN: 0470216409, January 1990, 148 Pages, also New Age International (P) Ltd. , New Delhi, ISBN: 812240880X, 2002, 158 pages.

[41] Chihara, Theodore Seio, ***An Introduction to Orthogonal Polynomials***, Gordon and Breach Publications, New York, **ISBN**: 0677041500, April 1978, 250 pages.

[42] Chistova, E.A., ***Tables of Bessel Functions of the True Argument and of Integrals Derived From Them***, Pergamon Press, 1959.

[43] Cholewinski, Frank M., ***The Finite Calculus Associated With Bessel Functions***, Contemporary Mathematics, Vol. 75, American Mathematical Society, ISBN: 0821850830, November 1988. 122 pages.

[44] Davis, H.T., ***Tables of Higher Mathematical Functions***, Principia Press, Bloomington, IN, 1935.

[45] Davis, P. J., "Leonhard Euler's Integral: A Historical Profile of the Gamma Function", ***Amer. Math. Monthly* 66**, 1959, pp. 849-869.

[46] Dieudonne, Jean A., Special Functions and Linear Representations of Lie Groups, Amer Mathematical Society, ISBN: 0821816926, May 1980, 59 pages.

[47] Dunkl, Charles and Xu, Yuan, ***Orthogonal Polynomials of Several Variables***, ISBN: 0-521-80043-9, April 2001, 400 Pages.

[48] Dunkl, Charles, Ismail, Monral and Wong, Roderick (Eds.), ***Proceedings of the International Workshop Special Functions***, Hong Kong, 21-25 June 1999, Singapore, World Scientific, **ISBN**: 9810243936, 2000, 438 pages.

[49] Dwork, Bernard M., ***Generalized Hypergeometric Functions***, Oxford Mathematical Monographs, Clarendon Pr., December, 1990, ISBN 0198535678, 188 pages.

[50] Erdélyi, Arthur, Magnus, W., Oberhettinger, F. and Tricomi, F.G., *Higher Transcendental Fuctions* vols. 1-3, McGraw-Hill, New York, 1953, also 5 vols., Krieger Publishing, Melbourne, Fla., ISBN: 0898742072, June 1981.

[51] Erdélyi, A.; Magnus, W., Oberhettinger, F. and Tricomi, F.G., *Tables of Integral Transforms*, Vols. 1-2, McGraw-Hill, New York, ISBN: 0070195501, 1954.

[52] Finch, S. R., *Mathematical Constants*, Cambridge University Press, Cambridge, England, 2003.

[53] Fletcher, A., Miller, J.C.P., Rosenhead, L. and Comrie, L.J. *An Index of Mathematical Tables*, vols. 1 and 2, 2nd ed., Addison-Wesley, Reading, Mass., 1962, 994 pages.

[54] Fox, L. and Parker, I.B., *Chebychev Polynomials in Numerical Analysis*, London, Oxford University Press, 1968, 205 pages.

[55] Freud, Geza, *Orthogonal Polynomials*, Birkhauser, Basel, 1969; English translation, Pergamon Press, Oxford, ISBN: 0080160476, 1971, 294 pages.

[56] Gautschi, W., Golub, G.H. and Opfer, G. (Eds.), *Applications and Computation of Orthogonal Polynomials*, Conference at the Mathematical Research Institute Oberwolfach, Germany, March 22-28, 1998, Basel, Switzerland, Birkhäuser, ISBN: 3764361379, 1999, 279 pages.

[57] Geronimus, Ya. L., *Polynomials Orthogonal on a Circle and Interval*, International Series of Monographs on Pure and Applied Mathematics, Vol. 18., Pergamon Press, Oxford-London-New York-Paris, 1961. 210 pages.

[58] Gormley, P.G., *The Zeros of Legendre Functions*, Proceedings of the Royal Irish Academy, Vol. XLIV, Sec. A, no. 4, Dublin 1937.

[59] Goudet, Georges, *Les fonctions de Bessel et leurs applications en physique*, Paris: Masson, 1943. 80 pages.

[60] Gray, Andrew and Mathews, G.B., *A Treatise on Bessel Functions*, Macmillan, 1922, 327 pages.

[61] Gray, Andrew and Mathiews, G.B., *A Treatise on Bessel Functions and Their Applications to Physics*, 2nd ed., New York, Dover Publications, 1966, 327 pages.

[62] Gumprecht, R.O. and Sliepcevich, C.M., *Functions of Partial Derivatives of Legendre Polynomials*, University of Michigan Press, Ann Arbor, MI, 1951, 310 pages.

[63] Gupta, B.D., *Mathematical Methods with Special Functions*, Stosius Inc/Advent Books Division, ISBN: 8122002684, 1992, 906 pages.

[64] Hardy, G.H., "A Chapter from Ramanujan's Note-Book", *Proc. Cambridge Philos. Soc.* **21**, 1923, pp. 492-503.

[65] Hardy, G.H., "Some Formulae of Ramanujan", *Proc. London Math. Soc.*, **22**, 1924.

[66] Hardy, G.H., *Ramanujan: Twelve Lectures on Subjects Suggested by His Life and Work, 3rd ed.*, Chelsea, New York, 1999.

References

[67] Havil, J., **Gamma: Exploring Euler's Constant**, Princeton University Press, Princeton, NJ, 2003.

[68] Henrici, Peter, **Applied and Computational Complex Analysis: Special Functions, Integral Transforms, Asymptotics, Continued Fractions**, John Wiley & Sons Inc., ISBN: 047154289X, March 1991, 672 pages.

[69] Higgins, J.R.; Bollobas, Bela (Editor); and Fulton, W., **Completeness And Basis Properties Of Sets Of Special Functions**, Cambridge University Press, ISBN: 0521604885, May 2004, 144 pages.

[70] Hochstadt, Harry, **Special Functions of Mathematical Physics**, Dover Publications, New York, ISBN: 0486652149, 1986, 332 pages.

[71] Iyanaga, Shokichi and Kawada, Yukiyosi, (Editors). **Encyclopedic Dictionary of Mathematics**. Cambridge, MA, MIT Press, ISBN: 0262590107, 1980.

[72] Jahnke, E., Emde, F. and Lösch, F. **Tables of Higher Functions**, McGraw-Hill, New York , 1960.

[73] Jones, W.W. (Editor), **Royal Society Shorter Mathematical Tables 1: A Short Table for Bessel functions**, ISBN: 0521061571, 1952.

[74] Johnson D.E. and Johnson, Johnny R., **Mathematical Problems in Engineering & Physics: Special Functions & Boundary Value Problems**, Ronald P., US, ISBN: 0826047904, December 1965.

[75] Kashiwara, Masaki, **Special Functions**, Proceedings of the Hayashibara Forum 1990, Held in Fujisaki Institute, Okayama, Japan, August 16-20, 1990, Springer Verlag, ISBN: 0387700854, September, 1991, 317 pages.

[76] Koelink, E. and Van Assche, W. (Editors), **Orthogonal Polynomials and Special Functions: Leuven 2002 (*Lecture Notes in Mathematics* 1817)**, Springer Verlag, ISBN: 3540403752, August 2003, 249 pages.

[77] Koepf, W., **Hypergeometric Summation: An Algorithmic Approach to Summation and Special Function Identities**, Vieweg, Braunschweig, Germany, 1998.

[78] Korenev, Boris G., **Bessel Functions and Their Applications**, Taylor and Francis, New York, ISBN: 041528130X, September, 2002, 280 pages.

[79] Krantz, S.G., **Handbook of Complex Variables**, Birkhäuser, Boston, MA, 1999.

[80] Lagrange, René. **Polynomes et fonctions de Legendre**. Paris, Gauthier-Villars, 1939.

[81] Le Lionnais, F., **Les nombres remarquables**, Hermann, Paris, 1983.

[82] Lebedev, A.V. and Fedorova, R.M. **A Guide to Mathematical Tables**, Pergamon Press, Oxford, 1960.

[83] Lebedev, Nikolai Nikolaevich, **Special Functions and Their Applications**, Prentice-Hall, Englewood Cliffs, New Jersey, **ASIN:** B0006BMX4E 1965, 308 pages; also revised English edition, New York, Dover Publications, ISBN: 0486606244, 1972, 308 pages.

[84] Legendre, A.M. "Sur l'attraction des Sphéroides." *Mém. Math. et Phys. présentés*

à l'Ac. r. des. sc. par divers savants **10**, 1785.

[85] Levin, A. and Lubinsky, D.S., "Christoffel Functions and Orthogonal Polynomials for Exponential Weights on [-1,1]", *Memoirs of the American Mathematical Society*, No.535, Vol. 11, 1994.

[86] Lowell, Herman H., *Tables of the Bessel-Kelvin Functions Ber, Bei, Ker, Kei, and Their Derivatives*, Washington, DC GPO 1959 NASA Technical Report R-32, 291 pages.

[87] Luke, Yudell L., *Integrals of Bessel Functions*, New York, McGraw-Hill, 1962, 419 pages.

[88] Luke, Yudell L., *The Special Functions and their Approximations*, Vols. 1 and 2, New York, Academic Press, 1969.

[89] Macdonald, I.G., *Symmetric Functions and Orthogonal Polynomials*, American Mathematical Society, ISBN: 0821807706, June 1998, 53 pages.

[90] MacRobert, Thomas Murray, *Spherical Harmonics*, Pergamon Press; 3rd edition, 1967, 349 pages.

[91] Magnus, W.; Oberhettinger, F.; and Soni, R.P., *Formulas and Theorems for the Special Functions of Mathematical Physics*, 3rd ed., Springer-Verlag, New York, ISBN: 0387035184, 1966, 508 pages.

[92] H.L Manocha, H.L.; Srinivasa Rao, K.; and Agarwal, R.P., *Selected Topics in Special Functions*, Allied, New Delhi, ISBN: 8177641697.

[93] Marcellan, Francisco, *Laredo Lectures on Orthogonal Polynomials and Special Functions*, Nova Science Pub. Inc., ISBN: 1594540098, June 2004, 210 pages.

[94] Marcum , J. I., *Tables of Hermite polynomials and the derivatives of the error functions*, Rand Corporation, 1948.

[95] Margenau, Henry and Murphy, G.M., *The Mathematics of Physics and Chemistry*, R. E. Krieger Pub. Co; 2d ed edition (1976), **ISBN:** 0882754238, 604 pages.

[96] Mathai, A.M., *A Handbook of Generalized Special Functions for Statistical and Physical Sciences*. Clarendon Press, ISBN: 0198535953, 1993, 235 pages.

[97] McLachlan, Norman William, *Bessel Functions for Engineers*, 2nd ed., Oxford University Press, London, 1955, 192 pages.

[98] McLachlan, Norman William, *Bessel Functions for Engineers*, 2nd ed. with corrections, Oxford, England Clarendon Press, 1961, 239 pages.

[99] Miller, Willard, Jr., *Lie Theory and Special Functions*, Academic Press, New York, **ASIN:** B0006BUXE6, June 1968, 338 pages.

[100] Morse, P.M. and Feshbach, H. *Methods of Theoretical Physics, Part I*, New York, McGraw-Hill, **ISBN:** 007043316X, June 1953, 997 pages.

[101] National Bureau of Standards**,** *Tables of Associated Legendre functions,* Columbia University Press 1945, 354 pages.

[102] National Bureau of Standards, *Tables of Bessel Functions of Fractional Order,* Columbia University Press, 1948, 413 pages; New York, Columbia

References

University Press, 1949 Volume II, 365 pages.

[103] National Bureau of Standards, *Table of the Bessel Functions $J_0(z)$ and $J_1(z)$ for Complex Arguments*, Columbia University Press, New York, 1947, 403 pages.

[104] Nemeth, Geza, *Mathematical Approximations of Special Functions: Ten Papers on Chebyshev Expansions*, Nova Science Pub. Inc., ISBN: 0941743950, 1991.

[105] Nevai, Paul G., *Orthogonal Polynomials*, American Mathematical Society, Reprint edition, **ISBN:** 0821822136, June 1980, 185 pages.

[106] Nevai, P.G. (Ed.): *Orthogonal Polynomials: Theory and Practice*, Proceedings of the NATO Advanced Study Institute on Orthogonal Polynomials and Their Applications, Colombus, Ohio, U.S.A., May 22-June 3, 1989, Kluwer Academic Publ., Dordrecht-Boston-London, **ISBN:** 0792305698, January 1990, 488 pages.

[107] Nikiforov, A.F. and Uvarov, V.B., *Special Functions of Mathematical Physics*, Translated from the Russian by R. P. Boas, Birkhauser, Basel, ISBN: 3764331836, 1988.

[108] Nikiforov, Arnold F.; and Uvarov, V.B., *Special Functions of Mathematical Physics: A Unified Introduction With Applications*, ISBN: 0817631836, Birkhauser Boston, February, 1988, 448 pages.

[109] Nikiforov, Arnold F.; Suslov, S.S.; and Uvarov, V.B., *Classical Orthogonal Polynomials of a Discrete Variable*, Springer-Verlag, Berlin-Heidelberg-New York, ISBN: 0387511237, 1992, 374 pages.

[110] Nikishin, E.M. and Sorokin, V.N., *Rational Approximations and Orthogonality*, translations of Mathematical Monographs 92, American Mathematical Society, Providence, Rhode Island, ISBN: 0821845454 1991. 221 pages.

[111] Onoe, Morio, *Tables of Modified Quotients of Bessel Functions of the First Kind for Real and Imaginary Arguments*, Columbia University Press, New York, 1958, 338 pages.

[112] Olver, Frank W.J., *Asymptotics and Special Functions*, AK Peters Ltd, 2nd edition, **ISBN:** 1568810695, June 1997, also Academic Press, ISBN: 012525850X, 1974.

[113] Olver, Frank W.J., Editor, Royal Society Mathematical Tables: Volume 7, *Bessel Functions, Part 3, Zeros and Associated Values*, Cambridge University Press, **ISBN:** 0521061539, January, 1960, 79 pages.

[114] Petiau, Gérard. *La théorie des fonctions de Bessel exposée en vue de ses applications à la Physique Mathématique*, Paris, CNRS, 1955. 477 pages.

[115] Press, W.H., Flannery, B.P., Teukolsky, S.A., and Vetterling, W.T., *Numerical Recipes in FORTRAN: The Art of Scientific Computing, 2nd ed.*, Cambridge, England, Cambridge University Press, 1992.

Special Functions and Orthogonal Polynomials

[116] Prudnikov, A.P., Brychkov, Yu. A. and Marichev, O.I., *Integrals and Series*, Vols. 1-3, T&F STM, ISBN: 2881247369, June, 1990, 2348 pages.

[117] Rainville, Earl David, *Special Functions*, Chelsea, New York, **ISBN**: 0828402582, June 1971, 365 pages.

[118] Rao Srinivasa K. et al, *Special Functions and Differential Equations*, Proceedings of a Workshop Held at the Institute of Mathematical Sciences, Chennai, India During 13-24 January 1997, Allied, New Delhi, ISBN: 8170237645, 1997.

[119] Relton, F.E., *Applied Bessel Functions*, Blackie, London, 1946.

[120] Ross, Bertram and Farrell, Orin J., *Solved Problems in Analysis: As Applied to Gamma, Beta, Legendre & Bessel Function*, Peter Smith Publisher Inc., ISBN: 0844600911, January 1984; also Dover Publications, ISBN: 0486627136, June, 1974, 410 pages.

[121] Sansone, Giovanni, *Orthogonal Functions,* rev. English ed., New York, Dover, 0486667308, June 1991, 411 pages.

[122] Saurer, Josef, *Bases of Special Functions and Their Domains of Convergence*, Vch Pub, October, 1993, **ISBN**: 3055016130, 158 pages.

[123] Sawyer, Charles, *Tables of the Bessel Functions Y0(x), Y1(x), K0(x), K1(x)*, National Bureau of Standards, Applied Mathematics Series-25, Department of Commerce 1952.

[124] Sedletskii, A.M., *Fourier Transforms and Approximations (Analytical Methods & Special Functions)*, Taylor & Francis, ISBN: 9056992341, 2000.

[125] Shohat, J., Théorie générale des polynomes orthogonaux de Tchebichef, Paris, Gauthier-Villars, 1934.

[126] Sibagaki, W., *Tables of Modified Bessel Functions with the Account of the Methods Used in the Calculation*, Tokyo, 1955.

[127] Siegel, K.M.; Brown, D.M.; H. E. Hunter, H.E. et al., *The Zeros of the associated legendre functions* $P_n^m(x)$ *of non- integral degree*, Ann Arbor, MI Willow Run Research Center, Engineering Research Institute, 1953, 30 pages.

[128] Singh P. and Denis, R.Y., *Special Functions and their Applications*, Dominant Publishers and Distributors, New Delhi, ISBN: 8187336919, 2001, 192 pages.

[129] Slater, L.J., *Confluent Hypergeometric Functions*, Cambridge University Press, London, ISBN: 0521064848, 1960, 260 pages.

[130] Salzer, Herbert E., *Table of the zeros and weight factors of the first twenty Hermite polynomials,* Journal of research of the National Bureau of Standards, 1952, 116 pages.

[131] Slavyanov, Sergei Yuryevitsh, and Lay, W., *Special Functions: A Unified Theory Based on Singularities*, Oxford University Press, **ISBN**: 0198505736, November 2000, 293 pages.

[132] Sneddon, Ian Naismith, *Special Functions of Mathematical Physics and Chemistry*, 2nd ed., Oliver and Boyd, Edinburgh, **ISBN**: 0582443962, 1961,

References

also Oliver and Boyd, 1956; also Longman Group United Kingdom, ISBN: 0582443962, 1980.

[133] Snow, Chester, *Hypergeometric and Legendre Functions with Applications to Integral Equations of Potential Theory.* Washington, DC. US Government Printing Office, 1952, 319 Pages.

[134] Society for Special Functions and their Applications (Jodhpur, India), *Special Functions: Selected Articles Proceedings of the First National Conference of the Society for Special Functions and Their Applications*, Scientific Publishers (India) ISBN: 817233267X, January 2001, 258 pages.

[135] Spanier, J. and Oldham, K.B., *An Atlas of Functions*, Hemisphere, Washington, DC.,1987.

[136] Stahl, Herbert and Totik, Vilmos, *General Orthogonal Polynomials*, Encyclopedia of Mathematics and Its Applications, Cambridge University Press, Cambridge, ISBN: 0521415349, April 1992, 264 pages.

[137] Sternberg, Wolfgang and Smith, Turner Linn, *The Theory of Potential and Spherical Harmonics*, 2nd ed., Toronto, University of Toronto Press, 1944, 312 pages.

[138] Stewart, Edward Wilson, *A Study of Hermite and Laguerre Polynomials*, University of Florida, 1960.

[139] Suetin, P.K., *Orthogonal Polynomials in Two Variables*, T&F STM, August, 1999, ISBN: 9056991671, 368 pages.

[140] Suslov, Sergei K., *Special Functions 2000: Current Perspective and Future Directions*, Kluwer Academic Pub., ISBN: 0792371208, 2001, 536 pages.

[141] Szegö, Gabor, *Orthogonal Polynomials,* Providence, RI, American Mathematical Society, Coll. Publ., vol. 23, 1939, New York, ISBN: 0821810235, 4th edition, 1975. 432 pages.

[142] Talman, James D., *Special Functions: A Group Theoretic Approach*, New York, W.A. Benjamin, 1968, 260 pages.

[143] Temme, Nico M., *Special Functions: An Introduction to the Classical Functions of Mathematical Physics*, New York, John Wiley & Sons Inc., ISBN: 0471113131, 1996. 374 pages.

[144] Thangavelu, S., *Lectures on Hermite and Laguerre Expansions*, Princeton University Press, ISBN: 0691000484, 1993, 214 pages.

[145] Titchmarsh, E.C., *The theory of the Riemann Zeta-function*, Oxford University Press, 2nd edition, ISBN: 0198533691, 1987, 380 pages.

[146] Todhunter, Isaac, *The Functions of Laplace, Lamé, and Bessel*, Macmillan and Co., London 1875, 2003, 362 pages.

[147] Tranter, C.J., *Bessel Functions with Some Physical Applications*, Hart Publishing, New York, ISBN: 0340049596, 1968, 148 pages.

[148] Truesdell, Clifford, *An Essay Toward a Unified Theory of Special Functions: Based upon the Functional Equation*, Princeton Univ Pr, ISBN: 0691095779, 1948, 182 pages.

[149] Van Assche, Walter, *Asymptotics for Orthogonal Polynomials*, Lecture Notes Math. 1265, Springer, Berlin-Heidelberg-New York, 1987.

[150] Van Assche, Walter, *Orthogonal Polynomials and Special Functions*, Leuven 2002, Springer Verlag, ISBN: 3540403752, 2003, 249 pages.

[151] Vilenkin, N.J., *Special Functions and the Theory of Group Representations*, American Mathematical Society, Revised edition, June, 1983, ISBN: 0821815725.

[152] Vilenkin, N.Ja. and Klimyk, A.U., *Representation of Lie Groups and Special Functions*, Vols. 1-3, and "Recent Advances", Kluwer Academic Publ., Dordrecht-Boston-London, ISBN: 0792314662, 1992-1995.

[153] Virchenko, Nina O., *Graphs of Elementary and Special Functions: Handbook*, Begell House, ISBN: 1567001564, April 2001.

[154] Virchenko, Nina O., *Generalized Associated Legendre Functions and Their Applications*, World Scientific Pub. Co. Inc., ISBN: 9810243537, 2001.

[155] Wang, Z.X., and Guo, D.R., *Special Functions*, Singapore, World Scientific Pub. Co., ISBN: 997150667X, October 1989. 710 pages.

[156] Watson, George Neville, *A Treatise on the Theory of Bessel Functions*, Cambridge University Press, ISBN: 0521483913, 1995, 812 pages.

[157] Wawrzynczyk, Antoni, *Group Representations and Special Functions*, D. Reidel Publishing Company, Dordrecht, Boston, Lancaster, ISBN: 9027712697, 1984, 708 pages.

[158] Whittaker, E.T. and Watson, G.N., *A Course in Modern Analysis, 4th ed.*, Cambridge University Press, Cambridge, England, 1990.

[159] Wong, Roderick; Dunkl, Charles; and Ismail, Monral, *Special Functions*, Proceedings of the International Workshop, Hong Kong 21 - 25 June 1999, World Scientific Pub. Co. Inc., ISBN: 9810243936, January 2000, 438 pages.

[160] Zhang, Shanjie and Jin, Jianming, *Computation of Special Functions*, New York, Wiley-Interscience; ISBN: 0471119636, July, 1996. 717p.

[161] Zhurina, M. I. and Karmazina, L. N, *Tables of the Legendre Functions: Parts 1 and 2*, New York, The Macmillan Company, 1964.

[162] Zwillinger, Daniel, (Ed.), *CRC Standard Mathematical Tables and Formulae*, Chapman & Hall/CRC, 31st edition, ISBN: 1584882913, November, 2002, 912 pages.